Springer Tracts in Mechanical Engineering

For further volumes:
http://www.springer.com/series/11693

Tianjian Lu · Fengxian Xin

Vibro-Acoustics of Lightweight Sandwich Structures

Tianjian Lu
Fengxian Xin
Xi'an Jiaotong University
Xi'an
China

ISSN 2195-9862 ISSN 2195-9870 (electronic)
ISBN 978-3-642-55357-8 ISBN 978-3-642-55358-5 (eBook)
DOI 10.1007/978-3-642-55358-5
Springer Heidelberg New York Dordrecht London

Jointly published with Science Press, Beijing
ISBN: 978-7-03-041322-2 Science Press, Beijing

Library of Congress Control Number: 2014942660

© Science Press Beijing and Springer-Verlag Berlin Heidelberg 2014
This work is subject to copyright. All rights are reserved by the Publishers, whether the whole or part of the material is concerned, specifically the rights of translation, reprinting, reuse of illustrations, recitation, broadcasting, reproduction on microfilms or in any other physical way, and transmission or information storage and retrieval, electronic adaptation, computer software, or by similar or dissimilar methodology now known or hereafter developed. Exempted from this legal reservation are brief excerpts in connection with reviews or scholarly analysis or material supplied specifically for the purpose of being entered and executed on a computer system, for exclusive use by the purchaser of the work. Duplication of this publication or parts thereof is permitted only under the provisions of the Copyright Law of the Publishers' locations, in its current version, and permission for use must always be obtained from Springer. Permissions for use may be obtained through RightsLink at the Copyright Clearance Center. Violations are liable to prosecution under the respective Copyright Law.
The use of general descriptive names, registered names, trademarks, service marks, etc. in this publication does not imply, even in the absence of a specific statement, that such names are exempt from the relevant protective laws and regulations and therefore free for general use.
While the advice and information in this book are believed to be true and accurate at the date of publication, neither the authors nor the editors nor the publishers can accept any legal responsibility for any errors or omissions that may be made. The publishers make no warranty, express or implied, with respect to the material contained herein.

Printed on acid-free paper

Springer is part of Springer Science+Business Media (www.springer.com)

Preface

The purpose of this book is to present the vibration and acoustical behavior of typical sandwich structures subject to mechanical and/or acoustical loadings, which actually form a class of structural elements of practical importance in huge amounts of engineering applications, such as aircraft fuselage, ship and submarine hulls. The contents of this book has grown out of the research activities of the authors in the field of sound radiation/transmission of/through lightweight sandwich structures.

The book is organized into six chapters: Chapter 1 deals with the vibro-acoustic performance of rectangular multiple-panel partitions with enclosed air cavity theoretically and experimentally, which has accounted for the simply supported and clamp supported boundary conditions. Chapter 2 concerns with the transmission of external jet-noise through a uniform skin plate of aircraft cabin fuselage in the presence of external mean flow. As an extension, Chap. 3 handles with the noise radiation and transmission from/through aeroelastic skin plates of aircraft fuselage stiffened by orthogonally distributed rib-stiffeners in the presence of convected mean flow. Chapter 4 develops a theoretical model for sound transmission through all-metallic, two-dimensional, periodic sandwich structures having corrugated core. Chapter 5 focuses on the sound radiation and transmission characteristics of periodically stiffened structures. Ultimately, Chap. 6 proposes the sound radiation and transmission behaviors of periodical sandwich structures having cavity-filling fibrous sound absorptive materials.

This book is involving multidisciplinary subjects especially including combined knowledge of vibration, aeroelastics and structural acoustics, which pays much attention on showing results and conclusions, in addition to mere theoretical modelling. Therefore this book should be of considerable interest to a wide range of readers in relevant fields. It is hoped that the content of the book will find application not only as a textbook for a wide audience of engineering students, but also a general reference for researchers in the field of vibrations and acoustics.

Xi'an, China
T.J. Lu
F.X. Xin

Acknowledgements

Although the contents of this book has grown out of the research activities of the authors, we would also like to deeply appreciate the reproduction permission for figures, tables etc. from Elsevier, AIAA, ASME, Acoustical Society of America and Taylor & Francis Ltd. These research activities are supported by the National Basic Research Program of China (Grant No. 2011CB610300), the National Natural Science Foundation of China (Grant Nos. 11102148, and 11321062) and the Fundamental Research Funds for Central Universities (xjj2011005).

Contents

1 Transmission of Sound Through Finite Multiple-Panel Partition 1
   1.1 Simply Supported Finite Double-Panel Partitions 2
        1.1.1 Introduction 2
        1.1.2 Vibroacoustic Theoretical Modeling 4
        1.1.3 Mathematic Formulation and Solution 5
        1.1.4 Convergence Check for Numerical Results 10
        1.1.5 Model Validation 11
        1.1.6 Effects of Air Cavity Thickness 13
        1.1.7 Effects of Panel Dimensions 17
        1.1.8 Effects of Incident Elevation Angle and Azimuth Angle 20
        1.1.9 Conclusions 23
   1.2 Clamped Finite Double-Panel Partitions 24
        1.2.1 Introduction 24
        1.2.2 Modeling of the Vibroacoustic Coupled System 26
        1.2.3 Model Validation 32
        1.2.4 Finite Versus Infinite Double-Panel Partition 34
        1.2.5 Effects of Panel Thickness on STL 35
        1.2.6 Effects of Air Cavity Thickness on STL 37
        1.2.7 Effects of Incident Angles on STL 37
        1.2.8 Conclusions 40
        1.2.9 Sound Transmission Measurements 41
        1.2.10 Relationships Between Clamped and Simply
                 Supported Boundary Conditions 47
        1.2.11 Conclusions 51
   1.3 Clamped Finite Triple-Panel Partitions 53
        1.3.1 Introduction 53
        1.3.2 Dynamic Structural Acoustic Formulation 56
        1.3.3 The Principle of Virtual Work 60
        1.3.4 Determination of Modal Coefficients 60
        1.3.5 Sound Transmission Loss 63
        1.3.6 Model Validation 63

		1.3.7	Physical Interpretation of STL Dips	64
		1.3.8	Comparison Among Single-, Double-, and Triple-Panel Partitions with Equivalent Total Mass	68
		1.3.9	Asymptotic Variation of STL Versus Frequency Curve from Finite to Infinite System	69
		1.3.10	Effects of Panel Thickness	70
		1.3.11	Effects of Air Cavity Depth	74
		1.3.12	Concluding Remarks	75
	Appendices			77
		Appendix A		77
		Appendix B		80
	References			83
2	**Vibroacoustics of Uniform Structures in Mean Flow**			87
	2.1	Finite Single-Leaf Aeroelastic Plate		88
		2.1.1	Introduction	88
		2.1.2	Modeling of Aeroelastic Coupled System	90
		2.1.3	Effects of Mean Flow in Incident Field	99
		2.1.4	Effects of Mean Flow in Transmitted Field	103
		2.1.5	Effects of Incident Elevation Angle in the Presence of Mean Flow on Both Incident Side and Transmitted Side	106
		2.1.6	Conclusions	108
	2.2	Infinite Double-Leaf Aeroelastic Plates		109
		2.2.1	Introduction	109
		2.2.2	Statement of the Problem	111
		2.2.3	Formulation of Plate Dynamics	112
		2.2.4	Consideration of Fluid-Structure Coupling	114
		2.2.5	Definition of Sound Transmission Loss	117
		2.2.6	Characteristic Impedance of an Infinite Plate	117
		2.2.7	Physical Interpretation for the Appearance of STL Peaks and Dips	119
		2.2.8	Effects of Mach Number	122
		2.2.9	Effects of Elevation Angle	127
		2.2.10	Effects of Azimuth Angle	128
		2.2.11	Effects of Panel Curvature and Cabin Internal Pressurization	129
		2.2.12	Conclusions	130
	2.3	Double-Leaf Panel Filled with Porous Materials		131
		2.3.1	Introduction	131
		2.3.2	Problem Description	133
		2.3.3	Theoretical Model	134
		2.3.4	Validation of Theoretical Model	140
		2.3.5	Influence of Porous Material and the Faceplates	141
		2.3.6	Influence of Porous Material Layer Thickness	143

		2.3.7	Influence of External Mean Flow	144
		2.3.8	Influence of Incident Sound Elevation Angle	147
		2.3.9	Influence of Sound Incident Azimuth Angle	148
		2.3.10	Conclusion	150
	Appendix			151
	Mass-Air-Mass Resonance			151
	Standing-Wave Attenuation			152
	Standing-Wave Resonance			152
	Coincidence Resonance			153
	References			154
3	**Vibroacoustics of Stiffened Structures in Mean Flow**			**159**
	3.1	Noise Radiation from Orthogonally Rib-Stiffened Plates		160
		3.1.1	Introduction	160
		3.1.2	Theoretical Formulation	162
		3.1.3	Effect of Mach Number	170
		3.1.4	Effect of Incidence Angle	172
		3.1.5	Effect of Periodic Spacings	173
		3.1.6	Concluding Remarks	175
	3.2	Transmission Loss of Orthogonally Rib-Stiffened Plates		176
		3.2.1	Introduction	176
		3.2.2	Theoretical Formulation	178
		3.2.3	Model Validation	187
		3.2.4	Effects of Mach Number of Mean Flow	189
		3.2.5	Effects of Rib-Stiffener Spacings	190
		3.2.6	Effects of Rib-Stiffener Thickness and Height	193
		3.2.7	Effects of Elevation and Azimuth Angles of Incident Sound	194
		3.2.8	Conclusions	197
	Appendices			198
	Appendix A			198
	Appendix B			201
	References			203
4	**Sound Transmission Across Sandwich Structures with Corrugated Cores**			**207**
	4.1	Introduction		207
	4.2	Development of Theoretical Model		209
	4.3	Effects of Core Topology on Sound Transmission Across the Sandwich Structure		214
	4.4	Physical Interpretation for the Existence of Peaks and Dips on STL Curves		215
	4.5	Optimal Design for Combined Sound Insulation and Structural Load Capacity		219
	4.6	Conclusion		220
	References			221

5 Sound Radiation, Transmission of Orthogonally Rib-Stiffened Sandwich Structures ... 225

- 5.1 Sound Radiation of Sandwich Structures ... 226
  - 5.1.1 Introduction ... 226
  - 5.1.2 Theoretical Modeling of Structural Dynamic Responses ... 228
  - 5.1.3 Solutions ... 234
  - 5.1.4 Far-Field Radiated Sound Pressure ... 238
  - 5.1.5 Validation of Theoretical Modeling ... 239
  - 5.1.6 Influences of Inertial Effects Arising from Rib-Stiffener Mass ... 240
  - 5.1.7 Influence of Excitation Position ... 242
  - 5.1.8 Influence of Rib-Stiffener Spacings ... 243
  - 5.1.9 Conclusions ... 244
- 5.2 Sound Transmission Through Sandwich Structures ... 245
  - 5.2.1 Introduction ... 245
  - 5.2.2 Analytic Formulation of Panel Vibration and Sound Transmission ... 248
  - 5.2.3 The Acoustic Pressure and Continuity Condition ... 257
  - 5.2.4 Solution of the Formulations with the Virtual Work Principle ... 258
  - 5.2.5 Virtual Work of Panel Elements ... 259
  - 5.2.6 Virtual Work of x-Wise Rib-Stiffeners ... 260
  - 5.2.7 Virtual Work of y-Wise Rib-Stiffeners ... 261
  - 5.2.8 Combination of Equations ... 261
  - 5.2.9 Definition of Sound Transmission Loss ... 264
  - 5.2.10 Convergence Check for Space-Harmonic Series Solution ... 265
  - 5.2.11 Validation of the Analytic Model ... 266
  - 5.2.12 Influence of Sound Incident Angles ... 267
  - 5.2.13 Influence of Inertial Effects Arising from Rib-Stiffener Mass ... 269
  - 5.2.14 Influence of Rib-Stiffener Spacings ... 270
  - 5.2.15 Influence of Airborne and Structure-Borne Paths ... 271
  - 5.2.16 Conclusions ... 272
- Appendices ... 273
  - Appendix A ... 273
  - Appendix B ... 278
- References ... 285

6 Sound Propagation in Rib-Stiffened Sandwich Structures with Cavity Absorption ... 289

- 6.1 Sound Radiation of Absorptive Sandwich Structures ... 290
  - 6.1.1 Introduction ... 290
  - 6.1.2 Structural Dynamic Responses to Time-Harmonic Point Force ... 292
  - 6.1.3 The Acoustic Pressure and Fluid-Structure Coupling ... 295

	6.1.4	Far-Field Sound-Radiated Pressure	300
	6.1.5	Convergence Check for Numerical Solution	300
	6.1.6	Validation of Theoretical Modeling	301
	6.1.7	Influence of Air-Structure Coupling Effect	303
	6.1.8	Influence of Fibrous Sound Absorptive Filling Material	305
	6.1.9	Conclusions	307
6.2	Sound Transmission Through Absorptive Sandwich Structure		308
	6.2.1	Introduction	308
	6.2.2	Analytic Formulation of Panel Vibration and Sound Transmission	309
	6.2.3	Application of the Periodicity of Structures	313
	6.2.4	Solution by Employing the Virtual Work Principle	316
	6.2.5	Model Validation	321
	6.2.6	Effects of Fluid-Structure Coupling on Sound Transmission	322
	6.2.7	Sound Transmission Loss Combined with Bending Stiffness and Structure Mass: Optimal Design of Sandwich	324
	6.2.8	Conclusions	327
Appendices			328
Appendix A			328
Appendix B			330
Appendix C			333
References			338

Chapter 1
Transmission of Sound Through Finite Multiple-Panel Partition

Abstract This chapter is organized as three parts: in the first part, the vibroacoustic performance of a rectangular double-panel partition with enclosed air cavity and simply mounted on an infinite acoustic rigid baffle is investigated analytically. The sound velocity potential method rather than the commonly used cavity modal function method is employed, which possesses good expandability and has significant implications for further vibroacoustic investigations. The simply supported boundary condition is accounted for by using the method of modal function, and double Fourier series solutions are obtained to characterize the vibroacoustic behaviors of the structure. Results for sound transmission loss (STL), panel vibration level, and sound pressure level are presented to explore the physical mechanisms of sound energy penetration across the finite double-panel partition. Specifically, focus is placed upon the influence of several key system parameters on sound transmission, including the thickness of air cavity, structural dimensions, and the elevation angle and azimuth angle of the incidence sound. Further extensions of the sound velocity potential method to typical framed double-panel structures are also proposed.

In the second part, the air-borne sound insulation performance of a rectangular double-panel partition clamp mounted on an infinite acoustic rigid baffle is investigated both analytically and experimentally, and compared with that of a simply supported one. With the clamped (or simply supported) boundary accounted for by using the method of modal function, a double series solution for the sound transmission loss (*STL*) of the structure is obtained by employing the weighted residual (Galerkin) method. Experimental measurements with Al double-panel partitions having air cavity are subsequently carried out to validate the theoretical model for both types of the boundary condition, and good overall agreement is achieved. A consistency check of the two different models (based separately on clamped modal function and simply supported modal function) is performed by extending the panel dimensions to infinite where no boundaries exist. The significant discrepancies between the two different boundary conditions are demonstrated in terms of the *STL* versus frequency plots as well as the panel deflection mode shapes.

In the third part, an analytical model for sound transmission through a clamped triple-panel partition of finite extent and separated by two impervious air cavities is formulated. The solution derived from the model takes the form of that for a clamp-supported rectangular plate. A set of modal functions (or more strictly speaking, the basic functions) are employed to account for the clamped boundary conditions, and the application of the virtual work principle leads to a set of simultaneous algebraic equations for determining the unknown modal coefficients. The sound transmission loss (STL) of the triple-panel partition as a function of excitation frequency is calculated and compared with that of a double-panel partition. The model predictions are then used to explore the physical mechanisms associated with the various dips on the STL versus frequency curve, including the equivalent "mass-spring" resonance, the standing-wave resonance, and the panel modal resonance. The asymptotic variation of the solution from a finite-sized partition to an infinitely large partition is illustrated in such a way as to demonstrate the influence of the boundary conditions on the soundproofing capability of the partition. In general, a triple-panel partition outperforms a double-panel partition in insulating the incident sound, and the relatively large number of system parameters pertinent to the triple-panel partition in comparison with that of the double-panel partition offers more design space for the former to tailor its noise reduction performance.

1.1 Simply Supported Finite Double-Panel Partitions

1.1.1 Introduction

Double-leaf partition structures have found increasingly wide applications in noise control engineering due to their superior sound insulation capability over single-leaf configurations. Typical examples include transportation vehicles, grazing windows and partition walls in buildings, aircraft fuselage shells, and so on [1–12].

Considerable efforts have been devoted to understanding and predicting the transmission of sound across single-leaf [13–15] and double-leaf [16–29] partitions. In fact, research about the former is often a prerequisite for studying the latter. For instance, Lomas [14] developed Green function solution for the steady-state vibration of an elastically supported rectangular plate coupled to a semi-infinite acoustic medium. An important feature of the investigation is the treatment of the elastic support boundary condition which was taken into account by assuming the rotational motion along the boundary controlled by distributions of massless rotary springs and by introducing the corresponding moments into the governing equations. The problem of sound radiation by a simply supported unbaffled panel was investigated by Laulagnet [13]. Both pressure jump and plate displacement in series of the simply supported plate models were developed.

Early sound transmission studies [16, 28–30] of double-panel structures with air cavity in between generally simplified the structure as infinite and hence did not account for the elastic boundary conditions on the periphery. For typical examples, Antonio et al. [17] gave an analytical evaluation of the acoustic insulation provided by double infinite walls and also did not take elastic boundary condition into account. Kropp et al. [19] addressed the optimization of sound insulation of double-panel constructions by dividing the frequency range into three cases, i.e., where the double wall resonance frequency is much higher (or closer or much lower) than the critical frequency of the total construction. Recently, Tadeu et al. [20] adopted an analytical method to assess the airborne sound and impact insulation properties of single- and double-leaf panels by neglecting the elastic boundary conditions. Bao and Pan [31] presented an experimental study on active control of sound transmission through double walls with different approaches, including cavity control, panel control, and room control.

For simply supported, finite rectangular double-panel structures, existing studies [3, 22–27, 32–37] concerned mainly with the loss of sound transmission across the structure, without detailed analysis about the energy transmission, the vibroacoustic coupling effects, and the physical mechanisms of sound transmission process across the structure. In particular, previous studies on double-panel partitions focus on either infinite extent or finite extent, without exploring the natural relationship between the two. The present study squarely addresses these deficiencies from the new perspectives of the integration analysis of STL, panel vibration level, and sound pressure level, with more details and the physical nature of sound penetration through double-panel partitions revealed. Since the rigid baffle bounds the cavity as well as the panel so that the cavity boundaries restrict the field to sinusoidal distributions parallel to the panel plane, analytical solutions in double Fourier series are proposed by applying the sinusoidal distributed sound velocity potential method. This method can be easily expanded to the vibroacoustic analysis of rib-stiffened double-panel structures, accounting for both the structure-borne route (i.e., structural connections between the two panels) and the airborne route (i.e., air cavity between the two panels), and hence can be regarded as an alternative of the cavity mode method in certain engineering applications. The model predictions are validated by comparing the analytical results with existing experimental data. The influences of key system parameters such as air cavity thickness, panel dimensions, and elevation angle and azimuth angle of incident sound on the sound insulation capability of the finite double-panel partition are systematically investigated. The results and conclusions of the present study should be referentially significant to others due to the similar physical nature of the vibroacoustic problem.

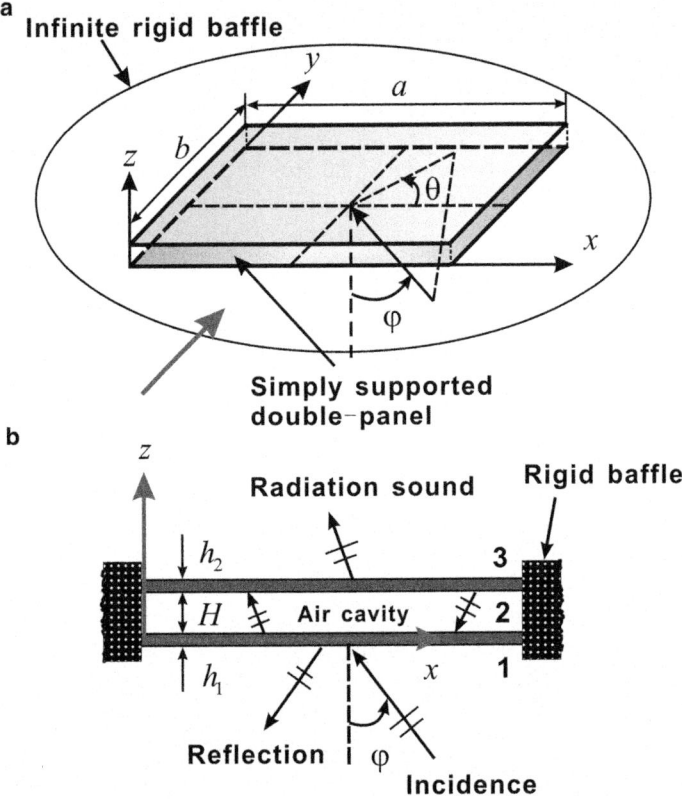

Fig. 1.1 Schematic of sound transmission through a baffled, rectangular, simply supported double-panel partition: (**a**) global view; (**b**) side view in the *arrow direction* in (**a**) (With permission from ASME)

1.1.2 Vibroacoustic Theoretical Modeling

The finite double-panel partition with enclosed air cavity is assumed to be rectangular, baffled, and simply supported along its boundaries, as shown in Fig. 1.1. The two panels are homogenous and isotropic and modeled as classical thin plate. The following geometrical dimensions are considered: the incident (bottom) panel and the radiating (top) panel have identical length a and width b, but may have different thicknesses h_1 and h_2 (Fig. 1.1b); the thickness of the air cavity is H (Fig. 1.1b). The whole configuration is mounted on an infinite acoustic rigid baffle which separates the space into two fields, i.e., sound incidence field ($z < 0$) and sound radiating field ($z > H$). A uniform plane sound wave varying harmonically in time is obliquely incident on the bottom panel, with incident elevation angle φ and azimuth angle θ (Fig. 1.1b). The vibration of the incident panel induced by the incident sound is transmitted through the enclosed air cavity to the radiating panel, which radiates sound into the acoustic medium. The vibroacoustic behaviors of the double-panel structure coupling with air cavity as well as sound transmission loss across the structure are to be solved analytically with the sound velocity potential method.

1.1.3 Mathematic Formulation and Solution

For an obliquely incident uniform plane sound wave varying harmonically in time, its acoustic velocity potential can be expressed as

$$\phi = I e^{-j(k_x x + k_y y + k_z z - \omega t)} \tag{1.1}$$

where I is the amplitude; $j = \sqrt{-1}$; ω is the angular frequency; and k_x, k_y, and k_z are the wavenumber components in the x-, y-, and z-directions, respectively:

$$k_x = k_0 \sin\varphi \cos\theta, \quad k_y = k_0 \sin\varphi \sin\theta, \quad k_z = k_0 \cos\varphi \tag{1.2}$$

Here, $k_0 = \omega/c_0$ is the acoustic wavenumber in air, with c_0 denoting the sound speed in air.

Due to the excitation of the incident sound wave, the double-panel configuration with enclosed air cavity vibrates and radiates sound. The vibroacoustic behaviors of the structure are governed by

$$D_1 \nabla^4 w_1 + m_1 \frac{\partial^2 w_1}{\partial t^2} - j\omega\rho_0 (\Phi_1 - \Phi_2) = 0 \tag{1.3}$$

$$D_2 \nabla^4 w_2 + m_2 \frac{\partial^2 w_2}{\partial t^2} - j\omega\rho_0 (\Phi_2 - \Phi_3) = 0 \tag{1.4}$$

where ρ_0 is the air density and (w_1, w_2), (m_1, m_2) and (D_1, D_2) are the transverse displacements, surface densities, and flexural rigidities of the incident and radiating panels, located at $z=0$ and $z=H$, respectively (Fig. 1.1). By introducing the loss factor of the panel material, the flexural rigidity of the panel D_i ($i=1,2$) can be written in terms of the complex Young's modulus $E_i(1+j\eta_i)$ as

$$D_i = \frac{E_i h_i^3 (1 + j\eta_i)}{12 (1 - v_i^2)} \tag{1.5}$$

The hard-walled cavity modal function $\phi_{mnl}^c = \cos(m\pi x/a)\cos(n\pi y/b)\cos(l\pi z/c)$ can only accurately model the sound field in a rigidly bounded cavity volume. It will therefore deviate somewhat from the precise results when the hard-walled cavity modal function is employed here to model the cavity bounded by two large flexural panels. In order to avoid this drawback, the sound velocity potential method is adopted, which is completely different from previous investigations based on cavity modal function. Let Φ_i ($i=1,2,3$) denote the velocity potentials of the three acoustic fields, i.e., sound incidence field, air cavity field, and structure radiating field (Fig. 1.1b), respectively. The velocity potential for the incident field can be defined as

$$\Phi_1(x,y,z;t) = I e^{-j(k_x x + k_y y + k_z z - \omega t)} + \beta e^{-j(k_x x + k_y y - k_z z - \omega t)} \quad (1.6)$$

where the first and second terms represent separately the velocity potential of the incident and the reflected plus radiating sound waves and I and β are the amplitudes of the incident (i.e., positive-going) and the reflected plus radiating (i.e., negative-going) waves, respectively. Similarly, the velocity potential in the air cavity can be written as

$$\Phi_2(x,y,z;t) = \varepsilon e^{-j(k_x x + k_y y + k_z z - \omega t)} + \zeta e^{-j(k_x x + k_y y - k_z z - \omega t)} \quad (1.7)$$

where ε is the amplitude of positive-going wave and ζ is the amplitude of negative-going wave. In the radiating field, there exist no reflected waves; thus, the velocity potential is only for radiating waves:

$$\Phi_3(x,y,z;t) = \xi e^{-j(k_x x + k_y y + k_z z - \omega t)} \quad (1.8)$$

where ξ is the amplitude of radiating (i.e., positive-going) wave. The local acoustic velocities and sound pressure are related to the velocity potentials by

$$\widehat{\mathbf{u}}_i = -\nabla \Phi_i, \quad p_i = \rho_0 \frac{\partial \Phi_i}{\partial t} = j\omega \rho_0 \Phi_i \quad (i=1,2,3) \quad (1.9)$$

For simply supported boundary condition, the transverse displacement and the transverse force are constrained to be zero at the periphery of the panel. Given that the double-panel structure is rectangular, the boundary conditions can be expressed as

$$x = 0, a: \quad w_1 = w_2 = 0, \quad \frac{\partial^2 w_1}{\partial x^2} = \frac{\partial^2 w_2}{\partial x^2} = 0 \quad (1.10)$$

$$y = 0, b: \quad w_1 = w_2 = 0, \quad \frac{\partial^2 w_1}{\partial y^2} = \frac{\partial^2 w_2}{\partial y^2} = 0 \quad (1.11)$$

At the air-panel interface, the normal velocity should be continuous, yielding the following velocity compatibility equations:

$$z = 0: \quad -\frac{\partial \Phi_1}{\partial z} = j\omega w_1, \quad -\frac{\partial \Phi_2}{\partial z} = j\omega w_1 \quad (1.12)$$

$$z = H: \quad -\frac{\partial \Phi_2}{\partial z} = j\omega w_2, \quad -\frac{\partial \Phi_3}{\partial z} = j\omega w_2 \quad (1.13)$$

1.1 Simply Supported Finite Double-Panel Partitions

For harmonic excitation of the finite double-panel system, the transverse displacements of the two panels can be written as

$$w_1(x,y,t) = \sum_{m=1}^{\infty}\sum_{n=1}^{\infty} \phi_{mn}(x,y) q_{1,mn}(t) \tag{1.14}$$

$$w_2(x,y,t) = \sum_{m=1}^{\infty}\sum_{n=1}^{\infty} \phi_{mn}(x,y) q_{2,mn}(t) \tag{1.15}$$

where the modal functions ϕ_{mn} and modal displacements $q_{i,mn}$ for simply supported boundary conditions (1.10) and (1.11) are given by

$$\phi_{mn}(x,y) = \sin\frac{m\pi x}{a} \sin\frac{n\pi y}{b} \tag{1.16}$$

$$q_{1,mn}(t) = \alpha_{1,mn} e^{j\omega t}, \quad q_{2,mn}(t) = \alpha_{2,mn} e^{j\omega t} \tag{1.17}$$

where $\alpha_{1,mn}$ and $\alpha_{2,mn}$ are the modal coefficients of the incident panel and the upper panel, respectively.

Since the rigid baffle bounds the cavity as well as the panel, the cavity boundaries restrict the field to sinusoidal distributions parallel to the panel plane. Therefore, the velocity potentials can be expressed in terms of the panel modal functions as

$$\Phi_1(x,y,z;t) = \sum_{m=1}^{\infty}\sum_{n=1}^{\infty} I_{mn} \phi_{mn}(x,y) e^{-j(k_z z - \omega t)}$$
$$+ \sum_{m=1}^{\infty}\sum_{n=1}^{\infty} \beta_{mn} \phi_{mn}(x,y) e^{-j(-k_z z - \omega t)} \tag{1.18}$$

$$\Phi_2(x,y,z;t) = \sum_{m=1}^{\infty}\sum_{n=1}^{\infty} \varepsilon_{mn} \phi_{mn}(x,y) e^{-j(k_z z - \omega t)}$$
$$+ \sum_{m=1}^{\infty}\sum_{n=1}^{\infty} \zeta_{mn} \phi_{mn}(x,y) e^{-j(-k_z z - \omega t)} \tag{1.19}$$

$$\Phi_3(x,y,z;t) = \sum_{m=1}^{\infty}\sum_{n=1}^{\infty} \xi_{mn} \phi_{mn}(x,y) e^{-j(k_z z - \omega t)} \tag{1.20}$$

where the unknown coefficients I_{mn}, β_{mn}, ε_{mn}, ζ_{mn}, and ξ_{mn} in (1.18), (1.19), and (1.20) can be determined by applying the orthogonality condition of the modal functions as

$$\tilde{\lambda}_{mn} = \frac{4}{ab} \int_0^b \int_0^a \tilde{\lambda} e^{-j(k_x x + k_y y)} \sin\frac{m\pi x}{a} \sin\frac{n\pi y}{b} dx dy \tag{1.21}$$

Here, the symbol $\tilde{\lambda}$ can be referred to any of the coefficients I, β, ε, ζ, and ξ. Note that the expressions in terms of either traveling wave or panel modal functions are completely equivalent in nature when they are both subjected to the same boundary conditions.

Substitution of Eqs. (1.18), (1.19), and (1.20) into Eqs. (1.12) and (1.13) leads to

$$\beta_{mn} = I_{mn} - \frac{\omega \alpha_{1,mn}}{k_z} \tag{1.22}$$

$$\varepsilon_{mn} = \frac{\omega \left(\alpha_{2,mn} e^{jk_z H} - \alpha_{1,mn} e^{2jk_z H}\right)}{k_z \left(1 - e^{2jk_z H}\right)} \tag{1.23}$$

$$\zeta_{mn} = \frac{\omega \left(\alpha_{2,mn} e^{jk_z H} - \alpha_{1,mn}\right)}{k_z \left(1 - e^{2jk_z H}\right)} \tag{1.24}$$

$$\xi_{mn} = \frac{\omega \alpha_2 e^{jk_z H}}{k_z} \tag{1.25}$$

Substituting Eqs. (1.14), (1.15), and (1.22), (1.23), (1.24), and (1.25) into Eqs. (1.3) and (1.4) and applying the orthogonal properties of modal functions, one gets

$$\ddot{q}_{1,kl}(t) + \omega_{1,kl}^2 q_{1,kl}(t) - \frac{j\omega\rho_0}{m_1} \left[(I_{kl} - \varepsilon_{kl}) e^{-j(k_z z - \omega t)} + (\beta_{kl} - \zeta_{kl}) e^{-j(-k_z z - \omega t)} \right] = 0 \tag{1.26}$$

$$\ddot{q}_{2,kl}(t) + \omega_{2,kl}^2 q_{2,kl}(t) - \frac{j\omega\rho_0}{m_2} \left[(\varepsilon_{kl} - \xi_{kl}) e^{-j(k_z z - \omega t)} + \zeta_{kl} e^{-j(-k_z z - \omega t)} \right] = 0 \tag{1.27}$$

where $\omega_{i,kl}$ is defined as

$$\omega_{i,kl}^2 = \frac{D_i \iint \nabla^4 \phi_{i,kl} \cdot \phi_{i,kl} \, dx \, dy}{m_i \iint \phi_{i,kl}^2 \, dx \, dy} \quad (i = 1, 2) \tag{1.28}$$

With Eq. (1.17), Eqs. (1.26) and (1.27) can be rewritten in matrix form as

$$\begin{bmatrix} Q_{11} & Q_{12} \\ Q_{21} & Q_{22} \end{bmatrix} \begin{Bmatrix} \alpha_{1,kl} \\ \alpha_{2,kl} \end{Bmatrix} = \begin{Bmatrix} F \\ 0 \end{Bmatrix} \tag{1.29}$$

where

$$Q_{11} = \omega_{1,kl}^2 - \omega^2 - \frac{j\omega\rho_0}{m_1} \frac{2\omega e^{2jk_z H}}{k_z \left(1 - e^{2jk_z H}\right)} \tag{1.30}$$

1.1 Simply Supported Finite Double-Panel Partitions

$$Q_{12} = \frac{j\omega\rho_0}{m_1} \frac{2\omega e^{jk_zH}}{k_z\left(1 - e^{2jk_zH}\right)} \tag{1.31}$$

$$Q_{21} = \frac{j\omega\rho_0}{m_2} \frac{2\omega e^{jk_zH}}{k_z\left(1 - e^{2jk_zH}\right)} \tag{1.32}$$

$$Q_{22} = \omega_{2,kl}^2 - \omega^2 - \frac{j\omega\rho_0}{m_2} \frac{2\omega e^{2jk_zH}}{k_z\left(1 - e^{2jk_zH}\right)} \tag{1.33}$$

$$F = \frac{2j\omega\rho_0 I_{kl}}{m_1} \tag{1.34}$$

The transmission coefficient of sound power is a function of the elevation angle φ and azimuth angle θ of the incidence sound, which can be expressed as

$$\tau(\varphi, \theta) = \frac{\sum_{m=1}^{\infty}\sum_{n=1}^{\infty} |\xi_{mn}|^2}{\sum_{m=1}^{\infty}\sum_{n=1}^{\infty} |I_{mn} + \beta_{mn}|^2} \tag{1.35}$$

For diffuse incident sound, due to the symmetry of the rectangular double-panel structure, the averaged transmission coefficient can be obtained by integration as

$$\tau_{\text{diff}} = \frac{\int_0^{\pi/4}\int_0^{\varphi_{\text{lim}}} \tau(\varphi, \theta) \sin\varphi \cos\varphi \sin\theta \cos\theta \, d\varphi d\theta}{\int_0^{\pi/4}\int_0^{\varphi_{\text{lim}}} \sin\varphi \cos\varphi \sin\theta \cos\theta \, d\varphi d\theta} \tag{1.36}$$

where φ_{lim} is the limiting angle defining the diffuseness of the incident field. Here, a limited incident angle φ_{lim} is introduced to carry out this integration, inasmuch as it is picked to give a good fit with experiments [23, 37–41].

In order to describe the vibration intensity of the two panels as well as the local distribution of sound pressure, two parameters are introduced below [42]:

(1) Averaged quadratic velocity $\overline{V}^2$:

$$\overline{V}^2 = \begin{cases} \overline{V}_1^2 = \dfrac{\omega^2}{2A_1} \displaystyle\int_{A_1} |w_1|^2 dA \\ \overline{V}_2^2 = \dfrac{\omega^2}{2A_2} \displaystyle\int_{A_2} |w_2|^2 dA \end{cases} \tag{1.37}$$

where the subscripts 1 and 2 denote the incident panel and upper panel, respectively. For the present numerical calculations, $\overline{V}^2$ will be plotted in decibel scale (dB) using a reference quadratic velocity of 2.5×10^{-15} m^2/s^2.

(2) Averaged quadratic sound pressure $\overline{P}^2$:

$$\overline{P}^2 = \begin{cases} \overline{P}_1^2 = \dfrac{1}{2A_1}\displaystyle\int_{A_1} |p_1|^2 \mathrm{d}A \\ \overline{P}_2^2 = \dfrac{1}{2A_2}\displaystyle\int_{A_2} |p_2|^2 \mathrm{d}A \end{cases} \qquad (1.38)$$

Here, only sound pressure in the near acoustic field (incidence field and radiating field) adjacent to each panel is considered, so the integration of Eq. (1.38) is taken over the panel area. Again, $\overline{P}^2$ will be plotted in dB scale, with reference to 4×10^{-10} Pa^2.

1.1.4 Convergence Check for Numerical Results

Numerical studies are performed to investigate the influence of relevant system parameters on the sound insulation property of simply supported double-panel partitions of finite extent, including the thickness of air cavity, panel dimensions, and the elevation and azimuth angles of incident sound. The material properties and structural dimensions of the panels are taken as follows. The two panels are made of aluminum, with Young's modulus $E = 70$ GPa, density $\rho = 2,700$ kg/m^3, Poisson ratio $v = 0.33$, and loss factor (damping) $\eta = 0.01$. The two rectangular panels have identical dimensions: $a = 1.2$ m in the x-direction and $b = 0.8$ m in the y-direction. Unless stated otherwise, the panels have thickness $h = 2$ mm, while the thickness of the air cavity is fixed at $H = 21.5$ mm. However, both h and H will be varied later to explore their influences on sound insulation. Air density is $\rho_0 = 1.21$ kg/m^3, sound speed in air is $c_0 = 343$ m/s, and the amplitude of the acoustic velocity potential for the incident sound is $I_0 = 1$ m^2/s.

Since the analytical solutions are presented in the form of double series, a sufficiently large number of terms must be adopted to ensure the solution convergence. It is admissible that once the solution is convergent at a given frequency, it is also convergent for all frequencies lower than that [43], so that the necessary number of terms is determined by the highest frequency of interest. Here, without loss of generality, $f = 6,000$ Hz is selected as the highest frequency for convergence checking. In the case when the incident sound is normal to the double-panel partition, the results shown in Fig. 1.2 demonstrate that the solution is rendered convergent if the single model number m (and n) has a value of 50 or larger. This implies that at least 2,500 terms (with both m and n ranging from 1 to 50) are needed in all the present calculations.

Moreover, note that the loss factor of the air cavity in between the two panels is too low (=0.001 approximately for air) to have a significant effect on STL especially when the depth of the air cavity is small. In other words, the discrepancy

1.1 Simply Supported Finite Double-Panel Partitions

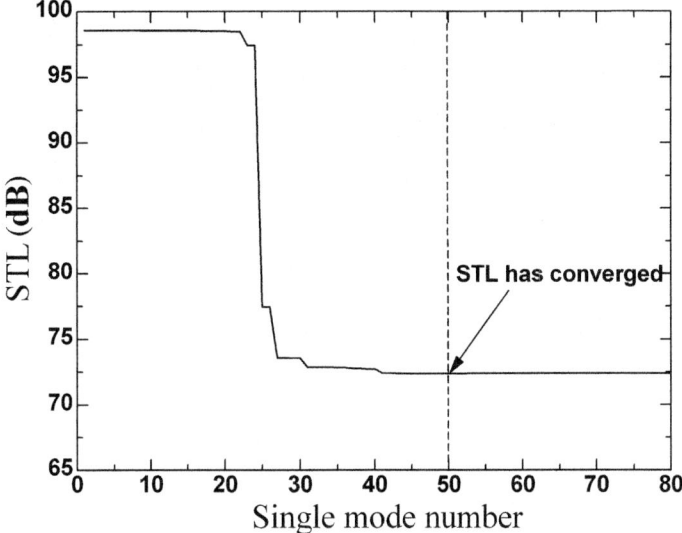

Fig. 1.2 Convergence of double Fourier series solution for sound transmission loss (*STL*) of a double-panel partition under the excitation of a normally incident sound at 6,000 Hz (With permission from ASME)

between the predicted STL curve with air damping and that without air damping is invisible, owing to the negligible damping of the air. The real crucial matter is the vibroacoustic coupling behavior of the air cavity with the two panels, thus which will be elucidated in the following section in terms of several key system parameters (i.e., air cavity thickness, panel dimensions, the elevation and azimuth angles of incident sound).

1.1.5 Model Validation

For validation, the analytical solutions are compared with the existing experiment results [22], as shown in Fig. 1.3. A diffuse incident sound is assumed for the calculation of structural STL, with the structural dimensions and panel material properties same as those used in the experiment [22]. As mentioned above, an empirical value of the limiting elevation angle $\varphi_{\lim}$ is usually assumed. Although the empirical limit on the angle of field plane wave incidence may still be controversial, more and more researchers [23, 37–41] have acknowledged that the limiting angle falls within the range of 65–80° and prefer to adopt the angle of 78° for numerical analyses. Here, the results calculated with $\varphi_{\lim} = 65°$ and 78°, respectively, are both presented in Fig. 1.3. It is seen from Fig. 1.3 that the theoretical predictions exhibit the same trend as the experimental measurements. Both the theoretical and experimental curves show two minima in the frequency range of concern, although

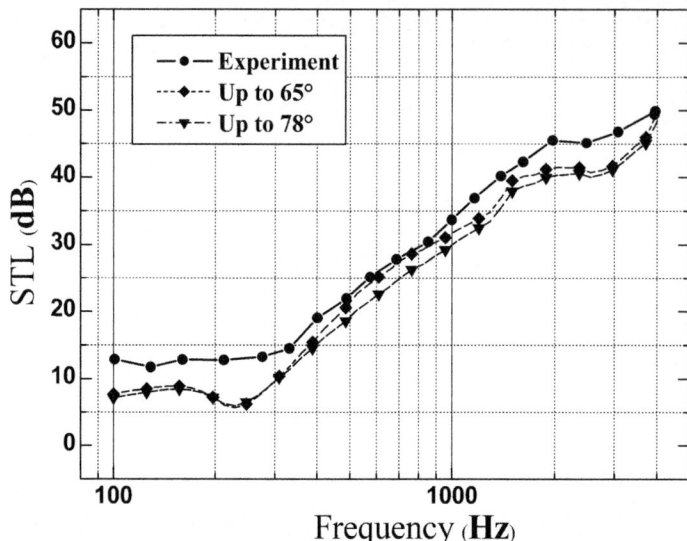

Fig. 1.3 STL plotted as a function of frequency for a double aluminum panel ($m_1 = m_2 = 0.239$ g/cm^2, H; diffuse incident sound): theory versus experiment [23] (With permission from ASME)

the first minimum on the experiment curve is only faintly observable. The first minimum at about 230 Hz corresponds to the mass-air-mass resonance related mainly to the low angles of incidence, while the second minimum appearing at the critical frequency of 2,546 Hz is mostly caused by the high angles of incidence, as mentioned by Villot et al. [23].

The experimentally measured STL values are consistently about 5 dB larger than the theoretically predicted values over the entire range of frequency studied (Fig. 1.3), which can be explained by the fact that glass fiber absorbent (several feet thick) was used around the edges of the double-panel partition in actual measurements. Otherwise, the theoretical predictions of Fig. 1.3 agree reasonably well with experiments.

To check the accuracy of the present model predictions further, another comparison is made with the theoretical and experimental results of Carneal et al. [26] for the case of normal incident sound (i.e., $\varphi = 0°$), as shown in Fig. 1.4. The double-panel partition considered consists of two identical aluminum plates (0.38 m by 0.30 m, 1.6 mm thickness), separated by a 0.048 m air cavity. Again, a close agreement between the present theoretical predictions and experiment measurements is observed. Due to the experiments that were carried out by using two clamped parallel panels, as an approximation, Carneal et al. [26] increased the stiffness of the simply supported plate by a factor of $\sqrt{2}$ for each boundary to approximate the clamped boundary condition; the present theoretical results in Fig. 1.4 adopts the same manner.

1.1 Simply Supported Finite Double-Panel Partitions

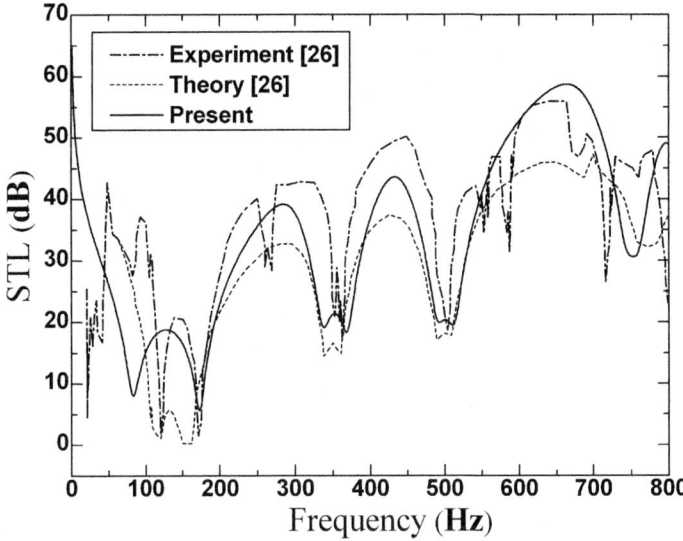

Fig. 1.4 STL plotted as a function of frequency for a double aluminum panel with dimensions $a = 0.38$ m, $b = 0.30$ m, and $H = 0.048$ m (normal incident sound): theory versus experiment [26] (With permission from ASME)

1.1.6 Effects of Air Cavity Thickness

Transmission of sound through a double-panel partition without any mechanical connection is due to the enclosed air between the two panels. Air in the cavity acts as springs, thus transmits the mechanical vibration of the incident panel to the radiating panel. The equivalent stiffness of the air between two parallel panels is given by Carneal and Fuller [26] and Fahy [30]:

$$K_a = \frac{\rho_0 c_0^2}{H} \tag{1.39}$$

The equivalent stiffness of the air is expected to have a significant effect on the sound transmission through the double-panel configuration. Therefore, it is necessary to investigate how the thickness of the air cavity influences the sound insulation capability of the structure.

With other geometrical dimensions of the structure fixed, the STL of the double-panel partition as a function of frequency is plotted in Fig. 1.5 for the case of normal incident sound and three selected values of air cavity thickness ($H = 5.275$ mm, 21.5 mm, and 86 mm). It is seen from Fig. 1.5 that the first resonance frequency corresponding to the first minimum on the STL versus frequency curve decreases as H is increased, which is expected because increase of air cavity thickness leads to reduced equivalent air stiffness (Eq. 1.39). Overall, the curve is shifted toward the left with increasing H (Fig. 1.5), indicating that the sound insulation property of the partition is improved.

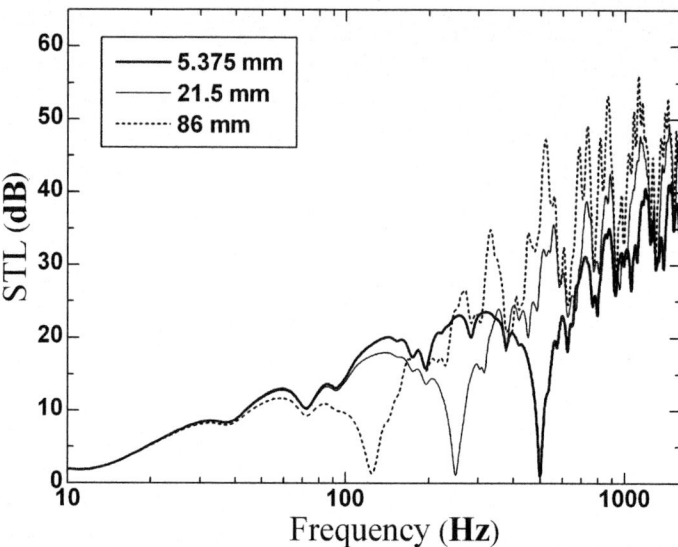

Fig. 1.5 STL plotted as a function of frequency for double-panel partitions with different thicknesses of enclosed air cavity (normal incident sound) (With permission from ASME)

The vibroacoustic performance of the double-panel structure is also investigated in terms of the averaged quadratic velocity of the two panels (Fig. 1.6) and the averaged quadratic sound pressure field in the close proximity of the two panels (Fig. 1.7). The double-panel partition addressed is identical to that of Fig. 1.5. The increase of air cavity thickness reduces the vibroacoustic coupling of the structure due to weakened air pumping effects (equivalent stiffness effects), which is reflected by the decline of distinctions between the averaged quadratic velocity levels of the incident panel (Fig. 1.6a) and the radiating panel (Fig. 1.6b) as H is increased. It is understandable that a stronger vibroacoustic effect of the air cavity (with relatively small thickness) leads to enhanced transmission of vibration energy from the incidence panel to the radiating panel, and hence, the vibration energies of the two panels will be in closer agreement because of the less energy expense in the transmission process. As the averaged quadratic velocity is directly related to the vibration energy of the panel, similar vibration energy levels of the two panels will be reflected by similar averaged quadratic velocity levels (Fig. 1.6). It should be pointed out that the first maximum in the averaged quadratic velocity curve of the radiating panel in Fig. 1.6b corresponds to the first minimum on the STL curve in Fig. 1.5. That is also expected because the intensive vibration of the radiating panel would radiate sound strongly, sharply decreasing the magnitude of the STL.

The double-panel partitions with different thicknesses of air cavity are excited by the same incident sound of unit amplitude, i.e., the input sound energy is identical for the three cases studied in Figs. 1.5, 1.6, and 1.7. However, the vibration energy

1.1 Simply Supported Finite Double-Panel Partitions

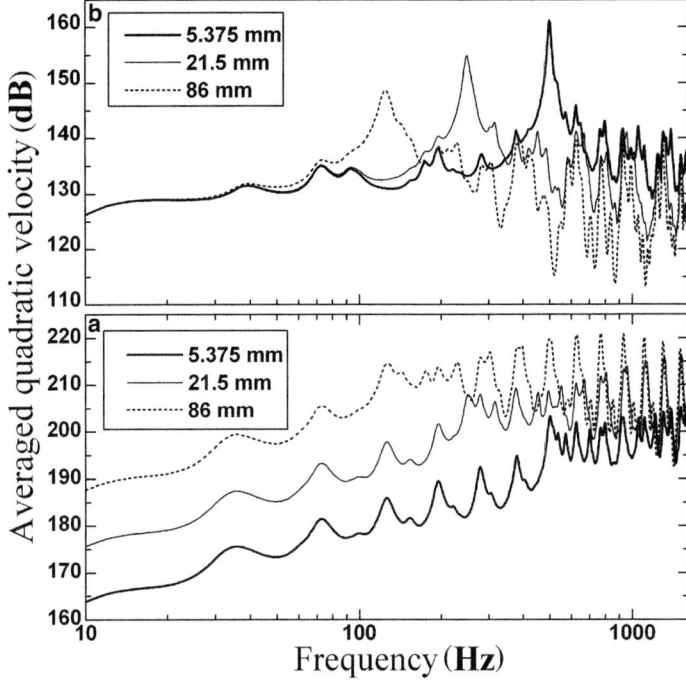

Fig. 1.6 Averaged quadratic velocity plotted as a function of frequency for double-panel partitions with different thicknesses of enclosed air cavity (normal incident sound): (**a**) incidence panel; (**b**) radiating panel (With permission from ASME)

of the incident panel (as reflected by the average quadratic velocity, Fig. 1.6a) increases with the increase of air cavity thickness while that of the radiating panel decreases (Fig. 1.6b). This also demonstrates that the increase of air cavity thickness weakens the vibroacoustic coupling effect of the structure, inducing less energy transmitting through the air cavity. Although subjected to the same sound excitation, the noticeable differences appearing in the vibration levels of the incident panel for the three cases (see Fig. 1.6a) are attributed to the different vibroacoustic performances of the backed air cavities having different depths, while the sound energy fluxes penetrating through the air cavity to the radiating panel do not deviate so much because of the larger vibration level of the incident panel with a weaker cavity coupling (e.g., the 86 mm case) and the smaller vibration level of the incident panel with a stronger cavity coupling (e.g., the 5.375 mm case). As a result, the vibration levels of the radiating panel are almost the same at low frequencies; see Fig. 1.6b. Since the sound pressure level in the transmitted field completely stems from the vibration of the radiating panel, it should remain nearly unchanged at low frequencies; see Fig. 1.7. Finally, given the definition of the STL, it should be almost the same for the considered three cases at low frequencies. Therefore, the noticeable differences in the vibration level of the incident panel in Fig. 1.6a at low frequencies

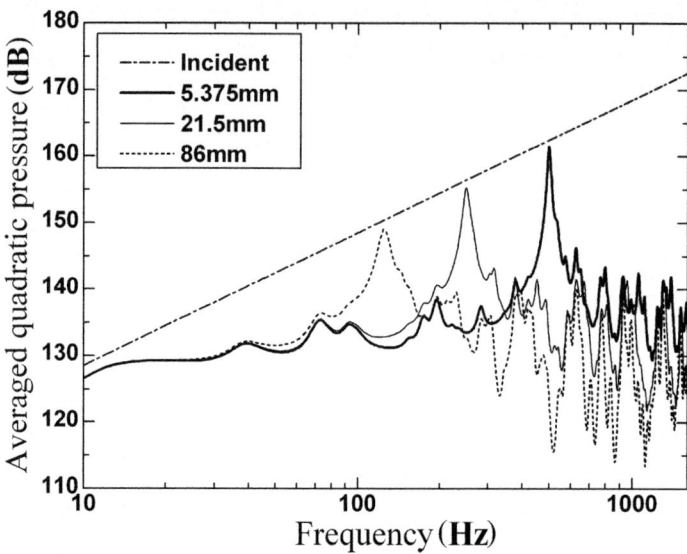

Fig. 1.7 Averaged quadratic sound pressure plotted as a function of frequency for double-panel partitions with different thicknesses of enclosed air cavity (normal incident sound) (With permission from ASME)

do not exist in Figs. 1.5, 1.6, and 1.7. Of course, the maxima and minima shown in these figures represent the modal behaviors of the two panels and the air cavity.

It is interesting to observe from Fig. 1.6 that the frequencies where most of the maxima and minima appear remain unchanged (or slightly shifted) as H is varied, although the actual values of the averaged quadratic velocity at these frequencies may change significantly. The reason is that the incident and radiating panels are made of the same material, have the identical dimensions, and are both simply supported on their edges, and hence, the two panels have the same natural resonance frequencies (which are related to the vibration energy maxima). The slight shifting of the maxima (or minima) at certain frequencies (Fig. 1.6) should be attributed to the influence of the vibroacoustic coupling effects of the air cavity.

The influence of air cavity thickness H on the averaged quadratic sound pressure is shown in Fig. 1.7 for both the incident field and the radiating field. Although the excitation intensity is the same, the sound pressure level of the radiating acoustic field varies considerably as H is changed. Note that, for a given H, the averaged quadratic sound pressure of the radiating acoustic field (Fig. 1.7) is almost the same as the averaged quadratic velocity of the radiating panel (Fig. 1.6b). This is expected because the sound energy adjacent to the radiating panel is generated directly by the panel. For the same reason, the first maximum on the averaged quadratic sound pressure versus frequency curve (Fig. 1.6b) corresponds to the first minimum appearing in the STL versus frequency curve of Fig. 1.5. Finally, it should be pointed out that the incident sound pressure not only varies with time but also depends upon the incident frequency, as shown by the dash-dot curve in Fig. 1.7.

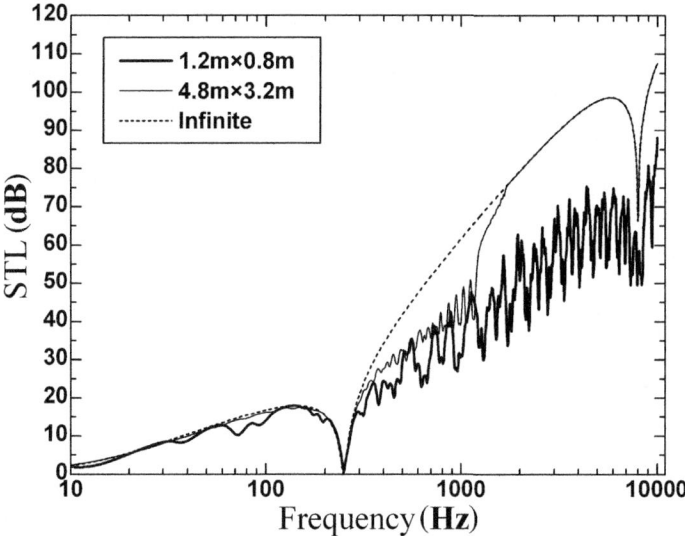

Fig. 1.8 STL plotted as a function of frequency for double-panel partitions having different in-plane dimensions (normal incident sound) (With permission from ASME)

1.1.7 Effects of Panel Dimensions

The influence of panel geometrical dimensions on the vibroacoustic performance of a finite double-panel partition is studied in terms of three parameters: STL (Fig. 1.8), averaged quadratic velocity (Fig. 1.9a, b), and averaged quadratic sound pressure (Fig. 1.10). Three different in-plane panel dimensions (1.2×0.8 m^2, 4.8×3.2 m^2, and infinite extent) are considered.

It can be observed from Fig. 1.8 that the STL curve becomes smooth with few maxima and minima when the panel dimensions are increased, which implies that the mode density of structures with smaller dimensions is larger than that with bigger dimensions due to the effect of the simply supported boundary conditions. The structural dimensions may have a significant effect on the sound insulation capability of the double-panel configuration, depending upon the frequency range of concern. Within the relatively low-frequency range of 10–300 Hz, the STL is not sensitive to the change of panel dimensions in the considered case here. As the sound frequency is increased beyond about 300 Hz, the STL values of structures with bigger dimensions are apparently larger than those with smaller dimensions. Furthermore, for frequencies larger than about 2,000 Hz, the STL curves corresponding to the two cases, 4.8×3.2 m^2 and infinite, fall onto one master curve, demonstrating that the infinitely large double-panel partition provides an upper bound STL for finite configurations.

A comparison of Fig. 1.5 with Fig. 1.8 suggests that the "mass-air-mass" resonance frequency associated with the first evident minimum on the STL versus

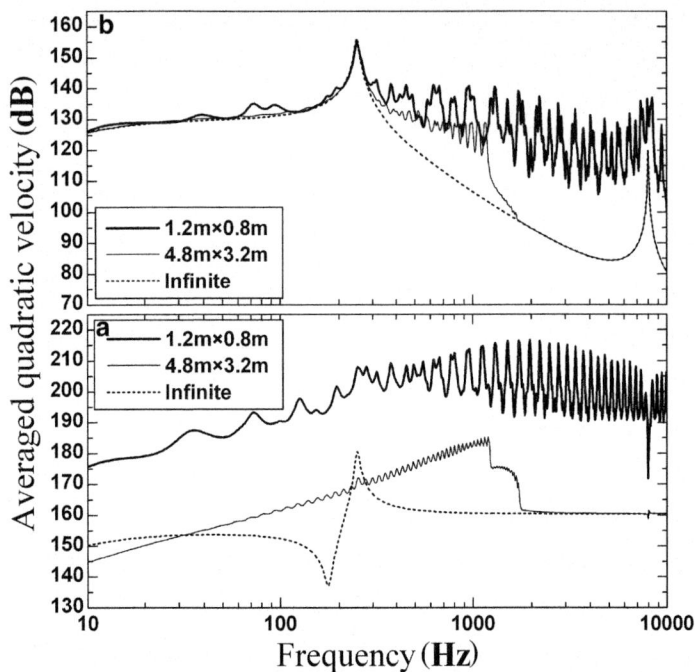

Fig. 1.9 Averaged quadratic velocity plotted as a function of frequency for double-panel partitions having different in-plane dimensions (normal incident sound): (**a**) incidence panel; (**b**) radiating panel (With permission from ASME)

frequency curve be independent of the in-plane panel dimensions as it is dominated by the thickness of the air cavity. Finally, a simple but worth noting fact is that a double-panel partition with enclosed air cavity has better sound insulation capabilities at relatively high frequencies than at low frequencies (Fig. 1.8).

Figure 1.9a and b presents the effects of in-plane panel dimensions on the averaged quadratic velocity of the incidence panel and the radiating panel, respectively. It is seen that structures with smaller dimensions possess higher mode density represented by the dense maxima and minima on the averaged quadratic velocity versus frequency curve than that having bigger dimensions. The averaged quadratic velocity is in fact directly related to the total vibration energy of the panel; see Eq. (1.37). The noticeable maximum at low frequency in Fig. 1.9b is associated with the "mass-air-mass" resonance of the double-panel system, at which the two panels vibrate in opposite phase, with the air cavity behaving like springs and the vibration level of the radiating panel sharply increasing to the maximum value. In comparison, the vibration level of the incident panel takes on different tendencies in different cases (see Fig. 1.9a) due to the complex interplay (destructive or constructive interference) between sound incidence and air cavity coupling.

1.1 Simply Supported Finite Double-Panel Partitions

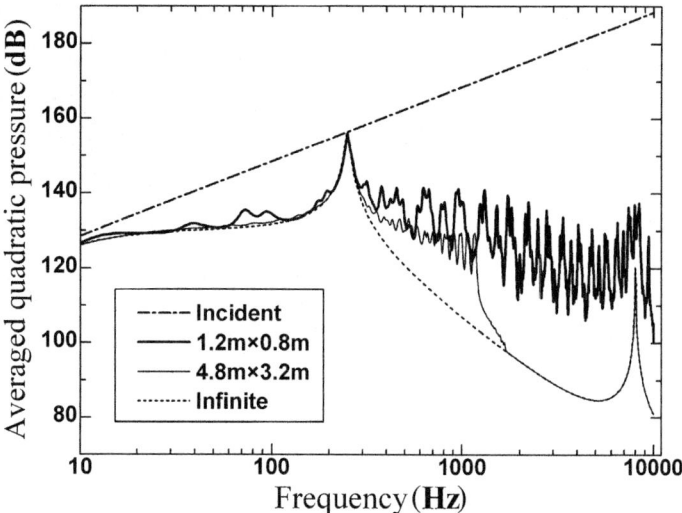

Fig. 1.10 Averaged quadratic sound pressure plotted as a function of frequency for double-panel partitions having different in-plane dimensions (normal incident sound) (With permission from ASME)

The averaged quadratic velocity of the infinite incidence panel maintains approximately a constant level (Fig. 1.9a), indicating the non-sensitivity of the structure with respect to excitation frequency, which is attributed to the small mode density of the infinite structure (Fig. 1.8). As the frequency is increased beyond the resonance frequency (~250 Hz), the vibration energy of the incidence panel reaches a higher level from 152 to 160 dB. As the in-plane dimensions of the panel are decreased, the vibration energy of the incidence panel is increased and becomes more sensitive to excitation frequency. Additionally, as aforementioned, the matching of the positions where the maxima and minima appear in Fig. 1.9a with those in Fig. 1.9b implies that the vibration energy of the radiating panel is strongly coupled to that of the incidence panel.

The effects of structural dimensions are also investigated by analyzing the averaged quadratic sound pressure; see Fig. 1.10. Again, the averaged quadratic sound pressures adjacent to the radiating panel are almost the same as the averaged quadratic velocities of the radiating panel. Note in particular that there exists a maximum (about 250 Hz) in the averaged quadratic sound pressure for the radiating acoustic field, which has nearly the same value as that for the incidence acoustic field (Fig. 1.10). The same phenomenon occurs in Fig. 1.7. This implies that the sound excitation energy transmits through the double-panel partition with little energy loss, as if the structure is nearly transparent at these particular frequencies in the considered case here (without considering air cavity damping). For practical sound insulation applications, these frequencies need to be identified and carefully avoided.

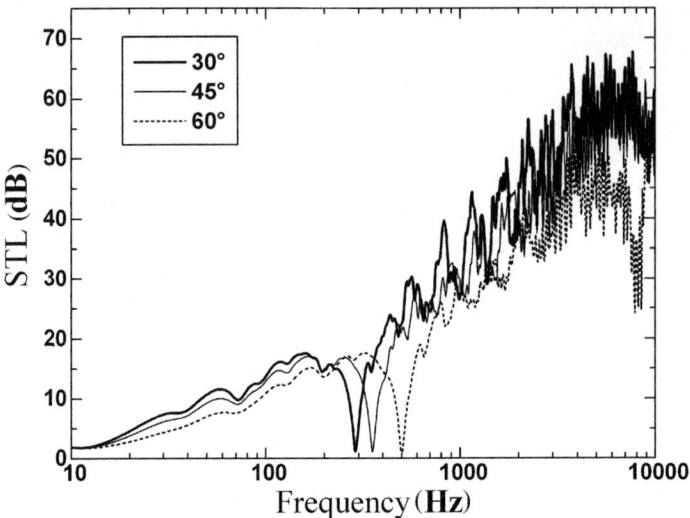

Fig. 1.11 STL plotted as a function of frequency for double-panel partitions excited by incident sound having different elevation angles and fixed azimuth angle ($\theta = 0°$) (With permission from ASME)

1.1.8 Effects of Incident Elevation Angle and Azimuth Angle

Figure 1.11 plots the predicted STL of the double-panel structure as a function of frequency for selected values of incident elevation angle, with the azimuth angle fixed at $\theta = 0°$. By increasing the elevation angle, it is seen that the "mass-air-mass" frequency increases, and it has been established that the results of Fig. 1.11 are consistent with the following expression for the "mass-air-mass" frequency of a double-panel structure with air cavity [8, 9, 12]:

$$f_a = \frac{1}{2\pi \cos \varphi} \sqrt{\frac{K_a (m_1 + m_2)}{m_1 m_2}} \tag{1.40}$$

Apart from changing the resonance frequencies, a smaller elevation incidence angle of sound also leads to enhanced sound insulation property of the structure over the considered frequency range (Fig. 1.11). It should be pointed out that although the above results are obtained by assuming $\theta = 0°$, the conclusions still hold for arbitrary values of the azimuth angle (from 0 to 2π).

The influence of incidence elevation angle on the averaged quadratic velocity of both the incidence and radiating panels is shown in Fig. 1.12. For the incidence panel, the averaged quadratic velocity varies significantly as the elevation angle is changed (Fig. 1.12a). The maxima on the averaged quadratic velocity versus frequency curves of Fig. 1.12a are mainly associated with the coincidence frequencies,

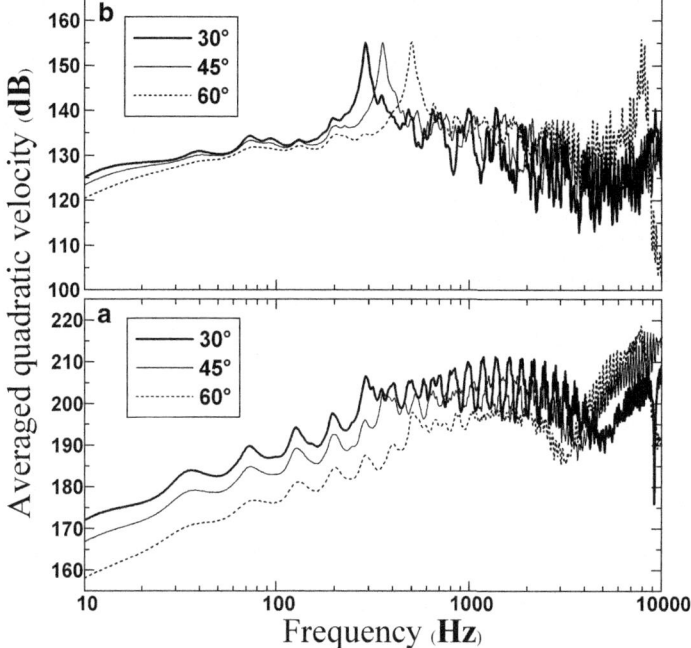

Fig. 1.12 Averaged quadratic velocity plotted as a function of frequency for double-panel partitions excited by incident sound having different elevation angles and fixed azimuth angle ($\theta = 0°$): (**a**) incidence panel; (**b**) radiating panel (With permission from ASME)

at which the wavelength of the flexural waves in the incidence panel matches with the trace wavelength of the incidence wave. Because the latter varies with the elevation angle, changes of the elevation angle affect the coincidence frequencies of the incidence panel, shifting the positions where these maxima appear (Fig. 1.12a).

It is interesting to see from Fig. 1.12 that sound with a smaller incidence elevation angle would excite the vibration of both the incidence panel and the radiating panel more intensely at frequencies below approximately 4,000 Hz, while the opposite is true for frequencies larger than 4,000 Hz. This may be attributed to the complex vibroacoustic coupling effects of the double-panel structure.

The sound energies in the radiating field for different elevation angles (for brevity, not shown here) are approximately the same as the vibration energies of the radiating panel shown in Fig. 1.12. The energy associated with a bigger elevation angle is slightly higher than that with a smaller elevation angle, indicating that the energy of sound is transmitted more easily through the double-panel configuration in the former case.

By using 3D (three-dimensional) overall view and contour map, Figs. 1.13 and 1.14 show the structural STL as a function of the elevation angle (or azimuth angle) and frequency of the incident sound. It is seen from Fig. 1.13 (with $\theta = 0°$) that the incident elevation angle has a significant effect on the STL of the considered

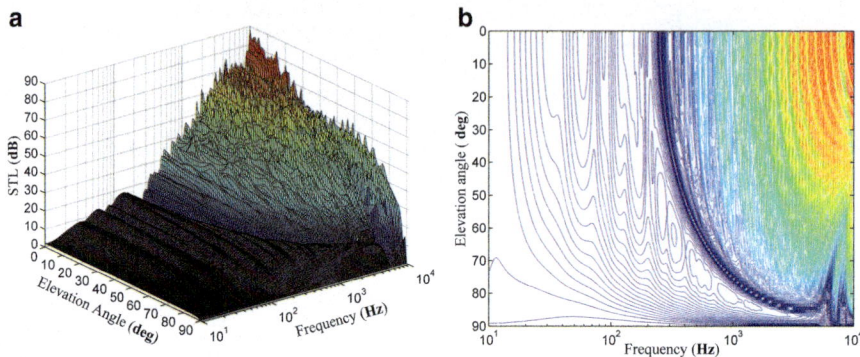

Fig. 1.13 Dependence of STL on incident sound elevation angel and frequency for fixed azimuth angle $\theta = 0°$: (**a**) global view; (**b**) contour map (With permission from ASME)

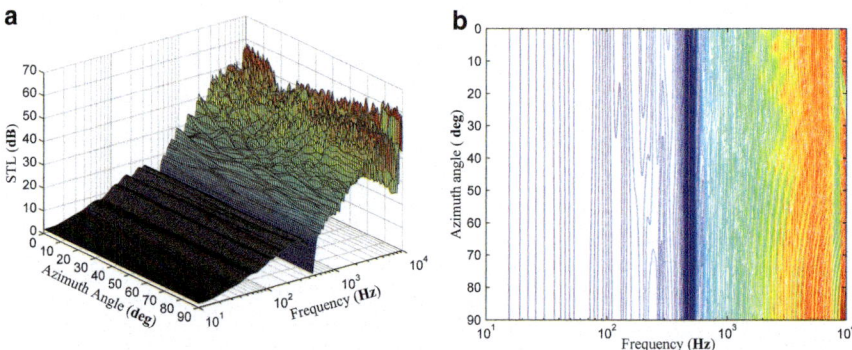

Fig. 1.14 Dependence of STL on incident sound elevation angel and frequency for fixed elevation angle $\varphi = 45°$: (**a**) global view; (**b**) contour map (With permission from ASME)

structure. Note also that a sharp valley exists on the STL curve, which is associated with the "mass-air-mass" resonance frequency and also the reason for the existence of the first evident minimum in Fig. 1.11 (Fig. 1.11 for three selected elevation angles is part of Fig. 1.13a). As previously discussed, at this "mass-air-mass" resonance frequency, the two panels resonate and radiate sound intensively due to the vibroacoustic coupling effects of the air cavity. When the frequency of the incidence sound surpasses the mass-air-mass resonance frequency, the value of STL increases significantly, as demonstrated by the maxima (or plateau) appearing in the 3D view of Fig. 1.13a.

The results of Fig. 1.14 (with the incidence elevation angle fixed at $\varphi = 45°$) suggest that the incident azimuth angle has a negligible influence on the vibroacoustic property of the double-panel structure, in sharp contrast with that of the elevation angle. This is attributed to the small aspect ratio of the rectangular panel considered. For a rectangular panel with small aspect ratio, the numerical calculation of STL for diffuse incident sound, i.e., Eq. (1.36), can be simplified as

$$\tau_{\text{diff}} = \frac{\int_0^{\pi/4} \int_0^{\varphi_{\text{lim}}} \tau(\varphi, \theta) \sin \varphi \cos \varphi \, d\varphi \, d\theta}{\int_0^{\pi/4} \int_0^{\varphi_{\text{lim}}} \sin \varphi \cos \varphi \, d\varphi \, d\theta} \qquad (1.41)$$

For the present panel of dimensions 1.2 m × 0.8 m, the results obtained with Eq. (1.41) are sufficiently accurate relative to Eq. (1.36).

Similar to Fig. 1.13, the sharp valley appearing in the 3D view and its contour map of Fig. 1.14 is related to the "mass-air-mass" resonance frequency. This valley is nonetheless parallel to the axis of azimuth angle, implying that the azimuth angle has negligible influence on the "mass-air-mass" frequency of the structure.

1.1.9 Conclusions

An analytical approach has been developed to investigate the sound transmission across simply supported rectangular double-panel partitions by introducing the sinusoidal distributed sound velocity potentials. The new approach is capable of describing more accurately the air cavity coupling than the commonly used cavity modal function method, because the rigid baffle bounds the cavity as well as the panel, so that the cavity boundaries restrict the field to sinusoidal distributions parallel to the panel plane. Consequently, the application of the sinusoidal distributed sound velocity potential for an air cavity is much closer to the physical nature than the cavity mode method. The precise handling of the air cavity coupling effect is ensured, as shown by the maxima and minima in the set of STL versus frequency plots.

With the method of modal function used to simulate the simply supported boundary conditions, analytical solutions are obtained in the form of double Fourier series. The truncated numbers of the double Fourier series are numerically estimated by ensuring the convergence of the solution. For model validation, the analytical predictions of structural STL are compared with existing experiment data, and close agreement is obtained. Subsequent numerical calculations focus on quantifying the effects of several key system parameters on the sound insulation capability of the structure, including the air cavity thickness, panel dimensions, and sound incident angles.

It is found that the overall vibroacoustic behavior of the double-panel partition can be significantly changed by altering the enclosed air cavity thickness without changing other geometrical dimensions of the structure. The elevation angle of the incident sound affects significantly the sound insulation capability of the finite double-panel structure, whereas the influence of incident azimuth angle is negligible. Due to the constraint of the simply supported boundary, the mode density of a double-panel partition with smaller in-plane dimensions is larger than that with bigger dimensions. This is consistent with the fact that the incidence panel of a smaller double-panel partition acquires more sound energy than that having bigger dimensions, even if the same sound pressure excitation is imposed. The STL of an infinitely large double-panel partition provides an upper bound estimate for finite configurations. It should be emphasized that there exist certain frequencies at which the sound excitation energy transmits through the double-panel partition with little energy loss, as if the structure is nearly transparent at these particular frequencies in the considered case here (without considering air cavity damping). For practical sound insulation applications, these frequencies need to be identified and carefully avoided.

As an extension of the vibroacoustic analysis of double-panel structures with air cavity, the pertinent governing equations for two typical framed structures, i.e., double panel with orthogonal parallel connections and that with one-dimensional parallel connections, of either infinite or finite extent, are presented. This serves to reveal the superiority and good expansibility of the sound velocity potential method over the commonly used cavity modal function method, in addition to providing significant implications for further vibroacoustic investigation of framed double-panel structures. The velocity potential method can be utilized not only for the vibroacoustic analysis of double-panel partitions without connections but also those with structural connections, while the cavity modal method cannot.

1.2 Clamped Finite Double-Panel Partitions

1.2.1 *Introduction*

With superior sound insulation properties over single-panel configurations, double-panel partitions have found a wide range of important applications in modern buildings, transportation vehicles, aerospace and aeronautical structures, etc. [3, 5–7, 9, 26]. To gain a fundamental understanding of the sound insulation mechanisms of double-panel partitions, the frequency characteristics of sound transmission loss (*STL*) are usually needed. In particular, how these are affected by different boundary conditions (i.e., clamped and simply supported) is of great theoretical and practical interest. The present investigation aims to address the significant differences between the two different boundary conditions for double-panel partitions containing air cavities, both theoretically and experimentally.

For decades the vibration responses of single- and double-panel constructions interacting with the surrounding fluid have been an attractive research topic. The approaches adopted to study the vibroacoustic behavior of both structures are similar, although there are more difficulties associated with the latter. Traditionally, the method of statistical energy analysis (SEA) advanced significantly by Maidanik [44] has been widely used to analyze the vibration response of a complex structure under force or sound excitation. However, the SEA method is less effective at relatively low frequencies on account of its pre-assumption that enough structural modes need to be excited. This is usually difficult to satisfy, causing statistical uncertainties that prevail in low-frequency force and sound excitations [35]. An

1.2 Clamped Finite Double-Panel Partitions

FEM (finite element method) model was developed by Ruzzene [45] to evaluate the acoustic characteristics of sandwich beams in terms of structural response and sound transmission reduction index, which is much effective for low frequencies but requires high computational cost for high frequencies [41]. For relatively simple structures, analytical solutions suited for a wide frequency range have been developed by various researchers [43, 46, 47].

As for the sound insulation properties of double-panel structures, the classical work of London [16] addressed an infinite double-panel structure, and hence, the influence of boundary conditions was not considered. Similarly, the study of Kropp et al. [19] and Antonio et al. [17] focused on the airborne sound insulation capability of infinite double wall constructions. The sound insulation property and radiation efficiency of an infinite double-plate connected by periodical studs have been investigated with the Fourier transform technique [48, 49] and with the space-harmonic expansion method [18, 50], respectively. Brunskog [51] examined the influence of finite cavities on the sound insulation properties of periodically framed infinite double-plate structures. To take into account the finite size of a double-panel structure, Villot et al. [23] developed an approximate technique based on the spatial windowing of plane waves. More recently, the problem of sound transmission through double-panel structures of finite extent was solved by considering simple boundary conditions [24, 35] on the basis of modal superposition theory from different viewpoints. In addition, to improve the sound isolation properties of double-panel partitions, various active control strategies have been proposed, both experimentally [26, 31] and theoretically [26, 52], which are significant from the viewpoint of practical noise control.

Although a persistent effort has been devoted to the studying of sound transmission through finite or infinite single- and double-leaf panels, many physical details remain an indistinct matter, especially for *clamped* double-panel configurations under oblique sound excitation. Often, the experimental measurements do not possess reproducibility, and different *STL* curves from different laboratories were obtained even though the same panels were compared [38]. According to Kim et al. [53], this may be attributed to the so-called tunneling effect, and the different mounting conditions adopted in different laboratories should be another key factor. On the other hand, to the authors' best knowledge, the problem of sound transmission through a double-panel partition with air cavity fully clamped (different from being simply supported) on its edges has not been analytically solved. The existing method [26, 46] for the influence of the clamped boundary condition is to modify the analytical solution for simply supported boundary conditions, which is only approximate for predicting the sound insulation properties of fully clamped finite single- or double-panel constructions. This chapter squarely addresses this deficiency. A step-by-step analysis for the vibroacoustic performance of a finite double-panel partition fully clamped on its edges under sound excitation is presented (note that aspects of the current theoretical formulations have been presented as a simplified outline in Ref. [6]). To gain full insight into the influence of the boundary condition on sound transmission across the structure, *STL* values obtained with the clamped boundary condition are compared with those obtained with the simply supported boundary condition. Experimental measurements for both types of boundary condition are subsequently carried out to validate the model predictions. The remarkable difference between the two different boundary conditions is highlighted. As sound transmission through a double-panel partition simply supported on its edges has been extensively studied, the present focus is placed upon exploring how this is different from that of a fully clamped double-panel partition.

1.2.2 Modeling of the Vibroacoustic Coupled System

Consider a rectangular double-panel partition with air cavity which is baffled and fully clamped (or simply supported) along its edges, as shown in Fig. 1.15a. The two panels are taken as homogenous, isotropic, and sufficiently thin. The geometrical dimensions of the structure are as follows: width of panel a, length of panel b, thickness of air cavity H, and thicknesses of incidence panel h_1 and radiating panel h_2 (Fig. 1.15b). The whole configuration is fully clamped (or simply supported) on an infinite acoustic rigid baffle which separates the space into two fields: sound incidence field ($z < 0$) and sound radiating field ($z > H$). Cartesian coordinates (x, y, z) are selected, as shown in Fig. 1.15.

An oblique plane sound wave varying harmonically in time is incident on the bottom panel with elevation angle φ and azimuth angle θ. The vibration of the bottom panel induced by the incident sound is transmitted through the hermetical air cavity to the upper panel, which radiates sound pressure waves into the upper acoustic domain (Fig. 1.15b). The model proposed below describes the vibroacoustic behavior of the double-panel structure, either fully clamped or simply supported, and its *STL* characteristics.

1.2.2.1 Theoretical Formulation and Solution

The acoustic velocity potential for a plane sound wave varying harmonically in time can be expressed as

$$\phi = I e^{-j(\mathbf{k}\cdot\mathbf{r}-\omega t)} \quad (1.42)$$

where I is the sound amplitude; ω is the angular frequency; $\mathbf{r}\ (= x\widehat{\mathbf{e}}_x + y\widehat{\mathbf{e}}_y + z\widehat{\mathbf{e}}_z)$ is the position vector (Fig. 1.15) with $\widehat{\mathbf{e}}_x, \widehat{\mathbf{e}}_y$, and $\widehat{\mathbf{e}}_z$ representing separately the unit vectors along x-, y-, and z-directions; and $\mathbf{k}\ (= k_x\widehat{\mathbf{e}}_x + k_y\widehat{\mathbf{e}}_y + k_z\widehat{\mathbf{e}}_z)$ is the wave vector with components k_x, k_y, and k_z. These wavenumbers are determined by the elevation angle φ and azimuth angle θ of the incidence sound wave as

$$k_x = k_0 \sin\varphi \cos\theta, \quad k_y = k_0 \sin\varphi \sin\theta, \quad k_z = k_0 \cos\varphi \quad (1.43)$$

where $k_0 = \omega/c_0$ is the acoustic wavenumber in air and c_0 is the acoustic speed in air.

1.2 Clamped Finite Double-Panel Partitions

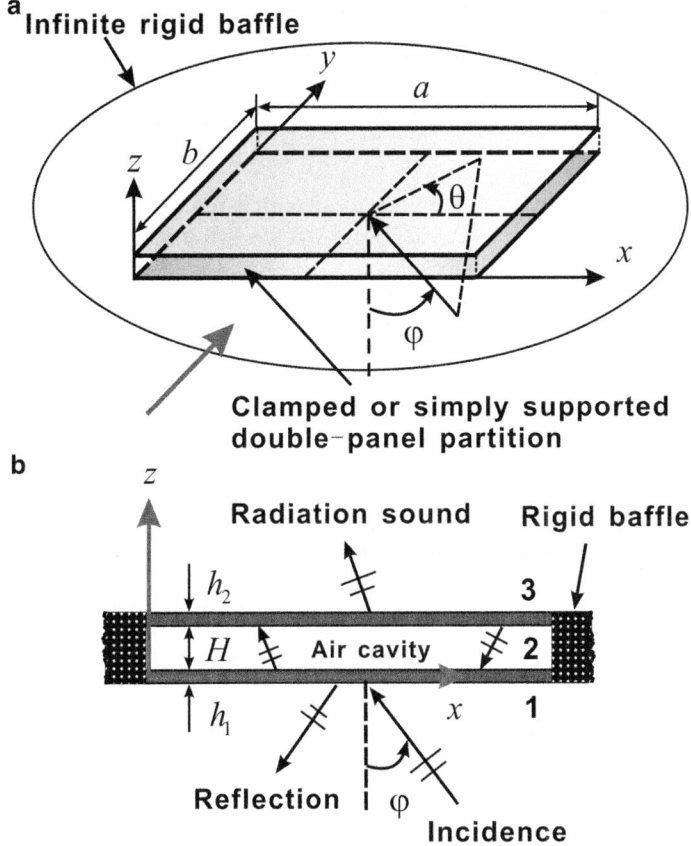

Fig. 1.15 Schematic illustration of sound transmission through a baffled double-panel partition which is clamped (or simply supported) on its edges: (**a**) global view; (**b**) side view in the arrow direction in (**a**) (With permission from Acoustical Society of America)

The flexural motions of a double-panel partition with air cavity induced by sound excitation (Fig. 1.15b) are governed by

$$D_1 \nabla^4 w_1 + m_1 \frac{\partial^2 w_1}{\partial t^2} - j\omega\rho_0 (\Phi_1 - \Phi_2) = 0 \quad (1.44)$$

$$D_2 \nabla^4 w_2 + m_2 \frac{\partial^2 w_2}{\partial t^2} - j\omega\rho_0 (\Phi_2 - \Phi_3) = 0 \quad (1.45)$$

where $\nabla^4 = (\partial^2/\partial x^2 + \partial^2/\partial y^2)^2$, ρ_0 is the air density, $j = \sqrt{-1}$, w_1 and w_2 are the transverse displacements, m_1 and m_2 are the mass per unit area, and D_1 and D_2 are the flexural rigidity of the bottom panel (panel 1) and the upper panel (panel 2), located at $z = 0$ and $z = H$, respectively. Damping of the panel material is taken into account by introducing the complex Young's modulus, $E_i(1+j\eta_i)$, where η is the loss factor. The flexural rigidity of the panel D_i ($i=1,2$) can thence be written as

$$D_i = \frac{E_i h_i^3 (1+j\eta_i)}{12(1-v_i^2)} \qquad (1.46)$$

Let Φ_i ($i=1,2,3$) denote the velocity potentials for the acoustic fields in the proximity of the two panels, corresponding to the sound incidence, the air cavity, and the structure radiating field (fields 1, 2, and 3 in Fig. 1.15b), respectively. The acoustic velocity potential in the incidence field (field 1, Fig. 1.15b) is defined as

$$\Phi_1(x,y,z;t) = I e^{-j(k_x x + k_y y + k_z z - \omega t)} + \beta e^{-j(k_x x + k_y y - k_z z - \omega t)} \qquad (1.47)$$

where the first term represents the velocity potential of the incident acoustic wave and the second term represents that of the reflected acoustic wave and I and β are the amplitudes of the incident (i.e., positive-going) and reflected (i.e., negative-going) waves, respectively. Similarly, the velocity potential in the air cavity (field 2, Fig. 1.15b) can be written as

$$\Phi_2(x,y,z;t) = \varepsilon e^{-j(k_x x + k_y y + k_z z - \omega t)} + \zeta e^{-j(k_x x + k_y y - k_z z - \omega t)} \qquad (1.48)$$

where ε is the amplitude of the positive-going wave and ζ is the amplitude of the negative-going wave. In the transmitting field (field 3, Fig. 1.15b) adjacent to the radiating upper panel, there exist no reflected waves, and therefore, the velocity potential is only for the transmitting (or radiating) waves, given as

$$\Phi_3(x,y,z;t) = \xi e^{-j(k_x x + k_y y + k_z z - \omega t)} \qquad (1.49)$$

where ξ is the amplitude of the radiating (i.e., positive-going) wave. These velocity potentials are related to the acoustic particle velocities by $\hat{u}_i = -\nabla \Phi_i$ and to the sound pressure by

$$p_i = \rho_0 \frac{\partial \Phi_i}{\partial t} = j\omega \rho_0 \Phi_i \quad (i=1,2,3) \qquad (1.50)$$

With the double-panel partition fully clamped onto a rigid baffle, the transverse deflection and the moment rotation of each panel are constrained to be zero along the edges. In view of the rectangular geometry of the double-panel structure, the boundary conditions can be expressed as

$$x=0,a \quad w_1 = w_2 = 0, \quad \frac{\partial w_1}{\partial x} = \frac{\partial w_2}{\partial x} = 0 \qquad (1.51)$$

$$y=0,b \quad w_1 = w_2 = 0, \quad \frac{\partial w_1}{\partial y} = \frac{\partial w_2}{\partial y} = 0 \qquad (1.52)$$

1.2 Clamped Finite Double-Panel Partitions

At the air-panel interface, the normal velocity is continuous, yielding the corresponding velocity compatibility condition equations:

$$z = 0 \qquad -\frac{\partial \Phi_1}{\partial z} = j\omega w_1, \quad -\frac{\partial \Phi_2}{\partial z} = j\omega w_1 \tag{1.53}$$

$$z = H \qquad -\frac{\partial \Phi_2}{\partial z} = j\omega w_2, \quad -\frac{\partial \Phi_3}{\partial z} = j\omega w_2 \tag{1.54}$$

Since the two panels are excited by a harmonic sound wave, their transverse displacements can be written as

$$w_1(x, y, t) = \sum_{m,n} \phi_{mn}(x, y) q_{1,mn}(t) \tag{1.55}$$

$$w_2(x, y, t) = \sum_{m,n} \phi_{mn}(x, y) q_{2,mn}(t) \tag{1.56}$$

where the modal functions (or, more strictly speaking, the basic functions) ϕ_{mn} and the modal displacements $q_{i,mn}$ take the following forms:

$$\phi_{mn}(x, y) = \left(1 - \cos\frac{2m\pi x}{a}\right)\left(1 - \cos\frac{2n\pi y}{b}\right) \tag{1.57}$$

$$q_{1,mn}(t) = \alpha_{1,mn} e^{j\omega t}, \quad q_{2,mn}(t) = \alpha_{2,mn} e^{j\omega t} \tag{1.58}$$

where $\alpha_{1,mn}$ and $\alpha_{2,mn}$ are the modal coefficients of the bottom panel and the upper panel, respectively. Note that the clamped modal function of Eq. (1.57) is different from the simply supported modal function $\phi^s_{mn} = \sin(m\pi x/a)\sin(n\pi y/b)$ used by previous researchers [2, 3, 24, 35, 54–56], because the former satisfies the boundary condition of zero moment rotation while the latter does not. Moreover, the clamped double-panel can transmit rotation on its edges, whereas the simply supported one cannot, which has been confirmed experimentally by Utley et al. [57].

By applying the modal functions for the clamped double-panel structure, the velocity potentials for the acoustic fields 1, 2, and 3 (Fig. 1.15b) can be expressed as

$$\Phi_1(x, y, z, t) = \sum_{m,n} I_{mn}\phi_{mn} e^{-j(k_z z - \omega t)} + \sum_{m,n} \beta_{mn}\phi_{mn} e^{-j(-k_z z - \omega t)} \tag{1.59}$$

$$\Phi_2(x, y, z, t) = \sum_{m,n} \varepsilon_{mn}\phi_{mn} e^{-j(k_z z - \omega t)} + \sum_{m,n} \zeta_{mn}\phi_{mn} e^{-j(-k_z z - \omega t)} \tag{1.60}$$

$$\Phi_3(x, y, z, t) = \sum_{m,n} \xi_{mn}\phi_{mn} e^{-j(k_z z - \omega t)} \tag{1.61}$$

where the coefficients I_{mn}, β_{mn}, ε_{mn}, ζ_{mn}, and ξ_{mn} are determined by

$$\tilde{\lambda}_{mn} = \frac{4}{ab} \int_0^b \int_0^a \tilde{\lambda} e^{-j(k_x x + k_y y)} \cos \frac{2m\pi x}{a} \cos \frac{2n\pi y}{b} dx dy \qquad (1.62)$$

Here, the symbol $\tilde{\lambda}$ can be referred to any of the coefficients I, β, ε, ζ, and ξ.

Substituting Eqs. (1.59), (1.60), and (1.61) into Eqs. (1.53) and (1.54) for the continuity of velocity at the air-panel interface and omitting the time factor $e^{j\omega t}$, one obtains

$$-I e^{-j(k_x x + k_y y)} + \sum_{m,n} \left[\beta_{mn} + \frac{\omega}{k_z} \alpha_{1,mn} \right] \times \phi_{mn}(x, y) = 0 \qquad (1.63)$$

$$\sum_{m,n} [k_z(-\varepsilon_{mn} + \zeta_{mn}) + \omega \alpha_{1,mn}] \times \phi_{mn}(x, y) = 0 \qquad (1.64)$$

$$\sum_{m,n} [k_z(-\varepsilon_{mn} e^{-jk_z H} + \zeta_{mn} e^{jk_z H}) + \omega \alpha_{2,mn}] \times \phi_{mn}(x, y) = 0 \qquad (1.65)$$

$$\sum_{m,n} [-k_z \xi_{mn} e^{-jk_z H} + \omega \alpha_{2,mn}] \times \phi_{mn}(x, y) = 0 \qquad (1.66)$$

According to the weighted residual (Galerkin) method, by setting the integral of a weighted residual of the modal function to zero, an arbitrarily accurate double series solution can be obtained. For the current double-leaf partition system, the integral equations are

$$\int_0^b \int_0^a \left[D_1 \nabla^4 w_1 + m_1 \frac{\partial^2 w_1}{\partial t^2} - j\omega \rho_0 (\Phi_1 - \Phi_2) \right] \phi_{mn}(x, y) dx dy = 0 \qquad (1.67)$$

$$\int_0^b \int_0^a \left[D_2 \nabla^4 w_2 + m_2 \frac{\partial^2 w_2}{\partial t^2} - j\omega \rho_0 (\Phi_2 - \Phi_3) \right] \phi_{mn}(x, y) dx dy = 0 \qquad (1.68)$$

Substituting Eqs. (1.55), (1.56), (1.59), (1.60), and (1.61) into (1.67) and (1.68) and then performing laborious but straightforward algebraic manipulations, one gets

$$4 D_1 \pi^4 ab \left\{ \left[3 \left(\frac{m}{a}\right)^4 + 3 \left(\frac{n}{b}\right)^4 + 2 \left(\frac{m}{a}\right)^2 \left(\frac{n}{b}\right)^2 \right] \alpha_{1,mn} + \sum_k 2 \left(\frac{n}{b}\right)^4 \alpha_{1,kn} + \sum_l 2 \left(\frac{m}{a}\right)^4 \alpha_{1,ml} \right\}$$
$$+ \frac{9ab}{4} Q_{1,mn} + \frac{3ab}{2} \sum_k Q_{1,kn} + \frac{3ab}{2} \sum_l Q_{1,ml} + ab \sum_{k,l} Q_{1,kl} = 2j\omega \rho_0 I f_{mn}(k_x, k_y)$$
$$\text{at } z = 0 \quad (k \neq m, l \neq n) \qquad (1.69)$$

1.2 Clamped Finite Double-Panel Partitions

$$4D_2\pi^4 ab\left\{\left[3\left(\tfrac{m}{a}\right)^4+3\left(\tfrac{n}{b}\right)^4+2\left(\tfrac{m}{a}\right)^2\left(\tfrac{n}{b}\right)^2\right]\alpha_{2,mn}+\sum_k 2\left(\tfrac{n}{b}\right)^4\alpha_{2,kn}+\sum_l 2\left(\tfrac{m}{a}\right)^4\alpha_{2,ml}\right\}$$
$$+\tfrac{9ab}{4}Q_{2,mn}\tfrac{3ab}{2}\sum_k Q_{2,kn}+\tfrac{3ab}{2}\sum_l Q_{2,ml}+ab\sum_{k,l}Q_{2,kl}=0$$
$$\text{at } z=H \quad (k\neq m, l\neq n) \tag{1.70}$$

where

$$Q_{1,mn}=-m_1\omega^2\alpha_{1,mn}+j\omega\rho_0\left(\frac{\omega}{k_z}\alpha_{1,mn}e^{jk_z z}+\varepsilon_{mn}e^{-jk_z z}+\zeta_{mn}e^{jk_z z}\right) \tag{1.71}$$

$$Q_{2,mn}=-m_2\omega^2\alpha_{2,mn}-j\omega\rho_0\left[(\varepsilon_{mn}-\xi_{mn})e^{-jk_z z}+\zeta_{mn}e^{jk_z z}\right] \tag{1.72}$$

In the above expressions, the abbreviated symbols $\sum_{k,l}$, $\sum_k$, and $\sum_l$ denote separately $\sum_{k=1}^\infty \sum_{l=1}^\infty$, $\sum_{k=1}^\infty$, and $\sum_{l=1}^\infty$ (the same can be said of $\sum_{m,n}$), and $f_{mn}(k_x,k_y)$ is a constant generated in the process of integration, expressed as

$$f_{mn}(k_x,k_y)$$
$$=\begin{cases} ab & \text{for } k_x=0 \text{ and } k_y=0 \\ \dfrac{4jn^2\pi^2 a\left(1-e^{-jk_y b}\right)}{k_y\left(k_y^2 b^2-4n^2\pi^2\right)} & \text{for } k_x=0 \text{ and } k_y\neq 0 \\ \dfrac{4jm^2\pi^2 b\left(1-e^{-jk_x a}\right)}{k_x\left(k_x^2 a^2-4m^2\pi^2\right)} & \text{for } k_x\neq 0 \text{ and } k_y=0 \\ -\dfrac{16m^2 n^2\pi^4\left(1-e^{-jk_x a}\right)\left(1-e^{-jk_y b}\right)}{k_x k_y\left(k_x^2 a^2-4m^2\pi^2\right)\left(k_y^2 b^2-4n^2\pi^2\right)} & \text{for } k_x\neq 0 \text{ and } k_y\neq 0 \end{cases}$$
$$\tag{1.73}$$

Together with Eqs. (1.63), (1.64), (1.65), and (1.66), Eqs. (1.69) and (1.70) form a set of infinite algebraic simultaneous equations for the unknown coefficients $\alpha_{1,mn}$ and $\alpha_{2,mn}$. For numerical calculation, it is necessary to take truncation at $1\leq m\leq M$ and $1\leq n\leq N$, leading to $2MN$ algebraic simultaneous equations. In matrix form, these can be grouped into

$$\begin{bmatrix} T_{11,kl} & T_{12,kl} \\ T_{21,kl} & T_{22,kl} \end{bmatrix}_{2MN\times 2MN} \begin{Bmatrix} \alpha_{1,kl} \\ \alpha_{2,kl} \end{Bmatrix}_{2MN\times 1} = \begin{Bmatrix} F_{kl} \\ 0 \end{Bmatrix}_{2MN\times 1} \tag{1.74}$$

where k and l take values from 1 to M and from 1 to N, respectively. Detailed derivations of Eq. (1.74) can be found in Appendix A. Once the unknowns $\alpha_{1,mn}$ and $\alpha_{2,mn}$ are determined by solving (1.74), the deflections (w_1, w_2) of the bottom and upper panels and the relevant parameters (β_{mn}, ε_{mn}, ζ_{mn}, ξ_{mn}) are also determined, enabling thence the analysis of sound transmission across the double-leaf configuration.

1.2.2.2 Definition of Sound Transmission Loss

The sound power of the relevant acoustic field can be defined as [9, 24, 36]

$$\Pi_i = \frac{1}{2}\text{Re} \iint_A p_i \cdot v_i^* dA, \quad (i = 1, 2, 3) \tag{1.75}$$

where the local volume velocity is associated with the sound pressure through the impendence of air as $v_i = p_i/(\rho_0 c_0)$. The superscript asterisk denotes the complex conjugate.

The power transmission coefficient that is a function of the incident angle (φ and θ) can be given by the ratio of the transmitted sound power to the incident sound power:

$$\tau(\varphi, \theta) = \frac{\Pi_3}{\Pi_1} \tag{1.76}$$

Then the sound transmission loss (*STL*) is defined as the inverse of the power transmission coefficient in decibels scale, given by

$$STL = 10 \log_{10}\left(\frac{1}{\tau}\right) \tag{1.77}$$

The *STL* index is commonly used as a measure of the effectiveness of the double-panel structure in isolating the incident sound.

1.2.3 Model Validation

The present analytical approach is validated by comparing the predicted sound transmission loss with the theoretical and experimental results of Carneal et al. [26], as shown in Fig. 1.16. The double-panel partition considered consists of two identical clamp-supported aluminum plates (0.38 m by 0.30 m, 1.6 mm thickness), separated by a 0.048 m air cavity. Other physical parameters used are listed in Table 1.1. The STL measurements were obtained by Carneal et al. under a normal incident sound excitation, who also provided theoretical results derived by modifying a theory developed originally for simply supported boundary condition. The results of Fig. 1.16 demonstrate clearly that the present analytical predictions are in closer agreement with the experimental measurements than the theoretical

1.2 Clamped Finite Double-Panel Partitions

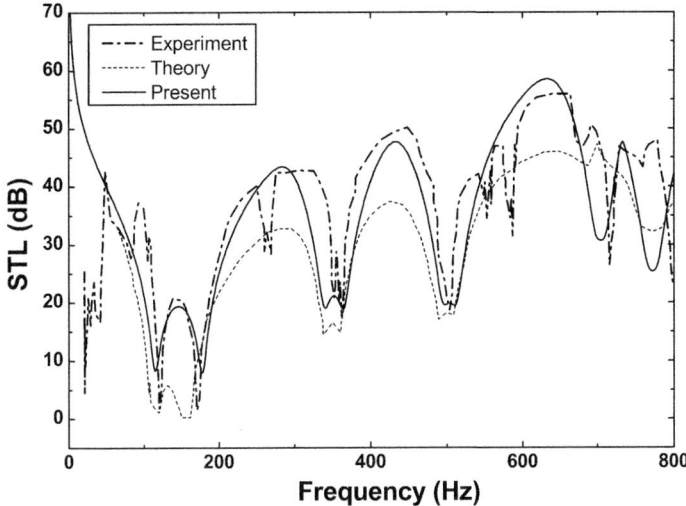

Fig. 1.16 STL of clamped aluminum double-panel partition under normal sound excitation: comparison of the present analytical predictions with the theoretical results and experimental measurements of Carneal et al. [26] (With permission from Acoustical Society of America)

Table 1.1 Structural dimensions and material properties

Panels	
Length	$a = 1$ m
Width	$b = 1$ m
Thickness	$h_1 = h_2 = 2$ (or 5, 10) mm
Young's modulus	$E = 70$ GPa
Density	$\rho = 2,700$ kg/m^3
Poisson ratio	$\nu = 0.33$
Loss factor	$\eta = 0.01$
Acoustic field	
Air cavity depth	$H = 0.08$ (or 0.04, 0.06) m
Density	$\rho_0 = 1.21$ kg/m^3
Sound speed	$c_0 = 343$ m/s
Initial amplitude	$I_0 = 1$ m^2/s

predictions of Carneal et al. Notice firstly that the experimental results are not reliable for frequencies below 50 Hz where the flanking paths of the test facility play a prominent role in measurements. Secondly, the extra dips in the experiment curve (compared with the theoretical curves) associated with (1, 2) ($\sim$260–280 Hz) and (4, 1) ($\sim$520–560 Hz) modes are attributed to the imperfect normal acoustic plane wave, unevenly damped plate, and/or the structural flanking path, as emphasized by Carneal et al. [26].

It should be pointed out that the present method is applicable not only to double-panel systems of finite extent but also to those of infinite extent. Furthermore, the method can be applied in both low- and high-frequency ranges, which is hardly

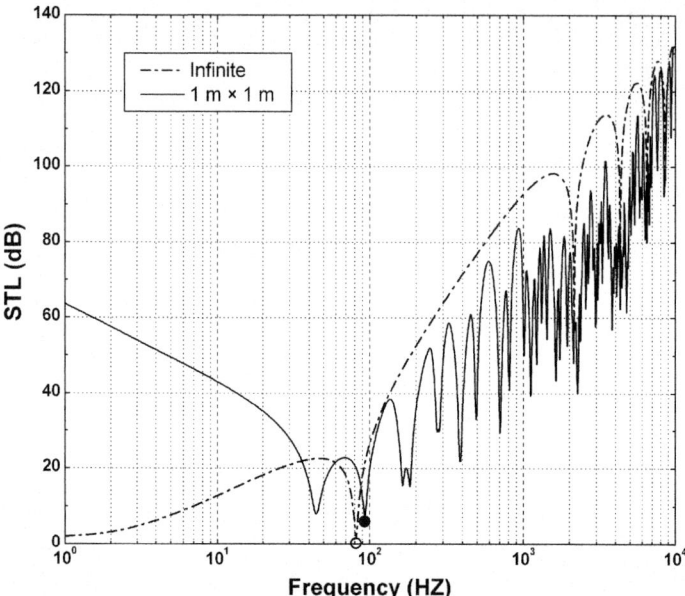

Fig. 1.17 Sound transmission loss plotted as a function of frequency for clamped double-panel partition with airproof cavity under normal sound excitation: *dash line* – infinite system ($H = 0.08$ m and $h_1 = h_2 = 0.005$ m); *solid line* – finite system with 1 m (length) × 1 m (width), $H = 0.08$ m and $h_1 = h_2 = 0.005$ m. ○: mass-air-mass resonance for infinite system; •: panel-cavity-panel resonance for finite system (With permission from Acoustical Society of America)

achievable by other approaches, e.g., FEM (finite element method), BEM (boundary element method), and SEA (statistical energy analysis) method.

1.2.4 Finite Versus Infinite Double-Panel Partition

Numerical calculations are performed in this section to quantify the effects of panel dimensions (length, width, and thickness) on STL of clamped double-panel structures.

Figure 1.17 compares the predicted structural STL of a double-panel system of finite extent (dash line) with that of an infinite double-panel system (solid line), both subjected to normal sound excitation, with $H = 0.08$ m and $h_1 = h_2 = 0.005$ m for both systems and 1 m (length) × 1 m (width) for the finite system. The two panels are identical and made of aluminum (Table 1.1).

The results of Fig. 1.17 exhibit analogous trend, with dips caused by the modal behavior of the double-panel system. For the finite double-panel system, the original modal behavior of the two panels (the radiating panel in particular) interacts strongly with the system behavior (including panel-cavity-panel resonance and standing-wave resonance) and hence plays a major role in the vibroacoustic performances of the whole system. For the infinite double-panel system, however, the original modal behavior of the panels has no influence on the overall system behavior. As a result, only dips related to system resonances show up in Fig. 1.17, with the first dip representing the mass-air-mass resonance and the remaining dips caused by the standing-wave resonance.

The intense peaks and dips in the STL versus frequency curve of the finite double-panel system occur because the finite system possesses a higher modal density than the infinite system over a wide range of frequency (approximately above 100 Hz, Fig. 1.17). Consequently, the infinite system provides an asymptotic maximum of STL for the finite system in this frequency range. Conversely, in the low-frequency range (less than about 100 Hz), there is no mode existing for the finite system. In other words, the infinite system is incapable of providing the right STL values at low frequencies for practical finite systems.

The dip denoted by a full black circle in Fig. 1.17 is associated with the so-called panel-cavity-panel resonance [36] for finite double-panel systems, while that represented by an open circle corresponds to the mass-air-mass resonance of infinite double-panel systems. The location of the former deviates slightly from the latter (the deviation increases with panel thickness; see Fig. 1.18 later), due to modal interactions of the finite system.

1.2.5 Effects of Panel Thickness on STL

To quantify the effects of panel thickness, the STL versus frequency curve is presented in Fig. 1.18a for an infinite system and in Fig. 1.18b for a finite system; the geometrical and material parameters are identical to those used to calculate Fig. 1.17, and normal sound excitation is imposed. Three values of panel thickness are selected: 2, 5, and 10 mm.

As expected and consistent with the well-known mass law, for both infinite and finite double-panel systems, it is seen from Fig. 1.18 that the STL values drastically increase as the panel thickness is increased. The influence of panel thickness on STL is particularly strong for finite systems at low frequencies (Fig. 1.18b), the significance of which should not be overlooked when designing clamped soundproof double-panel partitions of finite extent in practice.

For the infinite system, as the panel thickness is increased, the location of mass-air-mass resonance shifts to a lower frequency whereas the locations of standing-wave resonances remain unchanged (Fig. 1.18a) because they are only dependent on air cavity thickness and wavelength but independent of panel thickness.

Note that the STL predictions for the finite system (Fig. 1.18b) are only exhibited in the frequency range of 1–1,000 Hz as intense peaks and dips are present in this regime, much more complicated than those of the infinite system (Fig. 1.18a). This is attributed to the strong interaction of individual panel behavior with the overall system performance for the finite system, which is absent for the infinite system.

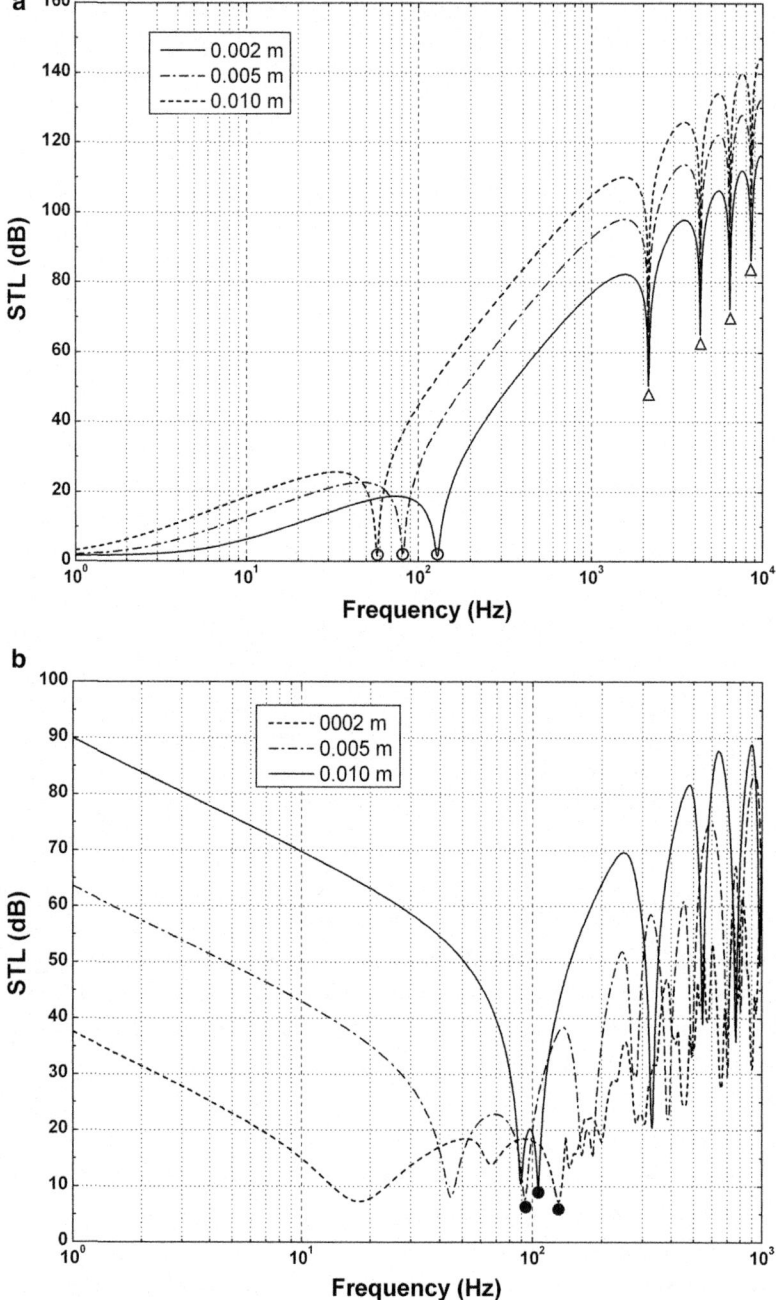

Fig. 1.18 Effects of panel thickness on STL of clamped double-panel partition under normal sound excitation: (**a**) infinite system ($H = 0.08$ m), ○: mass-air-mass resonance; △: standing-wave resonance; (**b**) finite system (1 m × 1 m, $H = 0.08$ m), ●: panel-cavity-panel resonance. Three different panel thicknesses (i.e., 2, 5, and 10 mm) were considered (With permission from Acoustical Society of America)

For the same reason, the positions of panel-cavity-panel resonances in Fig. 1.18b move to higher frequencies as the panel thickness is increased, in contrast to what is observed for the infinite system (Fig. 1.18a).

1.2.6 Effects of Air Cavity Thickness on STL

In order to explore the effects of air cavity thickness on STL and identify which mode (including panel mode and system coupling mode) is primarily responsible for each dip in the STL versus frequency curve of a clamped double-panel partition, the STLs are calculated for both infinite and finite systems with selected values of air cavity thickness ($H = 0.04, 0.06$, and 0.08 m), as shown in Fig. 1.19. Again, normal sound excitation is imposed, with 1 m (length) × 1 m (width) and $h_1 = h_2 = 0.01$ m for the finite system and $h_1 = h_2 = 0.002$ m for the infinite system. Both panels are made of aluminum (Table 1.1).

For the infinite system, as previously mentioned, the behavior of individual panels can be neglected, and only the system coupling effects (mass-air-mass resonance and standing-wave resonance) are of primary concern; see Fig. 1.19a. The positions of both mass-air-mass (or panel-cavity-panel) resonance and standing-wave resonances shift to lower frequencies with increasing air cavity thickness (Fig. 1.19a). Conversely, for the finite system, as can be seen from Fig. 1.19b, the position of the first dip is independent of air cavity thickness since it is completely addressed by the panel mode. However, the position of the second dip alters drastically as the air gap thickness is increased (in fact, moves to a lower frequency; Fig. 1.19b), owing to the fact that the system coupling effect (i.e., panel-cavity-panel resonance) plays a dominant role in this case. For the remaining dips exhibiting in Fig. 1.19b, their locations remain largely unchanged, implying that the panel mode and the system coupling effect are active in synchronization. Therefore, by tailoring the thickness of air cavity, it is possible to design finite double-panel partitions with better sound insulation properties over a wide frequency range (Fig. 1.19b). In comparison, the influence of air cavity on infinite systems is relatively small (Fig. 1.19a).

1.2.7 Effects of Incident Angles on STL

The influence of sound incident angles (elevation angle and azimuth angle) on the sound insulation property of a clamped double-panel partition of finite extent (1 m × 1 m, $H = 0.08$ m, and $h_1 = h_2 = 0.005$ m; both panels made of aluminum) is shown in Fig. 1.20.

The results of Fig. 1.20a demonstrate considerable influence of the incident elevation angle φ (with azimuth angle fixed at $\theta = 0°$) on the structural STL of the clamped finite system: the double-panel partition has marked selectivity for incident

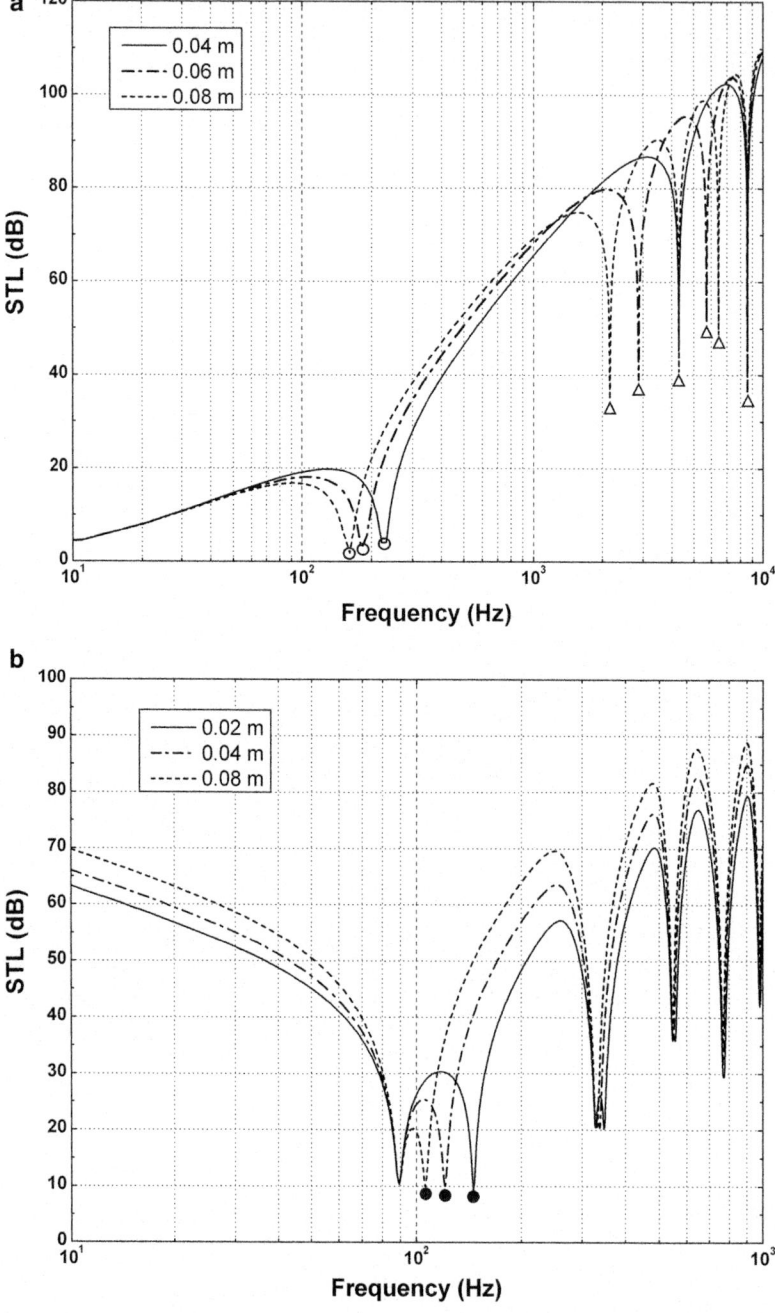

Fig. 1.19 Effects of air cavity thickness ($H = 0.04$, 0.06, and 0.08 m) on STL of clamped double-panel partition under normal sound excitation: (**a**) infinite system ($h_1 = h_2 = 0.002$ m), ○: mass-air-mass resonance; △: standing-wave resonance; (**b**) finite system (1 m × 1 m, $h_1 = h_2 = 0.010$ m), ●: panel-cavity-panel resonance (With permission from Acoustical Society of America)

1.2 Clamped Finite Double-Panel Partitions

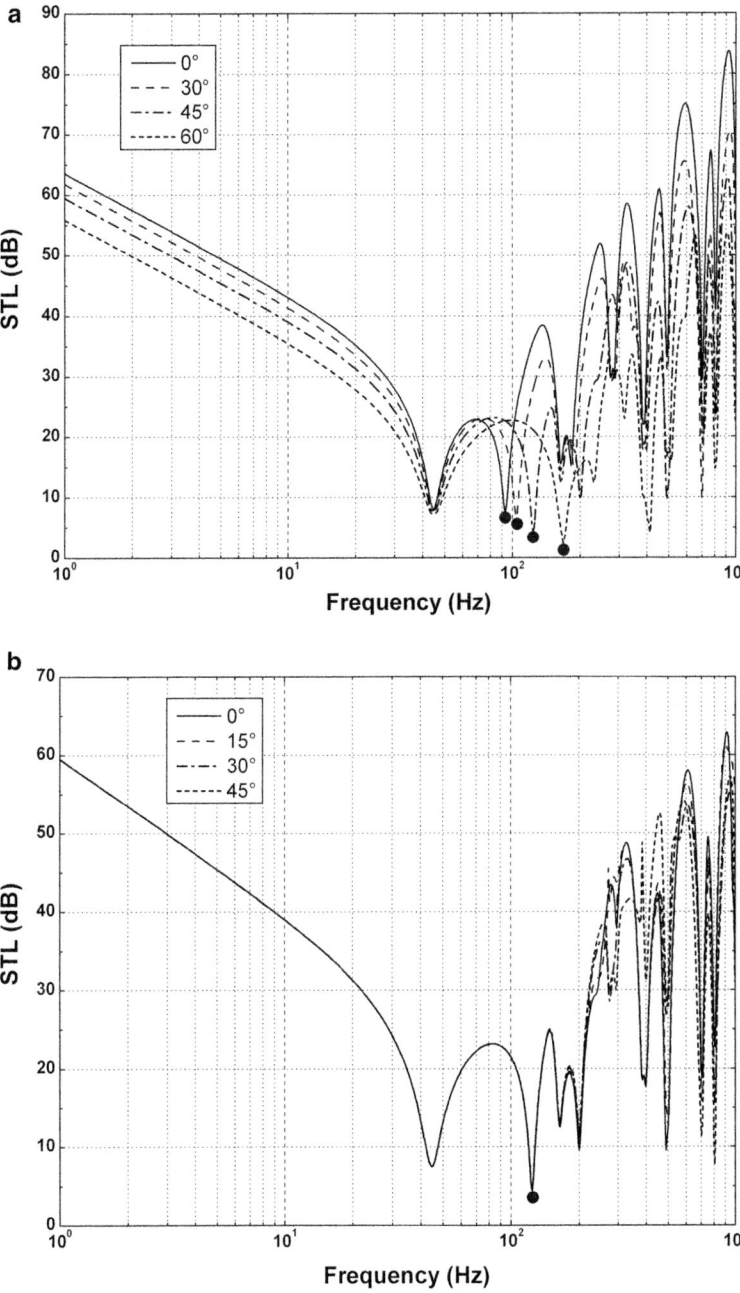

Fig. 1.20 STL plotted as a function of frequency for clamped double-panel partition of finite extent (1 m × 1 m, $H = 0.08$ m, $h_1 = h_2 = 0.005$ m) under sound excitation with: (**a**) varying elevation angle φ (azimuth angle fixed at $\theta = 0°$); (**b**) varying azimuth angle θ (elevation angle fixed at $\varphi = 45°$). ●: panel-cavity-panel resonance (With permission from Acoustical Society of America)

sound waves with different incident elevation angles. Generally speaking, incident sound waves with large elevation angles are easier to transmit through the double-panel structure than those with smaller elevation angles. For the case studied, the STL values decrease with increasing elevation angle for frequencies below about 43 Hz, while for frequencies above this value, the overall trend is similar apart from the complicated system modal behavior. In addition, it is worthwhile to note that the panel-cavity-panel resonance frequency increases with the elevation angle, which can be approximately described by a simple formula [25]:

$$f_\alpha = \frac{1}{2\pi \cos\varphi} \sqrt{\frac{\rho_0 c_0^2}{H} \frac{(m_1 + m_2)}{m_1 m_2}} \qquad (1.78)$$

The sound insulation properties of the clamped finite double-panel partition for selected incident azimuth angles ($\theta = 0°, 15°, 30°, 45°$) are compared in Fig. 1.20b, with the elevation angle fixed at $\varphi = 45°$. For frequencies below about 210 Hz, the STL curves for the four cases studied fall into one master curve. Beyond this frequency, small variations with varying azimuth angle are observed, caused by more complex structural modal behavior of the system at relatively high frequencies. Therefore, it may be concluded that the incident azimuth angle has negligible influence on the structural STL behavior of clamped finite systems. In other words, the selectivity of clamped double-panel partitions of finite extent for sound waves to different incident azimuth angles is limited.

1.2.8 Conclusions

An analytical model has been developed for studying the vibroacoustic behavior of a finite double-panel partition clamp mounted in an infinite acoustic rigid baffle. The method of modal function is used to simulate the clamped boundary conditions, and a double analytical series solution of the dynamic structural response is obtained using the weighted residual (Galerkin) method. The validity of model predictions is checked against existing experimental data, with good agreement achieved. The influence of several key system parameters on the sound insulation capability of clamped double-panel partitions is then systematically explored, including panel dimensions, thickness of air cavity, and elevation angle and azimuth angle of incidence sound.

For relatively high frequencies (above 100 Hz for the cases considered here), an infinite double-panel system sets an upper bound on the STL of clamped finite systems. For a finite double-panel system, the original modal behavior of individual panels (radiating panel in particular) interacts strongly with the system behavior (including panel-cavity-panel resonance and standing-wave resonance) and hence plays a major role in dictating the vibroacoustic performance of the whole system. For an infinite double-panel system, however, the original modal behavior of the

1.2 Clamped Finite Double-Panel Partitions

panels has no influence on the overall system behavior. And in the low-frequency range, the infinite system is incapable of providing the right STL values for practical finite systems.

Analogous to the mass law for single-leaf partitions, the increase of panel thickness (equivalent to increasing surface mass) considerably enhances the sound insulation property of clamped double-panel partitions. The influence of panel thickness on STL is particularly strong for finite systems at low frequencies, which is useful when designing clamped soundproof double-panel partitions.

As the thickness of air cavity is increased, the frequencies corresponding to the mass-air-mass (or panel-cavity-panel) resonance and standing-wave resonance change significantly. Moreover, by increasing the air cavity thickness, it is possible for the finite system to achieve better sound insulation capability in a wide frequency range. In comparison, air cavity thickness only has minimal influence on the infinite system.

Incident sound waves with bigger elevation angles are easier to transmit through a clamped double-panel structure than those with smaller elevation angles. The incident azimuth angle on the other hand has negligible effect on the structural STL of the system.

The proposed model is suitable for double-panel systems of finite or infinite extent and is applicable for both low- and high-frequency ranges. With these merits, it compares favorably with a number of other approaches, e.g., FEM (finite element method), BEM (boundary element method), and SEA (statistical energy analysis) method.

1.2.9 Sound Transmission Measurements

1.2.9.1 Experimental Setup

To validate the proposed theoretical model, *STL* measurements for fully *clamped* and simply supported double-panel partitions are separately carried out. The experimental setup is schematically illustrated in Fig. 1.21, while Fig. 1.22 presents more details for the clamped case. The transmission loss measurements are performed by utilizing a pressure method [58, 59]. Two condenser microphones (Knowles: FG-23742-150, diameter $d = 2.59$ mm) are located on the incident side and the radiated side, respectively, with the same distance 20 cm from the corresponding panel at the centerline of the structure. The microphone located on the radiated side can give a representative sound pressure level of the whole radiated field only when the microphone is placed normal to the planar structure [60]. The tested Al double panels are parallel and clamp (or simply) mounted on a large sandwich panel composed of two steel panels with thickness 2 mm each and heavy asbestos blanket in between as the core. As the steel sandwich panel has much superior sound insulation capability than the tested double-panel partition and is significantly larger than the Al double-panel in size, it may be regarded as an infinite acoustic rigid

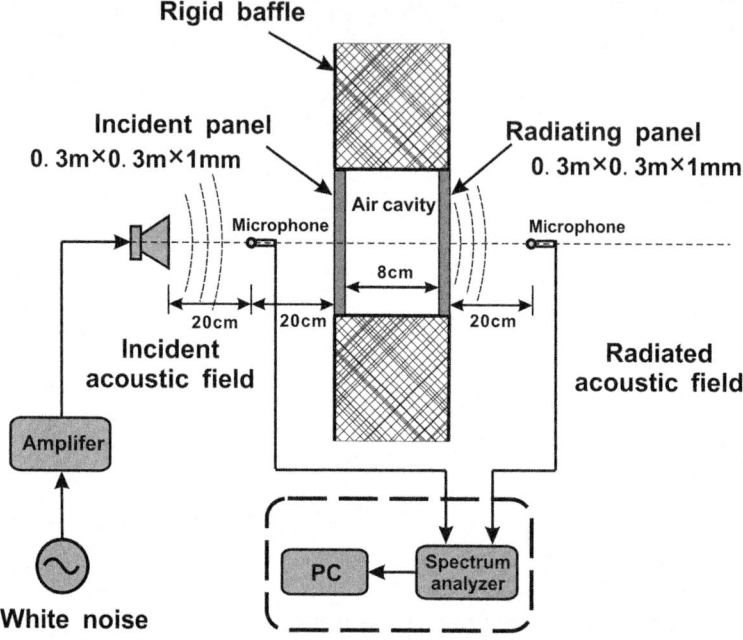

Fig. 1.21 Schematic of experimental setup for *STL* measurements of fully simply supported or fully clamped double-panel partition (With permission from Acoustical Society of America)

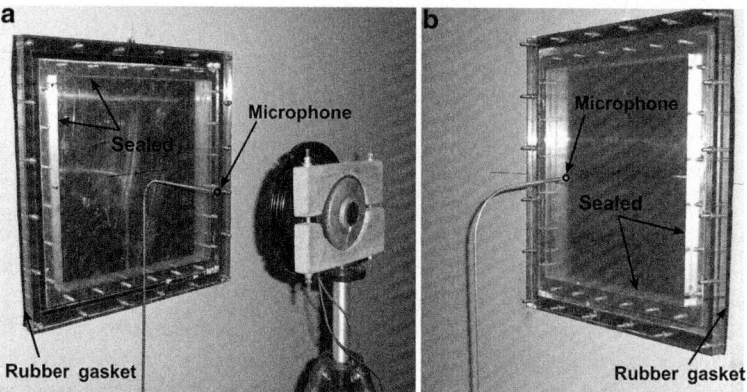

Fig. 1.22 Experimental setup for *STL* measurements of fully clamped double-panel partition: (**a**) incident side; (**b**) transmitted side (With permission from Acoustical Society of America)

baffle. The source room and the receiving room are both semi-anechoic, enabling as ideal a normal incident sound as possible in the source room and the measuring of the sound pressure completely radiated from the tested structure in the receiving room.

1.2 Clamped Finite Double-Panel Partitions

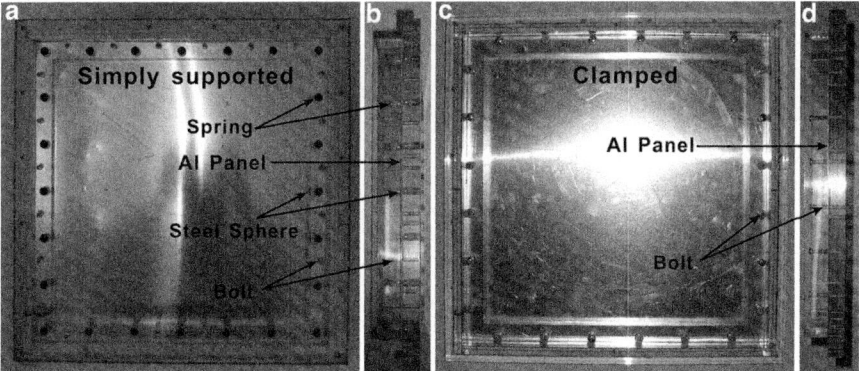

Fig. 1.23 Two typical mounted cases: simply supported case and clamped case, (**a**) front view of the simply supported fixture; (**b**) side view of (**a**); (**c**) front view of the clamped fixture; (**d**) side view of (**c**) (With permission from Acoustical Society of America)

The practical implementation of the two different boundary conditions is demonstrated in Fig. 1.23. Both the simply supported and the clamped fixtures for mounting the test panels are plexiglass made and are firmly fastened onto the rigid sandwich panel with bolts around the perimeters; rubber gaskets are used as intermediates to minimize sound leaking during the measurements (see Fig. 1.22). Figure 1.23a and b give the front view and the side view of the simply supported fixture, while Fig. 1.23c and d present the front view and the side view of the clamped fixture, respectively.

To implement the simply supported boundary condition, the Al panel is mounted around its perimeter on both sides by a set of steel spheres backed by elastic springs, and the two plexiglass frames holding the panel in between are directly bolted, as shown in Fig. 1.23a, b. Since the steel spheres can freely rotate without any restraint, the edges of the panel can also freely rotate but are restrained in the radial direction of the spheres, which represents the simply supported boundary condition. For the clamped boundary condition, each Al panel is directly held in between the two plexiglass frames, with each frame firmly fastened by bolts. All the cracks (if any) around the perimeter are sealed using adhesive glass cement (which for narrow cracks provides a seal equivalent to caulking), as shown in Fig. 1.22.

Sound transmission loss of the double-panel partition system is measured for two different cases, i.e., fully clamped and simply supported. The two identical Al panels with length $a = 0.3$ m, width $b = 0.3$ m, and thickness $h = 1$ mm are separated by an air cavity of depth $H = 8$ cm, as shown in Fig. 1.21. With a loudspeaker (diameter 20 cm) generating a white noise normally incident on the incident panel, the sound pressures on both sides of the partition system are measured by condenser microphones and analyzed by a dual-channel dynamical signal analyzer (HP: 35670A).

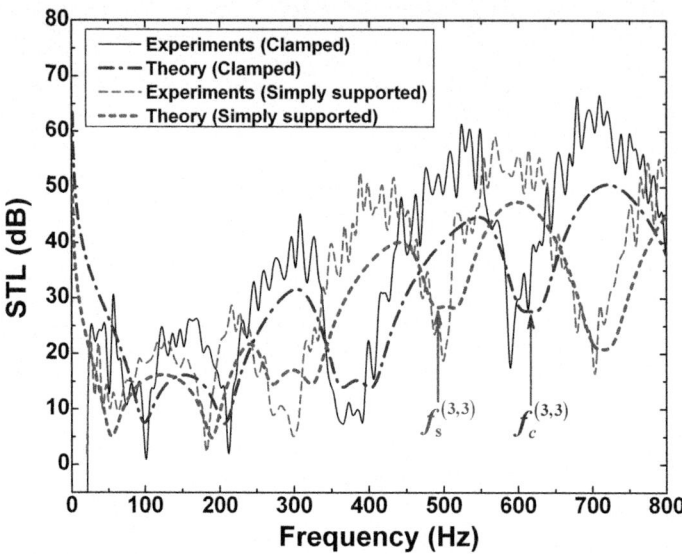

Fig. 1.24 *STL* plotted as a function of frequency for fully clamped and fully simply supported Al double-panel partitions in the case of normal sound incidence. The (3, 3) mode resonance dips for the simply supported double panel and the clamped double panel are particularly denoted by arrows at $f_s^{(3,3)} = 489$ Hz and $f_c^{(3,3)} = 619$ Hz, respectively (With permission from Acoustical Society of America)

1.2.9.2 Experimental Results and Model Validation

Figure 1.24 presents the measured *STL* value as a function of incident frequency (0–800 Hz) for both fully clamped and fully simply supported boundary conditions; for comparison, theoretical predictions obtained with the present model for each type of boundary condition are also included.

It is seen from Fig. 1.24 that, as far as the *STL* tendency with varying frequency is concerned, the agreement between the theoretical predictions and experimental measurements is good. The discernible discrepancies between the theory and experiment can be attributed to a number of factors, such as the imperfect normal plane sound wave, the uneven panel thickness, and the inevitable structural flanking transmission paths [26]. Note also that the experimental results at frequencies below 50 Hz are not reliable because the flanking transmission paths of the test facility play a prominent role in this frequency range [5, 6, 26].

The results of Fig. 1.24 clearly demonstrate the significant influence of boundary conditions on the transmission loss of a double-panel partition. The intense peaks and dips in the *STL* versus frequency curve reflect the inherent modal behaviors of the double-panel system. The *STL* dips in the simply supported case are shifted to lower frequencies in comparison with those of the *clamped* case, implying the fact that the natural frequencies of the simply supported system are lower than their

1.2 Clamped Finite Double-Panel Partitions

counterparts of the *clamped* system. It should be pointed out that the *STL* dips (apart from the second dips in the two theoretical curves) are dominated by the modal behavior of the radiating panel. It has been established that the second dips are associated with the "plate-cavity-plate" resonance [9, 36], which is insensitive to the imposed boundary conditions.

The research by Carneal and Fuller [26] on transmission loss across double-panel partitions should be mentioned here. The proposed theoretical model by Carneal and Fuller was established on simply supported boundary condition, whereas their experimental measurements were performed on two clamped plates. In order to experimentally verify the model, the stiffness of the simply supported plate was increased artificially by a factor of $\sqrt{2}$ for each boundary to approximate the clamped boundary condition. This assumption was also used to predict the natural frequencies of the clamed double-panel system.

Although the simply supported mode shapes may be a reasonable approximation of the *clamped* mode shapes [61], the natural frequencies associated with the simply supported boundary condition are lower, as the clamped condition provides a more rigorous constraint on panel vibration. To account for this increased constraint, simply increasing the panel stiffness by a factor of $\sqrt{2}$ as suggested by Carneal and Fuller may not be widely feasible, because the increased panel stiffness is determined by many parameters, such as panel dimensions, material properties, and incident sound frequency. In order to give a straightforward understanding of the distinctions between the different boundary conditions (i.e., clamped, simply supported, and modified simply supported by increasing panel stiffness), a comparison of the *STL* versus frequency curves obtained theoretically using the three different boundary conditions is presented in Fig. 1.25. The *STL* values predicted by the modified simply supported model are closer to the clamped model predictions when $f > 250$ Hz, while they agree better with the simply supported model predictions when $f < 250$ Hz. This suggests that by increasing the panel stiffness based on simply supported boundary conditions to emulate the clamped boundary condition may be feasible only when the frequency is sufficiently high.

Therefore, in subsequent studies of sound transmission through double-panel partitions, the modal functions of Eq. (1.57) that directly satisfy the clamped boundary condition (i.e., $w = 0$ and $\partial w/\partial n = 0$, which is different from the simply supported condition $w = 0$ and $\partial^2 w/\partial n^2 = 0$, where n denotes the outward vector on the edges) are applied. In a closely related study on sound transmission across double-panel partitions of finite extent [9], we have demonstrated that the correct implementation of the clamped boundary condition has noticeable superiority than the approximate approach adopted by Carneal and Fuller in view of the accurate prediction of the *STL* versus frequency curves and the natural frequencies.

To explore the boundary effects further, the typical (3, 3) mode behavior of a fully clamped double-panel partition is compared in Fig. 1.26 with that of a simply supported one. The (3, 3) mode natural frequency of the fully simply supported double-panel system occurs at $f_s^{(3,3)} = 489$ Hz, while it shifts to $f_c^{(3,3)} = 619$ Hz

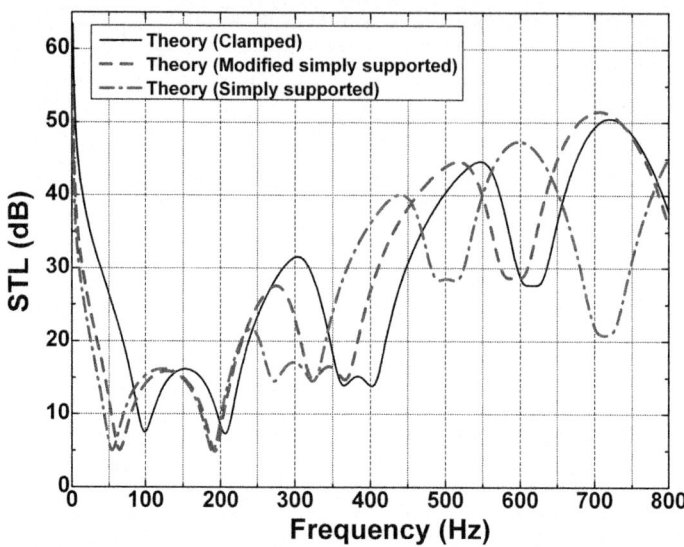

Fig. 1.25 Comparison of *STL* values obtained theoretically for three theoretical boundary conditions (i.e., clamped, simply supported, and modified simply supported) in the case of normal sound incidence (With permission from Acoustical Society of America)

when fully clamped. The corresponding (3, 3) mode shapes of the incident and radiating panels are presented in Fig. 1.26a1 and a2 for the fully simply supported case, while those for the fully clamped case are shown in Fig. 1.26b1 and b2. First, it is observed that the incident panel and the radiating panel vibrate in a symmetrical way (out-of-phase) for both cases; see Fig. 1.26a1 and b1, or Fig. 1.26b1 and b2. Symmetric motion (with respect to the symmetry plane running through the center of the partition) means that the panels move in breathing motion, both in or both out at a given position and a given time. Secondly, although the panel mode shapes at different boundary conditions exhibit similar forms (see Fig. 1.26a1 and b1, or Fig. 1.26a2 and b2), discernible discrepancies can be observed at panel edges, especially for the counterparts of Fig. 1.26a2 and b2. These differences at the panel edges reflect the boundary effects, i.e., the requirement that $\partial w/\partial n = 0$ for the clamped condition and that $\partial^2 w/\partial n^2 = 0$ for the simply supported condition.

To better differentiate the clamped case from the simply supported one, the predicted fundamental frequency coefficient $\Omega_{00} = \sqrt{\rho h/D}\omega_{00}a^2$ of a single-leaf panel either fully clamped or simply supported around its edges in vacuum is compared in Table 1.2. The significant alteration of the fundamental frequency coefficient from one case to another confirms once more the remarkable differences of the two types of boundary condition.

1.2 Clamped Finite Double-Panel Partitions

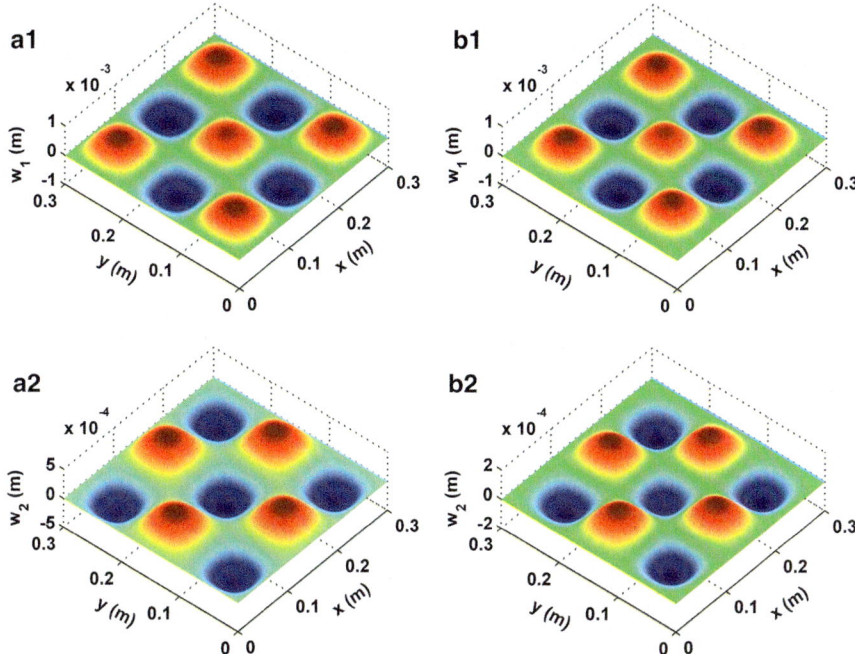

Fig. 1.26 Panel deflection mode shapes under normal sound excitation at frequency $f_s^{(3,3)} = 489$ Hz for fully simply supported case and $f_c^{(3,3)} = 619$ Hz for fully clamped case, where the responses are controlled by the (3, 3) natural mode: (**a1**) simply supported incident panel; (**a2**) simply supported radiating panel; (**b1**) clamped incident panel; (**b2**) clamped radiating panel (With permission from Acoustical Society of America)

Table 1.2 Comparison of the fundamental frequency coefficient $\Omega_{00} = \sqrt{\rho h/D}\omega_{00}a^2$ for a fully clamped single-leaf panel with that of fully simply supported

Boundary condition	$\lambda = a/b$	Leissa [62]	Laura et al. [63]	Present study
S-S-S-S[a]	1.0	19.73	19.74	19.74
C-C-C-C[a]	1.0	35.99	35.99	35.99

[a]"S-S-S-S" and "C-C-C-C" represent fully simply supported and fully clamped on four edges of the panel, respectively

1.2.10 Relationships Between Clamped and Simply Supported Boundary Conditions

Numerical studies are performed in this section to explore further the relationships between the two different boundary conditions and the significant influence of the boundary condition on the sound insulation properties of double-panel structures in terms of frequency characteristic curves (i.e., *STL* versus frequency curves) and panel vibration behaviors.

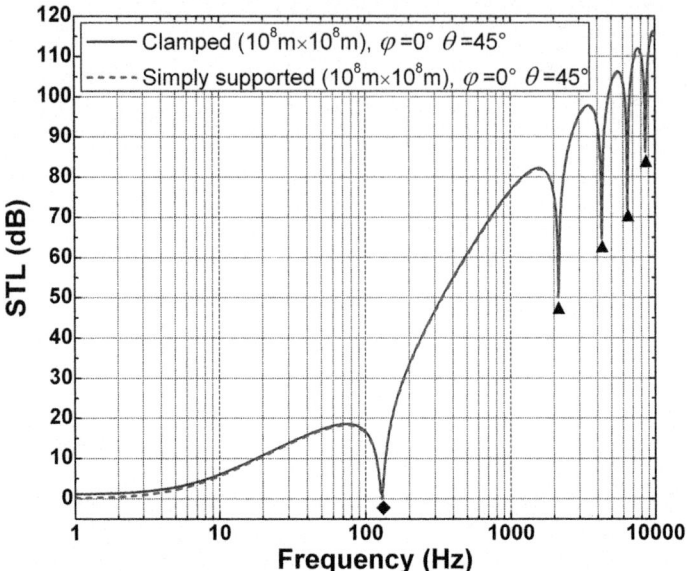

Fig. 1.27 *STL* of infinitely large double-panel partition theoretically obtained separately with clamped boundary and simply supported boundary. ♦: mass-air-mass resonance; ▲: standing-wave resonance (With permission from Acoustical Society of America)

1.2.10.1 Consistency of Clamped Model and Simply Supported Model

To check the consistency of the clamped model and the simply supported model, the panel dimensions are extended to infinite so that the constraint effect of the boundary conditions vanishes. The two models based on the clamped modal function and the simply supported modal function, respectively, are also applicable for infinitely large structures if sufficiently large values of panel length a and panel width b are taken in numerical calculations. The corresponding *STL* results for the two models with $a = 10^8$ m and $b = 10^8$ m assumed are shown in Fig. 1.27. As anticipated, an excellent agreement is achieved between the two models, as the effect of boundary condition becomes negligible (in other words, no boundaries exist) when the structure is extended to infinite.

The mass-air-mass resonance dip in Fig. 1.27 is marked by the symbol ♦, which is a unique phenomenon owned by the double-panel system and can be approximately predicted by the formula [9]

$$f_\alpha = \frac{1}{2\pi \cos\varphi} \sqrt{\frac{\rho_0 c_0^2}{H} \frac{(m_1 + m_2)}{m_1 m_2}} \qquad (1.79)$$

The standing-wave resonance dips labeled by the symbol ▲ occur when the depths of the air gap in between the two panels are integer numbers of half

1.2 Clamped Finite Double-Panel Partitions

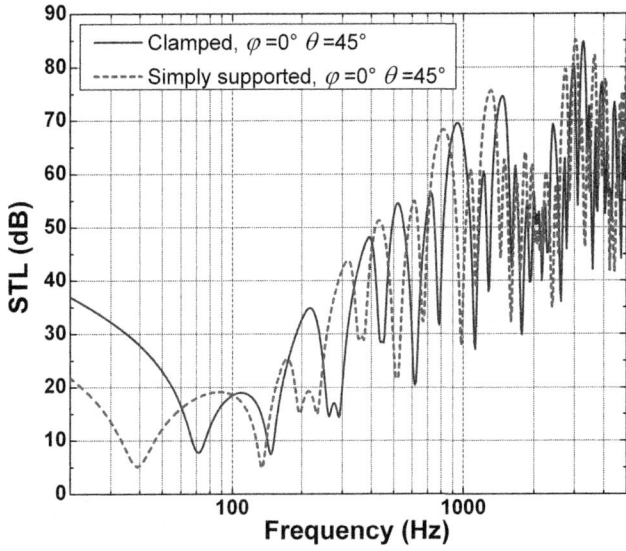

Fig. 1.28 Predicted *STL* of finite (0.5 m × 0.5 m) double-panel partition plotted as a function of frequency for incident sound with elevation angle $\varphi = 0°$ and azimuth angle $\theta = 45°$ (With permission from Acoustical Society of America)

wavelength of the incident sound. The corresponding resonance frequencies can be obtained by Wang et al. [18]:

$$f_{s,n} = \frac{nc_0}{2H} \quad (n = 1, 2, 3\ldots) \tag{1.80}$$

The theoretical predictions given in Fig. 1.27 agree excellently well with Eqs. (1.79) and (1.80).

1.2.10.2 Effects of Different Boundary Conditions on *STL*

In this section, the *STL* behaviors of the rectangular double-panel partition predicted by the two models based separately on the clamped boundary condition and the simply supported boundary condition are compared for different cases, e.g., sound incident with different elevation angles ($\varphi = 0°, 30°, 60°$) and fixed azimuth angle of $\theta = 45°$. The results are presented in Figs. 1.28, 1.30, and 1.32. Here, the incident azimuth angle is fixed at 45° for the purpose of fully exciting the relevant panel vibration modes.

The typical (3, 3) modal shapes of panel deflection associated with the cases of Figs. 1.28, 1.30, and 1.32 are presented in Figs. 1.29, 1.31, and 1.33, respectively. Note that although the mode shapes in Fig. 1.29 look similar to those shown in Fig. 1.26, different natural frequencies ($f_s^{(3,3)} = 352$ Hz for the simply supported

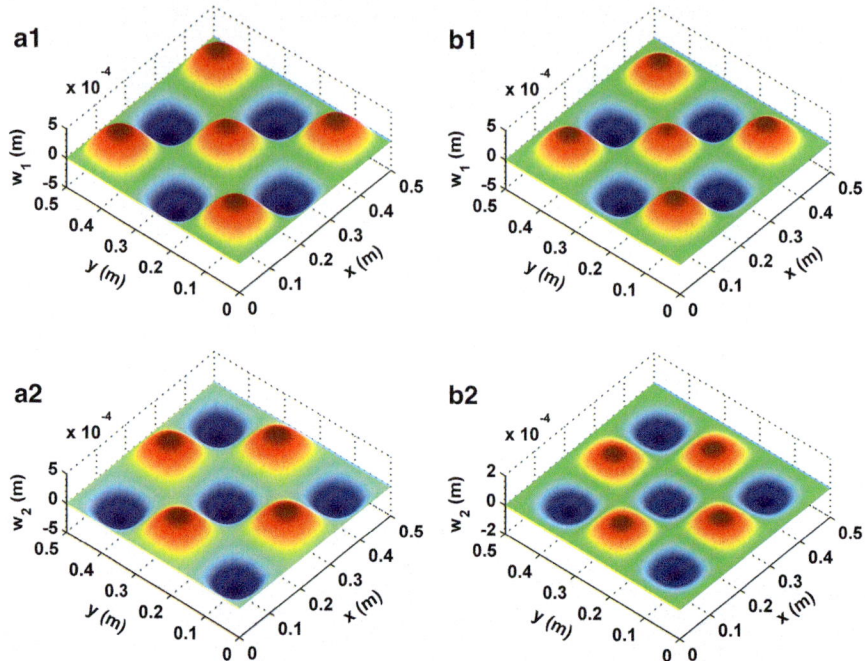

Fig. 1.29 Panel deflection mode shapes under normal sound excitation ($\varphi = 0°$, $\theta = 45°$) at frequency $f_s^{(3,3)} = 352$ Hz for fully simply supported case and $f_c^{(3,3)} = 446$ Hz for fully clamped case, where the responses are controlled by the (3, 3) natural mode: (**a1**) simply supported incident panel; (**a2**) simply supported radiating panel; (**b1**) clamped incident panel; (**b2**) clamped radiating panel (With permission from Acoustical Society of America)

case and $f_c^{(3,3)} = 446$ Hz for the clamped case) are obtained due to the different panel dimensions considered here (i.e., 0.5 m × 0.5 m × 2 mm).

As mentioned above, except for the second dip (i.e., the "plate-cavity-plate" resonance) which is insensitive to boundary conditions as seen in Figs. 1.28, 1.30, and 1.32, other dips determined mainly by the natural vibration of the radiating panel are significantly shifted when the boundary conditions are changed. Additionally, for the three considered cases, it can be seen that the *STL* values of the clamped system are distinctly higher than those of the simply supported system in the lower frequency range. For the higher frequency range, however, all the three cases show different trends. For example, for the case of elevation angle $\varphi = 0°$, the *STL* values obtained with the two different boundary conditions have overall the same order of magnitude, although the resonance dips are not in accord with each other. By comparing the three plots in succession, it can be seen that as the elevation angle is increased, the discrepancies between the *STL* values obtained with different boundary conditions increase.

Taking a whole view of the results shown in Figs. 1.29, 1.31, and 1.33, one can see that the mode shapes of the simply supported panels can be approximated

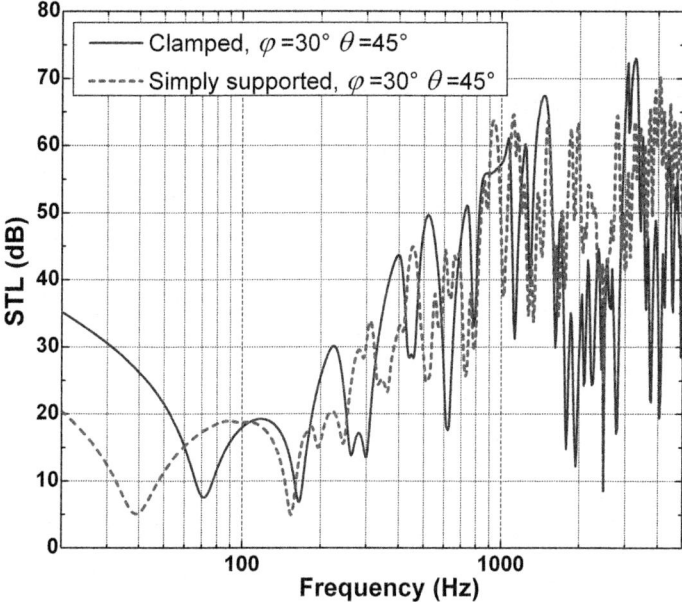

Fig. 1.30 Predicted *STL* of finite (0.5 m × 0.5 m) double-panel partition plotted as a function of frequency for incident sound with elevation angle $\varphi = 30°$ and azimuth angle $\theta = 45°$ (With permission from Acoustical Society of America)

as the corresponding clamped panel mode shapes only for the case of normal sound incidence (see Fig. 1.29). For oblique sound incidence, the mode shapes of the simply supported panels are dramatically different from those of the clamped panels (see Figs. 1.31 and 1.33). The asymmetric sound incidence (i.e., oblique sound incidence, $0° < \varphi < 90°$) induces asymmetric mode shapes of the simply supported panels (symmetric only about the incident plane), while the mode shapes of the clamped panels remain highly symmetric, which confirms the more rigorous restraint of the fully clamped condition on the movement of panel edges than that of the simply supported condition.

1.2.11 Conclusions

An analytical approach has been developed to investigate the influence of boundary constraints (fully clamped versus fully simply supported) on the sound insulation performance of a finite double-panel structure containing an air cavity. The theory is built upon the vibration responses of the two panels coupled by the air cavity. Experimental measurements are subsequently performed to validate the theoretical predictions, with good overall agreement achieved for both types of the boundary

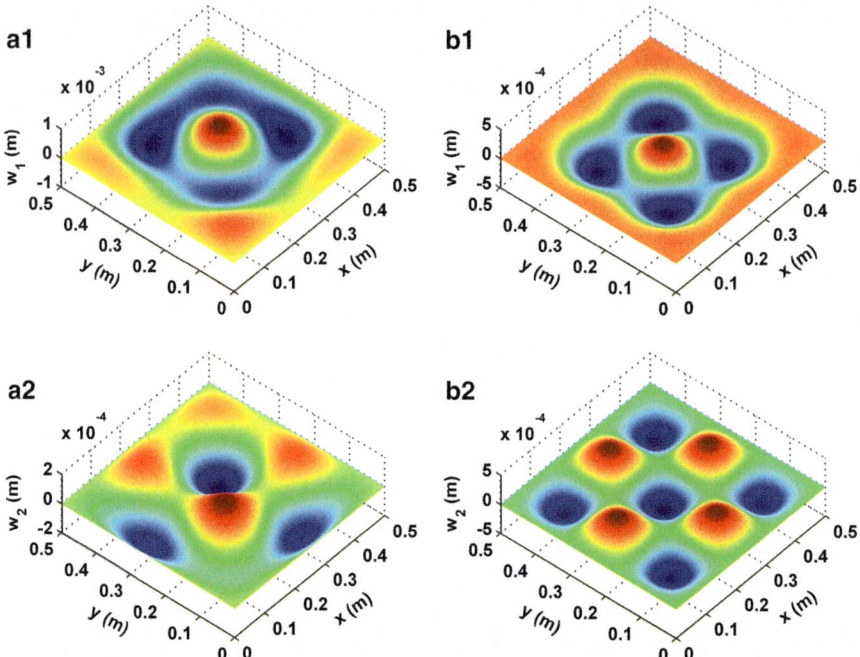

Fig. 1.31 Panel deflection mode shapes under oblique sound excitation ($\varphi = 30°, \theta = 45°$) at frequency $f_s^{(3,3)} = 352$ Hz for fully simply supported case and $f_c^{(3,3)} = 446$ Hz for fully clamped case, where the responses are controlled by the (3, 3) natural mode: (**a1**) simply supported incident panel; (**a2**) simply supported radiating panel; (**b1**) clamped incident panel; (**b2**) clamped radiating panel (With permission from Acoustical Society of America)

condition. The inherent consistency of the two models based separately on the clamped boundary condition and the simply supported boundary condition is confirmed when the panel dimensions become infinitely large. The model is then used to systematically explore the effects of different boundary conditions on the sound isolation capability of double-panel partitions, in terms of both the *STL* versus frequency plots and the (3, 3) mode shapes of panel vibration.

The comparison of the *STL* versus frequency plots obtained with the two different boundary conditions suggests that the natural frequencies of a fully clamped double-panel partition are higher than those of a fully simply supported one (except for the "plate-cavity-plate" resonance frequency). This is attributed to the more rigorous constraint provided by the clamped condition than that by the simply supported condition, which is equivalent to increasing the panel stiffness. However, to account for this increased panel stiffness by an artificial factor (e.g., $\sqrt{2}$) is not widely feasible, as the increased panel stiffness is determined by many parameters such as panel dimensions, material properties, and incident sound frequency.

Obtained results at oblique sound incidence for the two different boundary condition cases suggest that, at the lower frequency range, the *STL* values show

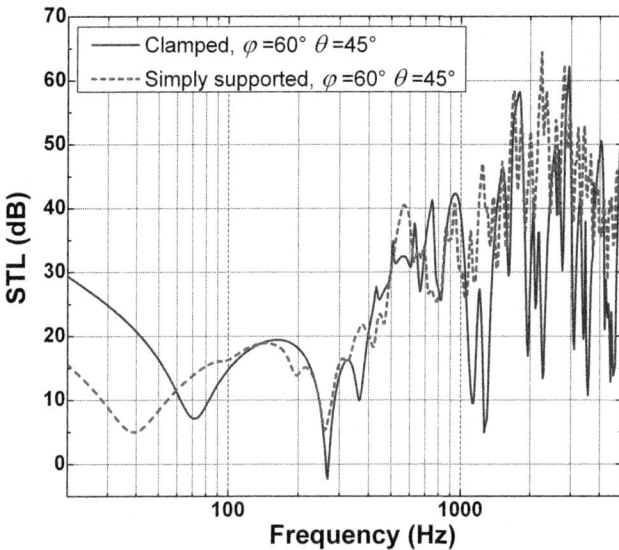

Fig. 1.32 Predicted *STL* of finite (0.5 m × 0.5 m) double-panel partition plotted as a function of frequency for incident sound with elevation angle $\varphi = 60°$ and azimuth angle $\theta = 45°$ (With permission from Acoustical Society of America)

noticeable discrepancies for the two cases, while at the higher frequency range, the discrepancies depend on the incident elevation angle. While the vibration mode shapes of a simply supported panel can be approximated as those of its clamped counterpart only in the case of normal sound incidence, dramatic distinctions exist between them in the case of oblique sound incidence.

1.3 Clamped Finite Triple-Panel Partitions

1.3.1 Introduction

Recent developments in building, transportation, environmental, and other engineering applications have prompted research on finding innovative ways for noise reduction. The transmission loss characteristics of a customarily used single panel follow in general the mass law. The traditional methods for low-frequency noise reduction require therefore the use of heavy damping materials, leading to weight penalty and hence offsetting the performance gains of the panel. As a result, double-panel partitions are extensively used in modern buildings, transportation vehicles, aerospace/aeronautical fuselages, etc., which have superior sound insulation properties than single-panel partitions [3–7, 11, 64–67]. Since the introduction of an additional panel can significantly enhance the transmission loss performance of a

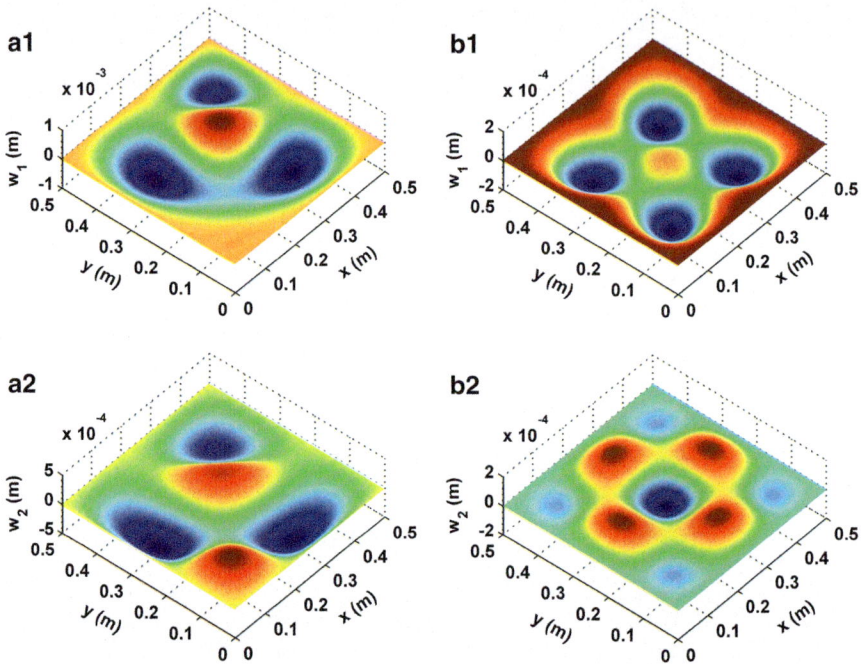

Fig. 1.33 Panel deflection mode shapes under oblique sound excitation ($\varphi = 60°, \theta = 45°$) at frequency $f_s^{(3,3)} = 352$ Hz for fully simply supported case and $f_c^{(3,3)} = 446$ Hz for fully clamped case, where the responses are controlled by the (3, 3) natural mode: (**a1**) simply supported incident panel; (**a2**) simply supported radiating panel; (**b1**) clamped incident panel; (**b2**) clamped radiating panel (With permission from Acoustical Society of America)

single panel, one would expect that a triple-panel partition constructed by adding another panel to the double-panel partition may lead to further gains in noise reduction. As a matter of fact, triple-panel partitions with separated air cavities have already been introduced as the standard configuration for glazing windows in some high-class buildings. We aim therefore in this research to develop an analytical model to investigate the sound transmission loss (STL) performance of a finite-sized triple-panel partition, which is clamp mounted on a rigid baffle and separated by two enclosed air cavities, and compare it with that of a clamped double-panel partition.

A significant amount of research has been devoted to developing accurate theoretical models of *STL* characteristics for single-panel [14, 15, 30, 39, 43, 46, 56, 68] and double-panel structures [3, 8, 16, 18, 28, 48, 69, 70]. Extensive experimental study [5, 6, 26, 34] has also been carried out and active control strategies [71–74] for noise reduction proposed.

Early research on interior noise reduction concentrated primarily on infinite structures, because the precise characterization of sound transmission across a finite-sized structure requires complex physical-mathematical treatment of the boundary

1.3 Clamped Finite Triple-Panel Partitions

conditions as well as the fluid-structure coupling effects, and the evaluation of the finite system response is more difficult due to the presence of violent peaks and dips on *STL* versus frequency curves.

As far as infinite structures are concerned, Beranek and Work [28] developed an early model of sound transmission through multiple structures containing flexible blankets based on the progressive impedance method. London [16] proposed a theory to deal with the transmission of reverberant sound through a double wall, also using the impedance method. The extension of Beranek's model to a random incidence field was carried out by Mulholland et al. [75]. The abovementioned research established on the progressive impedance method failed to fully account for the coincidence effect, which was later overcame by the analytical model of Antonio et al. [17]. An analysis method was developed by Lee and Kim [43] to study the sound transmission characteristics of a thin plate stiffened by equally spaced line stiffeners in terms of the space-harmonic approach. Wang et al. [18] further applied the space-harmonic approach to periodically connected double-leaf partitions. The transmission of structure-borne sound through a double-leaf structure with a porous absorptive layer inserted in the cavity was studied theoretically as well as experimentally by Yairi et al. [76], with the porous absorptive layer described using an electroacoustical equivalent model.

As for the sound transmission across finite-sized structures, several approaches have been employed to account for the edge conditions, such as FEM (finite element method) [36, 45, 77], BEM (boundary element method) [36, 78], SEA (statistical energy analysis) [44, 68, 79–81], modal expansion method [3, 35, 56], spatial windowing technique [23], and patch-mobility method [24]. In general, FEM is applied together with BEM to deal with the boundary and interface conditions. For example, for sound transmission through finite multilayer systems containing poroelastic materials, Panneton and Atalla [36] used the classical elastic and fluid elements to model the elastic and fluid media, but a BEM approach to account for the fluid-structure coupling effects. Sgard et al. [78] also employed a variational BEM approach to cope with the fluid loading effects. Based on two-dimensional FEM models, Del Coz Diaz et al. [77] established a methodology to predict the airborne sound insulation of building elements, which agreed well with their experimental measurements. Although FEM and BEM are well suited for low-frequency transmission loss calculations, they require high computation efforts when calculations over a wide frequency range (high frequencies in particular) are needed [41] and provide few physical insights. While SEA can be used as an alternative of FEM and BEM, for it is substantially more effective in providing sound transmission estimates for complex systems at high frequencies, it is seldom reliable at low frequencies because of the statistical uncertainties that prevail when there are only a few resonant modes present in the subsystems [9, 35]. Furthermore, analogous to FEM and BEM, SEA cannot provide detailed physical understanding of singular phenomena such as resonance. The modal expansion method based on simply supported edge constraints has been conventionally adopted to solve the vibroacoustic problem of finite systems on account of its excellent capability to deal with resonant modes. However, for clamped boundary conditions, this method

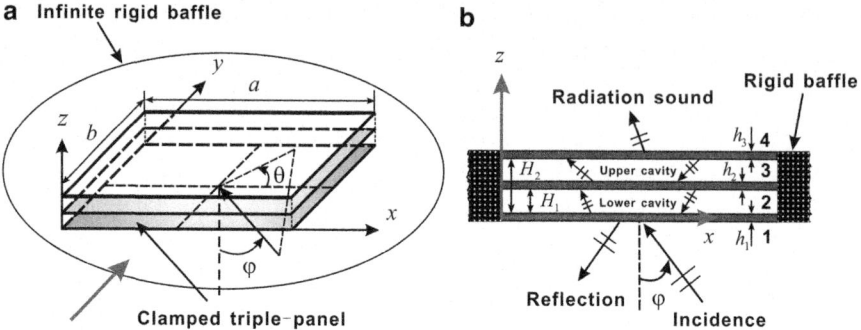

Fig. 1.34 Schematic of sound transmission through a clamp-mounted triple-panel partition: (**a**) overall view; (**b**) side view in the arrow direction in (**a**) (With permission from Elsevier)

is only approximate and needs special treatment for numerical convergence. The spatial windowing technique also suffers from the disadvantage of indeterminate boundary condition disposal and few physical insight gains. The patch-mobility method adopted by Chazot et al. [24] is essentially built upon the modal expansion method.

Although extensive theoretical research on sound transmission through finite and infinite multilayer systems has been carried out, there exists no analytical modeling of sound transmission across a finite-sized triple-panel construction with boundary constraints. A thorough literature search revealed that a few experimental studies [6, 82, 83] have concerned the transmission of sound through triple glazing windows. To squarely address this deficiency, the present research aims to develop an analytical model to investigate the detailed process of sound transmission across a clamped triple-panel partition of finite extent and separated by two impervious air cavities. The model is then used to validate one's expectation that a triple-panel partition would possess superior sound insulation properties than the widely used double-panel partition. Specifically, the "mass-spring" resonance, the standing-wave resonance, the modal behavior of the panel, and other interesting vibroacoustic phenomena are identified and interpreted on physical grounds, the asymptotic variation of the transmission loss from finite to infinite extent is illustrated, and a systematic parametrical study regarding the effects of panel thickness and air cavity depth is carried out.

1.3.2 Dynamic Structural Acoustic Formulation

We model theoretically the reflection, transmission, and radiation of a sound pressure wave propagating through a clamp-mounted rectangular triple-panel configuration (Fig. 1.34), with fluid-structure interaction and coupling effects duly accounted for. The structure is composed of three homogenous and isotropic panels

1.3 Clamped Finite Triple-Panel Partitions

and separated by two enclosed air cavities. A right-handed Cartesian coordinate system (x, y, z) is applied, with the x- and y-axes horizontally located on the surface of the incidence panel (bottom panel) and the z-axis pointing vertically upward, as shown in Fig. 1.34. In the defined coordinate system, the three panels are indexed as bottom, middle, and upper panels having thickness h_1, h_2, and h_3 and located at $z = 0$, H_1, and H_2, respectively (Fig. 1.34). The rectangular panels have identical dimensions $a \times b$ in the $x - y$ plane. The rigid baffle mounting the triple-panel partition is assumed to be infinitely large so as to exclude possible diffraction of sound around the baffle from the sound source side ($z < 0$) to the sound radiation/transmission side ($z > H_2$).

Under typical fluid-structure interface conditions, the uniform plane sound pressure varying harmonically in time constitutes a basic sound wave incidence on the bottom panel, characterized by the incident elevation angle φ and the incident azimuth angle θ, as shown in Fig. 1.34. Part of this disturbance is reflected back by the bottom panel, and the rest propagates consecutively through the bottom panel, the bottom cavity, the middle panel, and the top cavity to the upper panel and then is emitted by the vibrating upper panel. The model proposed for the above physical process is based upon the classical vibration theorem for thin flexural plates, which enables the three panels to possess different physical (e.g., Young's modulus, Poisson ratio, and loss factor) and geometrical (e.g., length and width) parameters, but requires the panels to be sufficiently thin relative to their length and width so that the shear displacement of a panel is much smaller than its lateral displacement.

The acoustic velocity potential for an obliquely incident uniform plane sound wave varying harmonically in time can be expressed as

$$\varphi = I e^{-j(k_x x + k_y y + k_z z - \omega t)} \tag{1.81}$$

where I is the amplitude, $j = \sqrt{-1}$, and ω is the angular frequency, and wavenumber components of the incident sound in x-, y-, and z-directions can be separately calculated as

$$k_x = k_0 \sin \varphi \cos \theta, \quad k_y = k_0 \sin \varphi \sin \theta, \quad k_z = k_0 \cos \varphi \tag{1.82}$$

where $k_0 = \omega/c_0$ is the wavenumber in air and c_0 is the sound speed in air.

As noted above, the three panels are modeled as classical thin plates so that, in terms of the thin plate vibration theorem, their motions under sound excitation are governed by

$$D_1 \nabla^4 w_1 + m_1 \frac{\partial^2 w_1}{\partial t^2} - j\omega \rho_0 (\Phi_1 - \Phi_2) = 0 \tag{1.83}$$

$$D_2 \nabla^4 w_2 + m_2 \frac{\partial^2 w_2}{\partial t^2} - j\omega \rho_0 (\Phi_2 - \Phi_3) = 0 \tag{1.84}$$

$$D_3 \nabla^4 w_3 + m_3 \frac{\partial^2 w_3}{\partial t^2} - j\omega\rho_0 (\Phi_3 - \Phi_4) = 0 \qquad (1.85)$$

where D_i, m_i, and w_i denote separately the panel flexural rigidity, surface density, and deflection, with $i = 1, 2, 3$ representing separately the bottom, middle, and top panels, ρ_0 is the air density, and Φ_α ($\alpha = 1, 2, 3, 4$) are the velocity potentials in the incident field, in the bottom and top air cavities, and in the transmitted field (Fig. 1.34), respectively.

In the context of harmonic sound excitation, the transverse displacements of the structure should also be time harmonic, which can be written as

$$w_i (x, y; t) = \sum_{m,n} \alpha_{i,mn} \varphi_{mn} (x, y) e^{j\omega t} \qquad (i = 1, 2, 3) \qquad (1.86)$$

where the modal functions (or, more strictly speaking, the basic functions) φ_{mn} take the following forms [12, 61, 84]:

$$\varphi_{mn} (x, y) = \left(1 - \cos \frac{2m\pi x}{a}\right) \left(1 - \cos \frac{2n\pi y}{b}\right) \qquad (1.87)$$

The velocity potentials of the acoustic field are associated with the local velocity by $\hat{u}_\alpha = -\nabla \Phi_\alpha$ and related to the acoustic pressure by $p_\alpha = \rho_0 \Phi_{\alpha,t} = j\omega\rho_0 \Phi_\alpha$. The velocity potentials can be described in terms of the modal function φ_{mn} as

$$\Phi_1 (\mathbf{r}; t) = \sum_{m,n} I_{mn} \varphi_{mn} (x, y) e^{-j(k_z z - \omega t)} + \sum_{m,n} \beta_{mn} \varphi_{mn} (x, y) e^{-j(-k_z z - \omega t)} \qquad (1.88)$$

$$\Phi_2 (\mathbf{r}; t) = \sum_{m,n} \varepsilon_{mn} \varphi_{mn} (x, y) e^{-j(k_z z - \omega t)} + \sum_{m,n} \zeta_{mn} \varphi_{mn} (x, y) e^{-j(-k_z z - \omega t)} \qquad (1.89)$$

$$\Phi_3 (\mathbf{r}; t) = \sum_{m,n} \xi_{mn} \varphi_{mn} (x, y) e^{-j(k_z z - \omega t)} + \sum_{m,n} \eta_{mn} \varphi_{mn} (x, y) e^{-j(-k_z z - \omega t)} \qquad (1.90)$$

$$\Phi_4 (\mathbf{r}; t) = \sum_{m,n} \mu_{mn} \varphi_{mn} (x, y) e^{-j(k_z z - \omega t)} \qquad (1.91)$$

where I_{mn} stands for the amplitudes of the incident sound wave; β_{mn} represents the amplitude of the reflected sound waves; ε_{mn} (ξ_{mn}) and ζ_{mn} (η_{mn}) denote the amplitudes of the positive- and negative-going waves in the bottom and top cavities, respectively; and μ_{mn} is the amplitudes of transmitted waves (i.e., the positive-going waves) in the transmitted domain (see Fig. 1.34).

1.3 Clamped Finite Triple-Panel Partitions

On the premise that the considered vibroacoustic problem involves fluid-structure interaction and mutual coupling, the velocity continuity condition (equivalent to the displacement continuity condition when the surrounding fluid is stationary with respect to the structure [85]) should be satisfied at the fluid-structure interface, so that

$$z = 0; \quad -\frac{\partial \Phi_1}{\partial z} = j\omega w_1, \quad -\frac{\partial \Phi_2}{\partial z} = j\omega w_1 \qquad (1.92)$$

$$z = H_1; \quad -\frac{\partial \Phi_2}{\partial z} = j\omega w_2, \quad -\frac{\partial \Phi_3}{\partial z} = j\omega w_2 \qquad (1.93)$$

$$z = H_2; \quad -\frac{\partial \Phi_3}{\partial z} = j\omega w_3, \quad -\frac{\partial \Phi_4}{\partial z} = j\omega w_3 \qquad (1.94)$$

Substitution of Eqs. (1.86) and Eqs. (1.88), (1.89), (1.90), and (1.91) into the above velocity continuity conditions leads to a set of simultaneous equations for the unknown coefficients β_{mn}, ε_{mn}, ζ_{mn}, ξ_{mn}, η_{mn}, and μ_{mn} as

$$-Ie^{-j(k_x x + k_y y)} + \sum_{m,n}\left[\beta_{mn} + \frac{\omega}{k_z}\alpha_{1,mn}\right] \times \varphi_{mn}(x,y) = 0 \qquad (1.95)$$

$$\sum_{m,n}[k_z(-\varepsilon_{mn} + \zeta_{mn}) + \omega\alpha_{1,mn}] \times \varphi_{mn}(x,y) = 0 \qquad (1.96)$$

$$\sum_{m,n}\left[k_z\left(-\varepsilon_{mn}e^{-jk_z H_1} + \zeta_{mn}e^{jk_z H_1}\right) + \omega\alpha_{2,mn}\right] \times \varphi_{mn}(x,y) = 0 \qquad (1.97)$$

$$\sum_{m,n}\left[k_z\left(-\xi_{mn}e^{-jk_z H_1} + \eta_{mn}e^{jk_z H_1}\right) + \omega\alpha_{2,mn}\right] \times \varphi_{mn}(x,y) = 0 \qquad (1.98)$$

$$\sum_{m,n}\left[k_z\left(-\xi_{mn}e^{-jk_z H_2} + \eta_{mn}e^{jk_z H_2}\right) + \omega\alpha_{3,mn}\right] \times \varphi_{mn}(x,y) = 0 \qquad (1.99)$$

$$\sum_{m,n}\left[-k_z\mu_{mn}e^{-jk_z H_2} + \omega\alpha_{3,mn}\right] \times \varphi_{mn}(x,y) = 0 \qquad (1.100)$$

Since the above simultaneous equations are valid at all values of x and y, they can be further simplified (except Eq. (1.95)) as

$$\varepsilon_{mn} = \frac{\omega\left(\alpha_{1,mn}e^{2jk_z H_1} - \alpha_{2,mn}e^{jk_z H_1}\right)}{k_z\left(e^{2jk_z H_1} - 1\right)} \qquad (1.101)$$

$$\zeta_{mn} = \frac{\omega\left(\alpha_{1,mn} - \alpha_{2,mn}e^{jk_z H_1}\right)}{k_z\left(e^{2jk_z H_1} - 1\right)} \qquad (1.102)$$

$$\xi_{mn} = \frac{\omega \left(\alpha_{3,mn} e^{jk_z H_1} - \alpha_{2,mn} e^{jk_z H_2}\right)}{k_z \left(e^{jk_z(H_1-H_2)} - e^{jk_z(H_2-H_1)}\right)} \quad (1.103)$$

$$\eta_{mn} = \frac{\omega \left(\alpha_{3,mn} e^{-jk_z H_1} - \alpha_{2,mn} e^{-jk_z H_2}\right)}{k_z \left(e^{jk_z(H_1-H_2)} - e^{jk_z(H_2-H_1)}\right)} \quad (1.104)$$

$$\mu_{mn} = \frac{\omega \alpha_{3,mn} e^{jk_z H_2}}{k_z} \quad (1.105)$$

1.3.3 The Principle of Virtual Work

To determine the unknown modal coefficients $\alpha_{1,mn}$, $\alpha_{2,mn}$, and $\alpha_{3,mn}$, the principle of virtual work is employed. The work exerted by the force on each particle that acts through an arbitrary virtual displacement (i.e., an arbitrary infinitesimal change in the position of the particle consistent with the constraints imposed on the motion of the particle) is given by

$$\delta w_i = \delta \alpha_{i,mn} \varphi_{mn}(x,y) \quad (i = 1, 2, 3) \quad (1.106)$$

Summing the above for the system gives the virtual work that must equal zero. For the present problem, the principle of virtual work in the weak form can be specified as

$$\iint_A \left[D_1 \nabla^4 w_1 + m_1 \frac{\partial^2 w_1}{\partial t^2} - j\omega \rho_0 (\Phi_1 - \Phi_2) \right] \cdot \delta w_1 dA = 0 \quad (1.107)$$

$$\iint_A \left[D_2 \nabla^4 w_2 + m_2 \frac{\partial^2 w_2}{\partial t^2} - j\omega \rho_0 (\Phi_2 - \Phi_3) \right] \cdot \delta w_2 dA = 0 \quad (1.108)$$

$$\iint_A \left[D_3 \nabla^4 w_3 + m_3 \frac{\partial^2 w_3}{\partial t^2} - j\omega \rho_0 (\Phi_3 - \Phi_4) \right] \cdot \delta w_3 dA = 0 \quad (1.109)$$

Upon substitution of Eqs. (1.86), (1.87), (1.88), (1.89), (1.90), (1.91), and (1.106) into Eqs. (1.107), (1.108), and (1.109) and with the help of Eqs. (1.95) and (1.101), (1.102), (1.103), (1.104), and (1.105), three infinite systems of equations are obtained, which can be solved simultaneously by truncation, as demonstrated below.

1.3.4 Determination of Modal Coefficients

Finally, following the procedures outlined in the preceding section and after some laborious but straightforward algebraic manipulations, three infinite systems of

1.3 Clamped Finite Triple-Panel Partitions

simultaneous algebraic equations for the unknown coefficients $\alpha_{1,mn}$, $\alpha_{2,mn}$, and $\alpha_{3,mn}$ can be obtained as

$$4D_1\pi^4 ab \left\{ \left[3\left(\frac{m}{a}\right)^4 + 3\left(\frac{n}{b}\right)^4 + 2\left(\frac{m}{a}\right)^2\left(\frac{n}{b}\right)^2 \right] \alpha_{1,mn} \right.$$

$$\left. + \sum_{k \neq m} 2\left(\frac{n}{b}\right)^4 \alpha_{1,kn} + \sum_{l \neq n} 2\left(\frac{m}{a}\right)^4 \alpha_{1,ml} \right\}$$

$$+ \frac{9ab}{4}\left[-m_1\omega^2 \alpha_{1,mn} + j\omega\rho_0 \left(\frac{\omega}{k_z}\alpha_{1,mn} + \varepsilon_{mn} + \zeta_{mn}\right)\right]$$

$$+ \frac{3ab}{2}\sum_{k \neq m}\left[-m_1\omega^2 \alpha_{1,kn} + j\omega\rho_0 \left(\frac{\omega}{k_z}\alpha_{1,kn} + \varepsilon_{kn} + \zeta_{kn}\right)\right]$$

$$+ \frac{3ab}{2}\sum_{l \neq n}\left[-m_1\omega^2 \alpha_{1,ml} + j\omega\rho_0 \left(\frac{\omega}{k_z}\alpha_{1,ml} + \varepsilon_{ml} + \zeta_{ml}\right)\right]$$

$$+ ab \sum_{k \neq m, l \neq n}\left[-m_1\omega^2 \alpha_{1,kl} + j\omega\rho_0 \left(\frac{\omega}{k_z}\alpha_{1,kl} + \varepsilon_{kl} + \zeta_{kl}\right)\right]$$

$$= 2j\omega\rho_0 I q_{mn}(k_x, k_y) \tag{1.110}$$

$$4D_2\pi^4 ab \left\{ \left[3\left(\frac{m}{a}\right)^4 + 3\left(\frac{n}{b}\right)^4 + 2\left(\frac{m}{a}\right)^2\left(\frac{n}{b}\right)^2 \right] \alpha_{2,mn} \right.$$

$$\left. + \sum_{k \neq m} 2\left(\frac{n}{b}\right)^4 \alpha_{2,kn} + \sum_{l \neq n} 2\left(\frac{m}{a}\right)^4 \alpha_{2,ml} \right\}$$

$$+ \frac{9ab}{4}\left\{-m_2\omega^2 \alpha_{2,mn} - j\omega\rho_0\left[(\varepsilon_{mn} - \xi_{mn})e^{-jk_zH_1} + (\zeta_{mn} - \eta_{mn})e^{jk_zH_1}\right]\right\}$$

$$+ \frac{3ab}{2}\sum_{k \neq m}\left\{-m_2\omega^2 \alpha_{2,kn} - j\omega\rho_0\left[(\varepsilon_{kn} - \xi_{kn})e^{-jk_zH_1} + (\zeta_{kn} - \eta_{kn})e^{jk_zH_1}\right]\right\}$$

$$+ \frac{3ab}{2}\sum_{l \neq n}\left\{-m_2\omega^2 \alpha_{2,ml} - j\omega\rho_0\left[(\varepsilon_{ml} - \xi_{ml})e^{-jk_zH_1} + (\zeta_{ml} - \eta_{ml})e^{jk_zH_1}\right]\right\}$$

$$+ ab \sum_{k \neq m, l \neq n}\left\{-m_2\omega^2 \alpha_{2,kl} - j\omega\rho_0\left[(\varepsilon_{kl} - \xi_{kl})e^{-jk_zH_1} + (\zeta_{kl} - \eta_{kl})e^{jk_zH_1}\right]\right\}$$

$$= 0 \tag{1.111}$$

$$4D_3\pi^4 ab \left\{ \left[3\left(\frac{m}{a}\right)^4 + 3\left(\frac{n}{b}\right)^4 + 2\left(\frac{m}{a}\right)^2\left(\frac{n}{b}\right)^2 \right] \alpha_{3,mn} \right.$$

$$\left. + \sum_{k \neq m} 2\left(\frac{n}{b}\right)^4 \alpha_{3,kn} + \sum_{l \neq n} 2\left(\frac{m}{a}\right)^4 \alpha_{3,ml} \right\}$$

$$+ \frac{9ab}{4} \left\{ -m_3\omega^2 \alpha_{3,mn} - j\omega\rho_0 \left[(\xi_{mn} - \mu_{mn}) e^{-jk_z H_2} + \eta_{mn} e^{jk_z H_2} \right] \right\}$$

$$+ \frac{3ab}{2} \sum_{k \neq m} \left\{ -m_3\omega^2 \alpha_{3,kn} - j\omega\rho_0 \left[(\xi_{kn} - \mu_{kn}) e^{-jk_z H_2} + \eta_{kn} e^{jk_z H_2} \right] \right\}$$

$$+ \frac{3ab}{2} \sum_{l \neq n} \left\{ -m_3\omega^2 \alpha_{3,ml} - j\omega\rho_0 \left[(\xi_{ml} - \mu_{ml}) e^{-jk_z H_2} + \eta_{ml} e^{jk_z H_2} \right] \right\}$$

$$+ ab \sum_{k \neq m, l \neq n} \left\{ -m_3\omega^2 \alpha_{3,kl} - j\omega\rho_0 \left[(\xi_{kl} - \mu_{kl}) e^{-jk_z H_2} + \eta_{kl} e^{jk_z H_2} \right] \right\}$$

$$= 0 \tag{1.112}$$

where the abbreviation $\sum_{k \neq m}$ (or $\sum_{l \neq n}$) has the conventional meaning that summation is intended with the index k (or l) taking integer values from 1 to $+\infty$ except for the specified value m (or n). Similarly, the notation $\sum_{k \neq m, l \neq n}$ denotes double summation about indices k and l from 1 to $+\infty$ apart from the prescribed values m and n. Additionally, a parameter associated with the generalized force $\boldsymbol{F}$ (see Appendix B for details) appears during the process of integration as

$$q_{mn}(k_x, k_y)$$
$$= \begin{cases} ab & k_x = 0 \text{ and } k_y = 0 \\ \dfrac{4jn^2\pi^2 a \left(1 - e^{-jk_y b}\right)}{k_y \left(k_y^2 b^2 - 4n^2\pi^2\right)} & k_x = 0 \text{ and } k_y \neq 0 \\ \dfrac{4jm^2\pi^2 b \left(1 - e^{-jk_x a}\right)}{k_x \left(k_x^2 a^2 - 4m^2\pi^2\right)} & k_x \neq 0 \text{ and } k_y = 0 \\ -\dfrac{16m^2 n^2 \pi^4 \left(1 - e^{-jk_x a}\right)\left(1 - e^{-jk_y b}\right)}{k_x k_y \left(k_x^2 a^2 - 4m^2\pi^2\right)\left(k_y^2 b^2 - 4n^2\pi^2\right)} & k_x \neq 0 \text{ and } k_y \neq 0 \end{cases}$$
$$\tag{1.113}$$

In terms of matrix formulation, Eqs. (1.110), (1.111), and (1.112) can be rewritten as a linear system of equations consisting of $3MN$ equilibrium equations by taking truncation of m from 1 to M and n from 1 to N as

1.3 Clamped Finite Triple-Panel Partitions

$$\begin{bmatrix} T_{11} & T_{12} & 0 \\ T_{21} & T_{22} & T_{23} \\ 0 & T_{32} & T_{33} \end{bmatrix}_{3MN \times 3MN} \begin{Bmatrix} \alpha_1 \\ \alpha_2 \\ \alpha_3 \end{Bmatrix}_{3MN \times 1} = \begin{Bmatrix} F \\ 0 \\ 0 \end{Bmatrix}_{3MN \times 1} \quad (1.114)$$

where T_{11} and T_{12} are the vectors derived from Eq. (1.110) that correspond to the unknown modal coefficient vector α_1; T_{21}, T_{22}, and T_{23} are the vectors arising from Eq. (1.111) that relate to the unknown modal coefficient vector α_2; and the vectors T_{32} and T_{33} are derived from Eq. (1.112) with respect to α_3. More details can be found in Appendix B.

1.3.5 Sound Transmission Loss

The sound power of the relevant acoustic fields ($\alpha = 1, 2, 3, 4$; see Fig. 1.34) can be defined as [24, 36]

$$\Pi_\alpha = \frac{1}{2} \text{Re} \iint_A p_\alpha \cdot v_\alpha^* dA \quad (1.115)$$

where the local volume velocity is related to the sound pressure through the impendence of air as $v_\alpha = p_\alpha/(\rho_0 c_0)$ and the superscript asterisk denotes complex conjugate.

The transmission coefficient is defined as the ratio of the transmitted sound power to the incident sound power:

$$\tau(\varphi, \theta) = \frac{\Pi_4}{\Pi_1} \quad (1.116)$$

which is dependent upon the incident angles φ and θ. The sound transmission loss (*STL*) is then customarily defined as the inverse of the power transmission coefficient in decibels scale as

$$STL = 10 \, \log_{10}\left(\frac{1}{\tau}\right) \quad (1.117)$$

Throughout the present study, *STL* is used as a measure of the effectiveness of a clamped triple-panel (or, for comparison, double-panel) configuration of finite extent in isolating the incident sound.

1.3.6 Model Validation

The validity and feasibility of the proposed theoretical model for sound transmission across triple-panel partitions is checked by comparing model predictions with

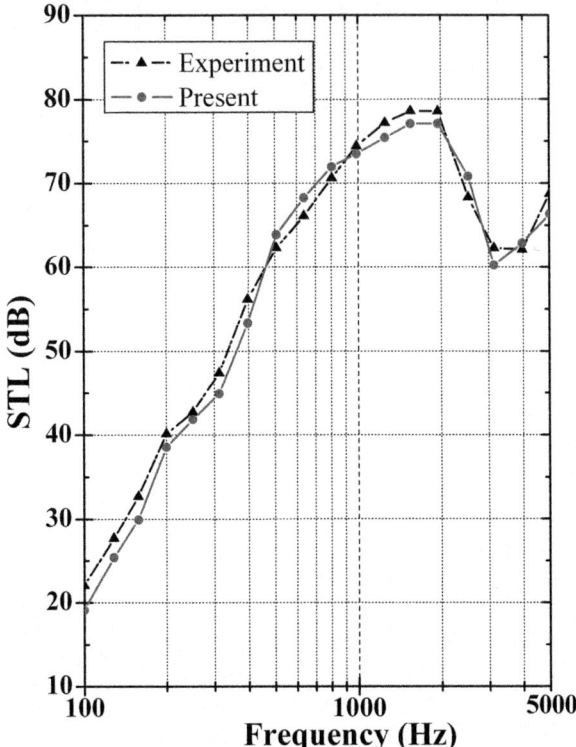

Fig. 1.35 Diffuse sound transmission loss (*STL*) plotted as a function of incident frequency: comparison between present model predictions with experimental measurements [83] (With permission from Elsevier)

existing experiment results [83], as shown in Fig. 1.35. In the present simulation, sound transmission loss (STL) is calculated in 1/3 octave bands with a diffuse field integration. Overall, as illustrated in Fig. 1.35, the present theoretical results agree excellently with those measured. The small discrepancies may be attributed to the fact that the mineral wool filled around the edges of the partition in factual measurement is not accounted for in the model.

1.3.7 Physical Interpretation of STL Dips

To exclude the panel modal behavior due to edge constraints, Fig. 1.36 shows separately the characteristics of normal incident transmission loss through a single-, double-, and triple-panel partition of infinite extent. As anticipated, the *STL* versus frequency curve of the single panel obeys the mass law, while a set of dips appear on the *STL* versus frequency curves of double- and triple-panel partitions at frequencies related to the system resonance mode. Note that similar plots for double-panel partitions have been presented in our previous work. With special focus placed upon the *STL* versus frequency curve of the triple-panel partition, it is observed that within

1.3 Clamped Finite Triple-Panel Partitions

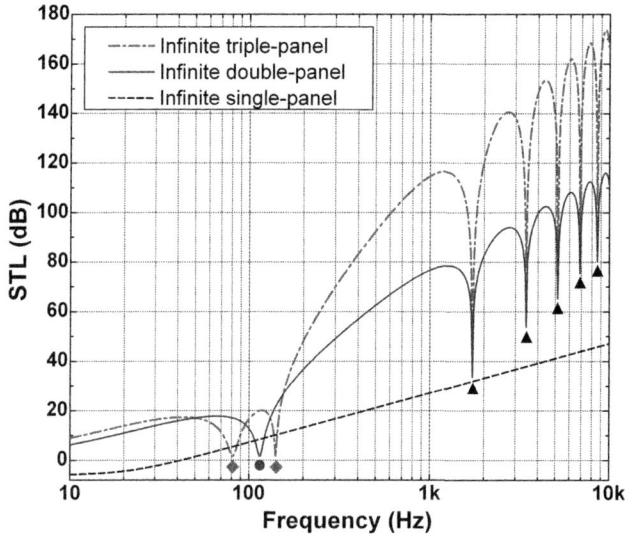

Fig. 1.36 Comparison of *STL* versus frequency curves among infinite single-, double-, and triple-panel partitions ($h^s = 0.002$ m for single panel; $\tilde{H} = 0.1$ m, $h_1^d = h_2^d = 0.002$ m for double panel; $\tilde{H}_1 = \tilde{H}_2 = 0.1$ m, $h_1^t = h_2^t = h_3^t = 0.002$ m for triple panel). Symbols: ● mass-air-mass resonance; ♦ mass-air-mass-air-mass resonance; ▲ standing-wave resonance (With permission from Elsevier)

the frequency range considered, two different kinds of resonance exist, i.e., those associated with the first and second dips labeled by the symbol ♦ and subsequent dips at higher frequencies labeled by the symbol ▲.

Analogous to the "mass-air-mass" resonance of a double-panel system, the triple-panel partition can also be simplified as a system of three lumped masses (m_1, m_2, and m_3) connected together by two springs with stiffness k_1 and k_2, respectively. In the case of sound incident normally to the partition, the eigenvalue equation of this simplified mass-spring system can be written as

$$|K - \omega^2 M| = \begin{vmatrix} k_1 - \omega^2 m_1 & -k_1 & 0 \\ -k_1 & k_1 + k_2 - \omega^2 m_2 & -k_2 \\ 0 & -k_2 & k_2 - \omega^2 m_3 \end{vmatrix} = 0 \quad (1.118)$$

from which the resonance frequencies of the equivalent mass-spring system can be obtained as

$$f_1 = \frac{\sqrt{2}}{4\pi} \sqrt{\frac{\lambda_1 + \lambda_2 - \sqrt{(\lambda_1 - \lambda_2)^2 + 4k_1 k_2 m_1^2 m_3^2}}{m_1 m_2 m_3}} \quad (1.119)$$

Table 1.3 Comparison between theory and closed-form formulae for resonance frequencies of infinite triple-panel partition

Order	Mass-spring resonance f_i (Hz)		Standing-wave resonance $f_{s,n}$ (Hz)	
	Theory	Eqs. (1.119) and (1.120)	Theory	Eq. (1.121)
1	81.20	81.72	1,719	1,715
2	140.89	141.54	3,436	3,430
3	\	\	5,146	5,145
4	\	\	6,863	6,860

$$f_2 = \frac{\sqrt{2}}{4\pi} \sqrt{\frac{\bar{\lambda}_1 + \bar{\lambda}_2 + \sqrt{(\bar{\lambda}_1 - \bar{\lambda}_2)^2 + 4k_1 k_2 m_1^2 m_3^2}}{m_1 m_2 m_3}} \qquad (1.120)$$

Here, $\bar{\lambda}_1 = k_1 m_3 (m_1 + m_2)$ and $\bar{\lambda}_2 = k_2 m_1 (m_2 + m_3)$, with $k_i = \rho_0 c_0^2 / \tilde{H}_i$ ($i = 1, 2$), are the equivalent stiffness of the lower air cavity and the upper air cavity, respectively. The introduction of the additional panel and air cavity induces more complicated fluid-structural coupling in the triple-panel system. As a result, one dip corresponding to the "mass-air-mass" resonance of the double-panel system is divided into two dips for the triple-panel system. The two formulae (1.119) and (1.120) can be used to estimate the resonance frequencies associated with the two dips, which have been specially labeled by symbol ♦ in Fig. 1.36.

At larger frequencies, the resonance dips denoted by symbol ▲ in Fig. 1.36 are associated with the standing-wave resonance phenomenon due to the interaction effect of successively reflected waves inside the air cavity. For such phenomenon to occur, the depth of the air cavity should be integer numbers of the half wavelength of the incident sound. The nth standing-wave resonances occur therefore at the frequency [18]

$$f_{s,n} = \frac{n c_0}{2 \tilde{H}} \quad (n = 1, 2, 3 \ldots) \qquad (1.121)$$

where $\tilde{H}$ is the depth of the air cavity having a value of 0.1 m for both the double- and triple-panel systems considered. In accordance with the prediction of (1.121), the standing-wave resonance frequencies should be the same for the two systems, which is confirmed by the excellent agreement shown in Fig. 1.36 for frequencies above 1 kHz.

The predictions of the present theory for the resonance frequencies of an infinite triple-panel partition are compared in Table 1.3 with those of the closed formulae, i.e., Eqs. (1.119), (1.120), and (1.121). Excellent agreement is achieved, suggesting that the theoretical modeling is consistent with the above stated physical nature of the *STL* dips.

1.3 Clamped Finite Triple-Panel Partitions

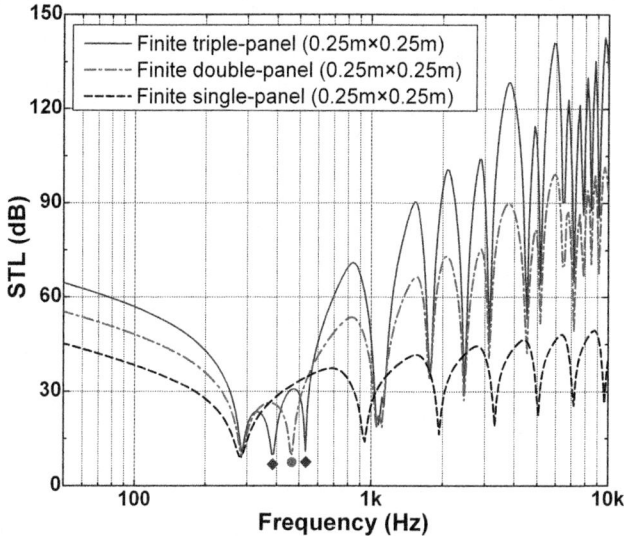

Fig. 1.37 Comparison of *STL* versus frequency curves among clamp-supported rectangular single-, double-, and triple-panel partitions ($h^s = 0.002$ m for single panel; $\tilde{H} = 0.010$ m, $h_1^d = h_2^d = 0.002$ m for double panel; $\tilde{H}_1 = \tilde{H}_2 = 0.010$ m, $h_1^t = h_2^t = h_3^t = 0.002$ m for triple panel). Symbols: ● mass-air-mass resonance; ◆ mass-air-mass-air-mass resonance (With permission from Elsevier)

For a clamp-supported triple-panel partition of finite extent, the modal behavior of the panel plays a dominant role in the appearance of numerous resonance dips on the *STL* versus frequency curve, as shown in Fig. 1.37. To clearly identify the resonance dips induced by the panel natural vibratory modes, the *STL* versus frequency curves of the single and double panels are plotted together with that of the triple panel. It is seen from Fig. 1.37 that the first dips associated with the three partitions agree well with each other, while other dips associated with the panel vibratory modes achieve good agreement only between the double- and triple-panel partitions, which deviate away from those of the single-panel partition. This is caused by the air cavity coupling effect that is absent in the single-panel system. Moreover, several additional dips appear on the *STL* frequency curves of double- and triple-panel partitions due to the mass-spring resonance and standing-wave resonance.

Built upon the results for simply supported boundary conditions, the frequencies corresponding to the *STL* dips arising from panel vibratory modes in clamp-supported double- and triple-panel partitions can be approximately estimated by

$$f_{mn} = \frac{\pi}{2} \left(\frac{m^2}{a^2} + \frac{n^2}{b^2} \right) \sqrt{\frac{\sqrt{2} E h^2}{12 \rho (1 - \nu^2)}} \qquad (1.122)$$

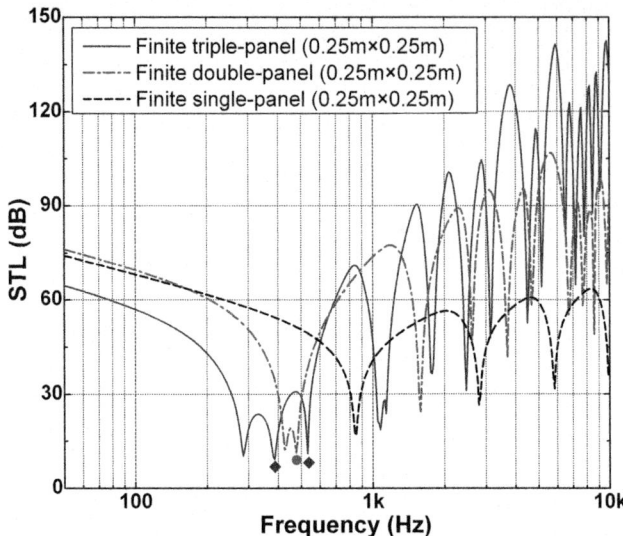

Fig. 1.38 Comparison of *STL* versus frequency curves among clamp-supported rectangular single-, double-, and triple-panel partitions ($h^s = 0.006$ m for single panel; $\tilde{H} = 0.020$ m, $h_1^d = h_2^d = 0.003$ m for double panel; $\tilde{H}_1 = \tilde{H}_2 = 0.010$ m, $h_1^t = h_2^t = h_3^t = 0.002$ m for triple panel). Symbols: • mass-air-mass resonance; ♦ mass-air-mass-air-mass resonance (With permission from Elsevier)

1.3.8 Comparison Among Single-, Double-, and Triple-Panel Partitions with Equivalent Total Mass

As illustrated in Fig. 1.37, for relatively high frequencies exceeding the mass-spring resonance frequency, improved sound insulation is demonstrated for triple-panel partitions over single- and double-panel partitions. However, an increase of STL would be expected simply from the mass increase resulting from the addition of a third panel. To provide a fair comparison, theoretical results for the three configurations (i.e., single-, double-, and triple-panel partitions) with equivalent total mass are plotted in Fig. 1.38.

In comparison, the *STL* dips induced by the mass-spring resonance deviate much among different structures, due mainly to the different panel masses and air cavity coupling effects. In the frequency range above the mass-spring resonance frequency, on the whole, a triple-panel partition provides larger STL than a double-panel partition and remarkably larger STL than a single-panel partition. This suggests that cavity coupling effects play a dominant role in this frequency range for double- and triple-panel partitions. In contrast, in the frequency range below the mass-spring resonance frequency, the triple-panel partition exhibits poorer sound insulation than both single- and double-panel partitions. In other words, multi-panel partitions do not provide improved soundproofing capability than single panels

1.3 Clamped Finite Triple-Panel Partitions

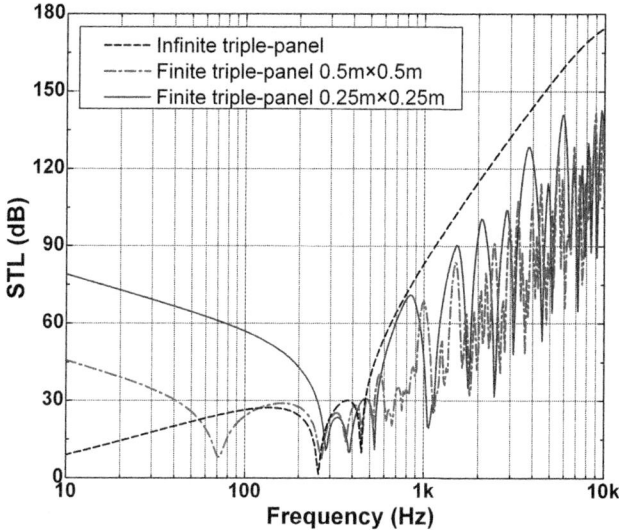

Fig. 1.39 Variation of sound transmission loss with panel dimensions for triple-panel partitions ($\tilde{H}_1 = \tilde{H}_2 = 0.010$ m, $h_1^t = h_2^t = h_3^t = 0.002$ m) (With permission from Elsevier)

with equivalent mass in frequencies below the mass-spring resonance frequency, implying that cavity coupling of multi-panel partitions has little effect on STL in the low-frequency range.

For frequencies below the cutoff frequency for the cavities, the present theoretical results as discussed above are completely in accordance with the experimental results of Brekke [83] for double- and triple-panel partitions.

1.3.9 Asymptotic Variation of STL Versus Frequency Curve from Finite to Infinite System

To examine the variation of the transmission loss characteristics of a triple-panel partition with varying geometrical dimensions, two selected finite cases (i.e., 0.25 m × 0.25 m and 0.5 m × 0.5 m) are compared in Fig. 1.39 with the infinite case, with the air cavity depth fixed at 0.010 m. It is seen from Fig. 1.39 that, as the panel dimensions increase, the panel mode-dominated *STL* dips are shifted toward the lower frequencies, which are consistent with the predictions of Eq. (1.122). Beyond the mass-spring resonance dip, the *STL* versus frequency curve of the infinite triple-panel structure sets upper bound for the finite-sized partitions, because the panel mode-dominated STL dips vanish in the infinite case. At frequencies below the mass-spring dip, however, the soundproofing performance of the infinite structure is significantly inferior to that of finite-sized structures due to boundary constraint effects. Actually, this trend is mainly affected by the (1, 1) panel mode resonance

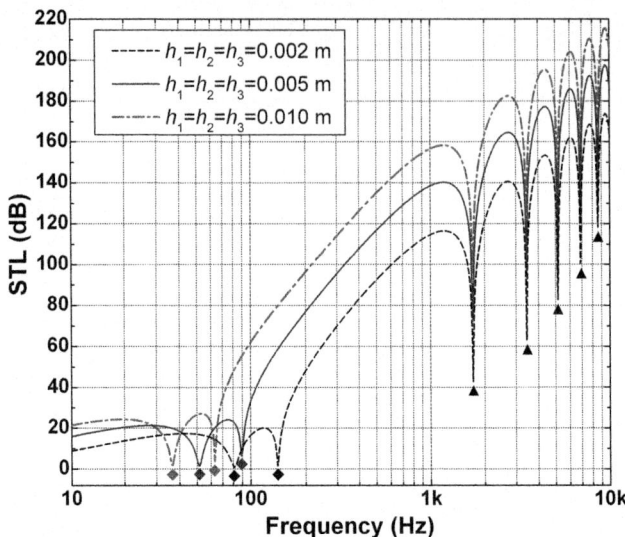

Fig. 1.40 Effects of panel thickness on *STL* for infinite triple-panel partition ($\tilde{H}_1 = \tilde{H}_2 = 0.1$ m) under normal sound excitation: ♦ mass-air-mass-air-mass resonance; ▲ standing-wave resonance (With permission from Elsevier)

dips for different panel geometrical dimensions. As can be seen in Eq. (1.122), the (1, 1) panel mode resonance frequency decreases with increasing panel dimensions, which are separately associated with the first dip for the two finite cases and the dip at 0 Hz for the infinite case. Also, for the same reason, the sound proofing capability of the triple-panel partition increases with decreasing panel dimensions. Since for many applications, noise reduction at the low-frequency range (<500 Hz) is of vital importance, this finding has significant implications on the practical design of triple-panel partitions such as the soundproofing windows installed in high-class buildings and aircraft fuselages.

1.3.10 Effects of Panel Thickness

To demonstrate how the sound transmission performance of a triple-panel partition varies with panel thickness, the *STL* versus frequency curves for the infinite case is plotted in Figs. 1.40 and 1.41, while those for the finite case are presented in Figs. 1.42 and 1.43. As shown in Fig. 1.40, the *STL* value is increased significantly as the panel thickness is increased, which is consistent with the mass law for a single panel but more noticeable due to cavity coupling effects. The mass-spring resonance dips shift downward as the panel thickness is increased, due to the increased surface density of the panel. The standing-wave resonance dips reside in their original locations, however, implying that these are independent of the panel thickness.

1.3 Clamped Finite Triple-Panel Partitions

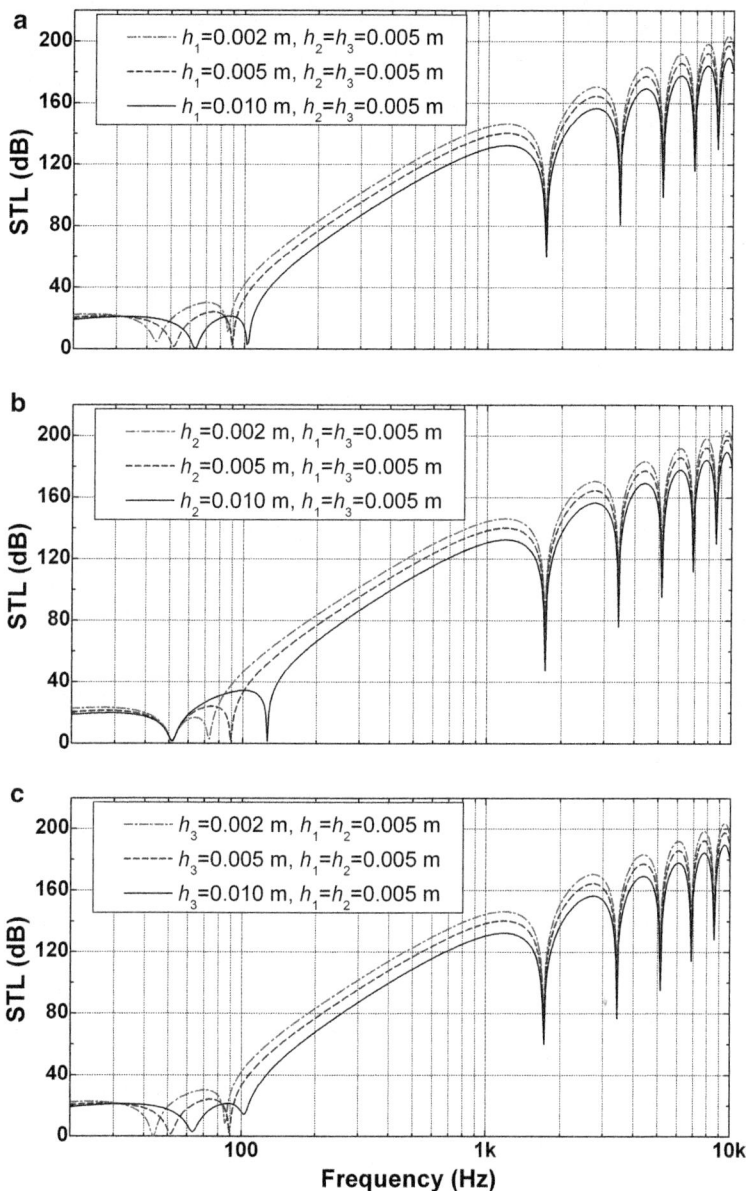

Fig. 1.41 Variation of *STL* for infinite triple-panel partition ($\tilde{H}_1 = \tilde{H}_2 = 0.1$ m) under normal sound excitation with the thickness of: (**a**) incident panel; (**b**) middle panel; (**c**) radiation panel (With permission from Elsevier)

To highlight the different roles played by the three panels in the sound transmission process, the thicknesses of arbitrarily selected two panels are fixed while that of the remaining one is systematically varied. The results for a triple-panel partition of infinite extent are firstly presented in Fig. 1.41, and it is seen that

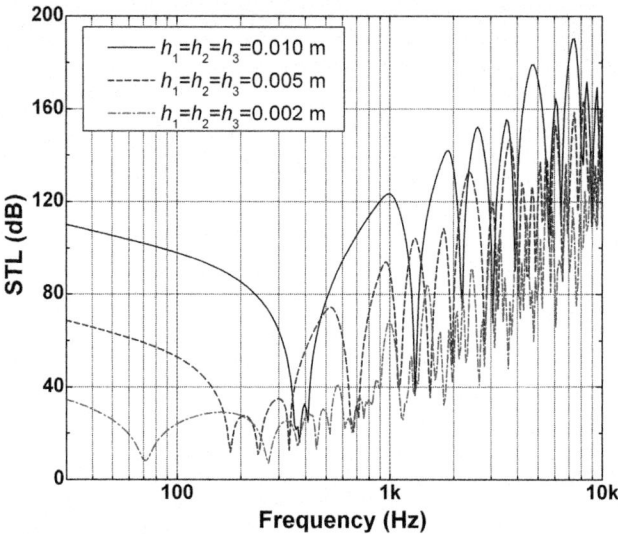

Fig. 1.42 Effects of panel thickness on *STL* for finite triple-panel partition (0.5 m × 0.5 m, $\tilde{H}_1 = \tilde{H}_2 = 0.01$ m) under normal sound excitation (With permission from Elsevier)

the increased thickness of any panel among the three leads to the increase in *STL*. The good agreement between Fig. 1.41a and c suggests the same role played by the incident panel and the radiation panel, which completely follows the acoustical reciprocal theorem. The mass-spring resonance dips shift in a distinct way as the thickness of the middle panel is increased (Fig. 1.41b), which is in accordance with the predictions of Eqs. (1.119) and (1.120).

For triple-panel partitions of finite extent, as the panel thickness is increased, two prominent features can be observed from Fig. 1.42: (1) remarkable increase of the *STL* value and (2) shifting of the resonance dips toward higher frequencies, which is attributed to the increased panel stiffness and surface density. The individual role of each panel in sound transmission is illustrated in Fig. 1.43a, b, and c for the incident panel, the middle panel, and the radiation panel, respectively. The most noticeable feature of the results shown in Fig. 1.43 is that increasing the incident panel thickness causes the *STL* value to increase more dramatically than increasing the thickness of the middle or radiation panel, especially at relatively low frequencies. This is because the middle or radiation panel does not significantly affect the coupling between the panels through air stiffness for frequencies below the cutoff frequency of the cavities [83]. This feature is also consistent with the experimental results of [82]. The pronounced deviation between the resonant dips for different cases shown in Fig. 1.43a implies the predominant role of the incident panel vibratory modes. In contrast, the good agreement between the resonant dips in the high-frequency range for different panel thicknesses in Fig. 1.43b and c suggests that the vibratory modes of the middle or radiation panel have negligible effects.

1.3 Clamped Finite Triple-Panel Partitions

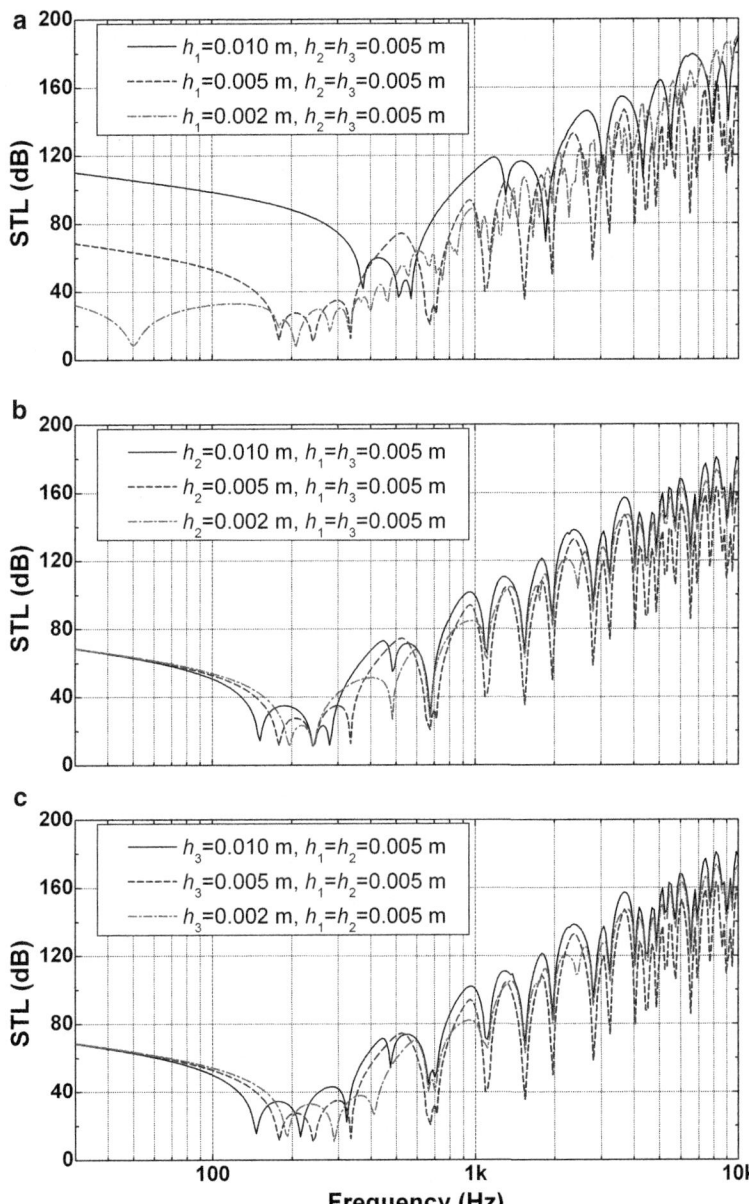

Fig. 1.43 Variation of *STL* for finite triple-panel partition (0.5 m × 0.5 m, $\tilde{H}_1 = \tilde{H}_2 = 0.01$ m) under normal sound excitation with the thickness of: (**a**) incident panel; (**b**) middle panel; (**c**) radiation panel (With permission from Elsevier)

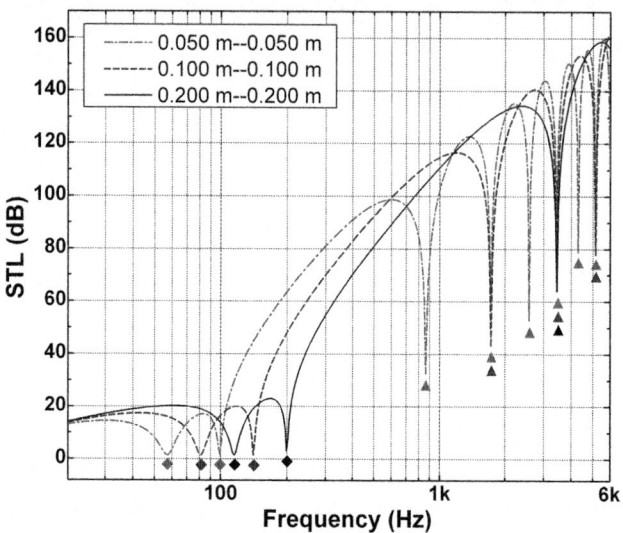

Fig. 1.44 Effects of air cavity thickness on *STL* for infinite triple-panel partition ($h_1^t = h_2^t = h_3^t = 0.002$ m) under normal sound excitation: ♦ mass-air-mass-air-mass resonance; ▲ standing-wave resonance (With permission from Elsevier)

At relatively low frequencies, however, the shifting of the resonant dips with varying panel thickness demonstrates the significant effects of each panel (see Fig. 1.43a, b, c). Note that the steeper dips for the case of ($h_1^t = h_2^t = h_3^t = 0.005$ m) than other cases are attributed to the noticeable enhancement of the same vibratory modes of the three identical panels.

1.3.11 Effects of Air Cavity Depth

For an infinitely large triple-panel partition under normal sound excitation, Fig. 1.44 plots the *STL* versus frequency curves for selected air cavity depths, with the thickness of each panel fixed at 0.002 m. As the air cavity depth is increased, the tendency of the *STL* versus frequency curve varies significantly. Within the frequency range between the mass-spring resonance and the first-order standing-wave resonance, increasing the air cavity depth leads to remarkable increase of the *STL* value. The mass-spring resonance dips shift downward with increasing air cavity depth, which is attributed to the decreased equivalent stiffness of the air cavities. According to Eq. (1.121), the natural frequency of each standing-wave resonance decreases as the air cavity depth is increased, which is consistent with the results of Fig. 1.44.

The effects of air cavity depth on sound transmission across a finite-sized triple-panel partition are shown in Figs. 1.45 and 1.46, again with the panel thickness fixed

1.3 Clamped Finite Triple-Panel Partitions

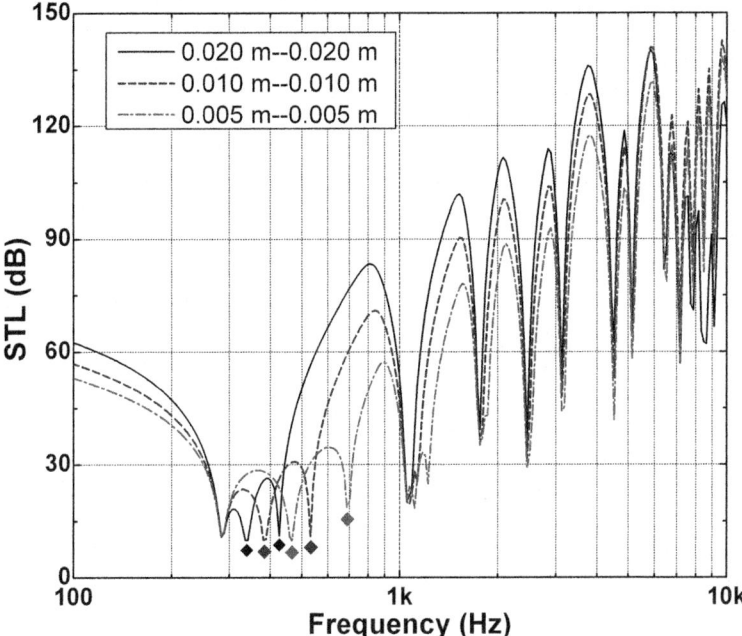

Fig. 1.45 Effects of air cavity thickness on *STL* for finite triple-panel partition (0.25 m × 0.25 m, $h_1^t = h_2^t = h_3^t = 0.002$ m) under normal sound excitation: ◆ mass-air-mass-air-mass resonance (With permission from Elsevier)

at 0.002 m. It is seen that while the increase of air cavity depth leads to enhanced soundproofing capability of the structure, the effects are particularly noticeable if the depth of the two cavities is increased simultaneously (see Fig. 1.45). The mass-spring resonance dips alter significantly, consistent with the predictions of Eqs. (1.119) and (1.120). The dips dominated by panel vibratory modes change little because the boundary condition plays a stronger effect than cavity coupling effects at these dips. In addition, the good agreement between Fig. 1.46a and b demonstrates the identical role of the two air cavities in the process of sound transmission through the finite triple-panel structure, which can be explained with the acoustical reciprocal theorem.

1.3.12 Concluding Remarks

A theory has been established that can be used to predict the sound transmission characteristics of a clamp-mounted triple-panel partition separated by two enclosed air cavities. A set of modal functions (basic functions) are employed to account for the clamped boundary conditions, and the application of the virtual work principle

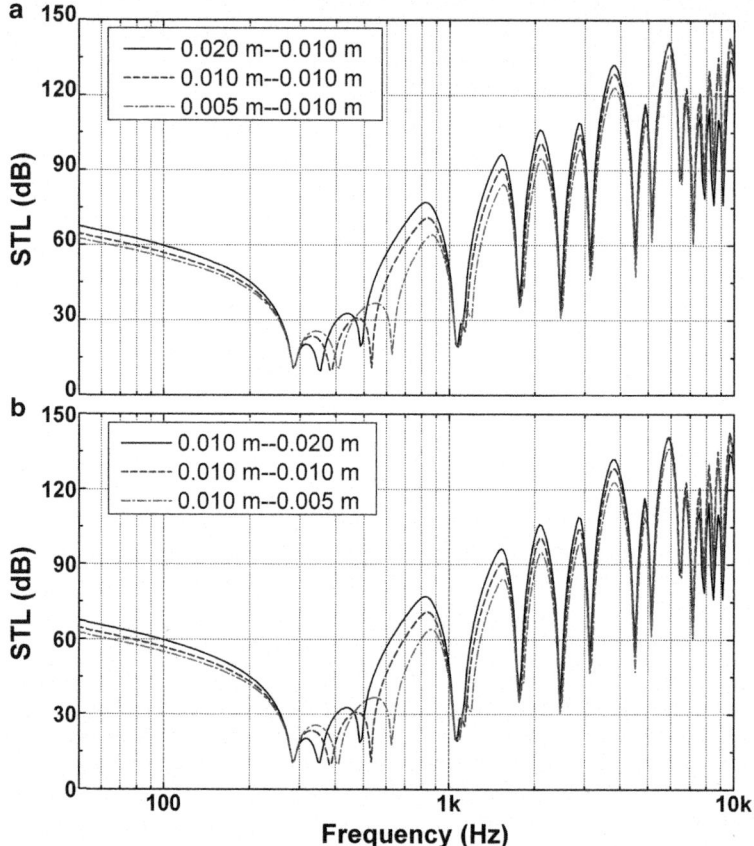

Fig. 1.46 Effects of air cavity thickness on *STL* for finite triple-panel partition (0.25 m × 0.25 m, $h_1^t = h_2^t = h_3^t = 0.002$ m) partition under normal sound excitation: (**a**) only varying the depth of the air cavity adjacent to the incident panel; (**b**) only varying the depth of the air cavity adjacent to the radiating panel (With permission from Elsevier)

leads to a set of simultaneous algebraic equations for determining the unknown modal coefficients. The present theory has the advantage of clearly showing the major vibroacoustic phenomena associated with the edge constraints and fluid-structure coupling, such as the equivalent mass-spring resonance, the standing-wave resonance, and the modal resonance of the system. Extensive numerical calculations are carried out to obtain the frequency characteristic curves of the transmission loss performance of the triple-panel structure, with detailed physical explanations given for the abovementioned resonance dips. Comparison of the triple-panel partition with the double-panel partition suggests that, for the purpose of maximizing the transmission loss, the former is a preferred alternative of the latter although the superiority is not remarkable when the total masses of the two are equivalent. Moreover, the relatively large number of system parameters owned by the triple-panel partition allows more design space for tailoring its noise reduction capability. Since for many applications, noise reduction at the low-frequency range (<500 Hz) is of vital importance, the finding that the sound proofing capability of a triple-panel partition increases with decreasing panel dimensions has significant implications on the design of soundproofing windows installed in high-class buildings and aircraft fuselages.

As a future work, an active control strategy to minimize the sound transmission across a clamp-mounted triple-panel structure will be analytically and experimentally developed based upon the proposed theory from the viewpoint of practical noise reduction.

Appendices

Appendix A

The deflection coefficients of the two panels are

$$\{\alpha_{1,kl}\} = \begin{bmatrix} \alpha_{1,11} & \alpha_{1,21} & \cdots & \alpha_{1,M1} & \alpha_{1,12} & \alpha_{1,22} & \cdots & \alpha_{1,M2} & \cdots & \alpha_{1,MN} \end{bmatrix}^{\mathrm{T}}_{MN \times 1} \quad (1.\mathrm{A}.1)$$

$$\{\alpha_{2,kl}\} = \begin{bmatrix} \alpha_{2,11} & \alpha_{2,21} & \cdots & \alpha_{2,M1} & \alpha_{2,12} & \alpha_{2,22} & \cdots & \alpha_{2,M2} & \cdots & \alpha_{2,MN} \end{bmatrix}^{\mathrm{T}}_{MN \times 1} \quad (1.\mathrm{A}.2)$$

The left-hand side of Eq. (1.74) represents the generalized force, where

$$\{F_{kl}\} = 2j\omega\rho_0 I \begin{bmatrix} f_{11} & f_{21} & \cdots & f_{M1} & f_{12} & f_{22} & \cdots & f_{M2} & \cdots & f_{MN} \end{bmatrix}^{\mathrm{T}}_{MN \times 1} \quad (1.\mathrm{A}.3)$$

$$\lambda^{*1}_{1,mn} = 3\left(\frac{m}{a}\right)^4 + 3\left(\frac{n}{b}\right)^4 + 2\left(\frac{m}{a}\right)^2\left(\frac{n}{b}\right)^2 \quad (1.\mathrm{A}.4)$$

$$\Delta^{*1}_1 = \begin{bmatrix} \lambda^{*1}_{1,11} & & & & & & & & \\ & \lambda^{*1}_{1,21} & & & & & & & \\ & & \ddots & & & & & & \\ & & & \lambda^{*1}_{1,M1} & & & & & \\ & & & & \lambda^{*1}_{1,12} & & & & \\ & & & & & \lambda^{*1}_{1,22} & & & \\ & & & & & & \ddots & & \\ & & & & & & & \lambda^{*1}_{1,M2} & \\ & & & & & & & & \ddots & \\ & & & & & & & & & \lambda^{*1}_{1,MN} \end{bmatrix}_{MN \times MN}$$

$$(1.\mathrm{A}.5)$$

$$\lambda_{1,n}^{*2} = \frac{2n^4}{b^4} \begin{bmatrix} 0 & 1 & 1 & \cdots & 1 \\ 1 & 0 & 1 & \cdots & 1 \\ 1 & 1 & 0 & \cdots & \cdots \\ \cdots & \cdots & \cdots & \ddots & 1 \\ 1 & 1 & \cdots & 1 & 0 \end{bmatrix}_{M \times N} \tag{1.A.6}$$

$$\Delta_1^{*2} = \begin{bmatrix} \lambda_{1,1}^{*2} & & & \\ & \lambda_{1,2}^{*2} & & \\ & & \ddots & \\ & & & \lambda_{1,N}^{*2} \end{bmatrix}_{MN \times MN} \tag{1.A.7}$$

$$\lambda_1^{*3} = \frac{2}{a^4} \begin{bmatrix} 1^4 & & & \\ & 2^4 & & \\ & & \ddots & \\ & & & M^4 \end{bmatrix}_{M \times N} \tag{1.A.8}$$

$$\Delta_1^{*3} = \begin{bmatrix} 0 & \lambda_1^{*3} & \lambda_1^{*3} & \cdots & \lambda_1^{*3} \\ \lambda_1^{*3} & 0 & \lambda_1^{*3} & \cdots & \lambda_1^{*3} \\ \lambda_1^{*3} & \lambda_1^{*3} & 0 & \cdots & \cdots \\ \cdots & \cdots & \cdots & \ddots & \lambda_1^{*3} \\ \lambda_1^{*3} & \lambda_1^{*3} & \cdots & \lambda_1^{*3} & 0 \end{bmatrix}_{MN \times MN} \tag{1.A.9}$$

$$\Delta_2^{*1} = \frac{9ab}{4} \begin{bmatrix} 1 & & & \\ & 1 & & \\ & & \ddots & \\ & & & 1 \end{bmatrix}_{MN \times MN} \tag{1.A.10}$$

$$\lambda_2^{*2} = \frac{3ab}{2} \begin{bmatrix} 0 & 1 & 1 & \cdots & 1 \\ 1 & 0 & 1 & \cdots & 1 \\ 1 & 1 & 0 & \cdots & \cdots \\ \cdots & \cdots & \cdots & \ddots & 1 \\ 1 & 1 & \cdots & 1 & 0 \end{bmatrix}_{M \times N} \tag{1.A.11}$$

$$\Delta_2^{*2} = \begin{bmatrix} \lambda_2^{*2} & & & \\ & \lambda_2^{*2} & & \\ & & \ddots & \\ & & & \lambda_2^{*2} \end{bmatrix}_{MN \times MN} \tag{1.A.12}$$

$$\lambda_2^{*3} = \frac{3ab}{2}\begin{bmatrix} 1 & & & & \\ & 1 & & & \\ & & \ddots & & \\ & & & & 1 \end{bmatrix}_{M\times N} \quad (1.A.13)$$

$$\Delta_2^{*3} = \begin{bmatrix} 0 & \lambda_2^{*3} & \lambda_2^{*3} & \cdots & \lambda_2^{*3} \\ \lambda_2^{*3} & 0 & \lambda_2^{*3} & \cdots & \lambda_2^{*3} \\ \lambda_2^{*3} & \lambda_2^{*3} & 0 & \cdots & \cdots \\ \cdots & \cdots & \cdots & \ddots & \lambda_2^{*3} \\ \lambda_2^{*3} & \lambda_2^{*3} & \cdots & \lambda_2^{*3} & 0 \end{bmatrix}_{MN\times MN} \quad (1.A.14)$$

$$\lambda_2^{*4} = ab\begin{bmatrix} 0 & 1 & 1 & \cdots & 1 \\ 1 & 0 & 1 & \cdots & 1 \\ 1 & 1 & 0 & \cdots & \cdots \\ \cdots & \cdots & \cdots & \ddots & 1 \\ 1 & 1 & \cdots & 1 & 0 \end{bmatrix}_{M\times N} \quad (1.A.15)$$

$$\Delta_2^{*4} = \begin{bmatrix} 0 & \lambda_2^{*4} & \lambda_2^{*4} & \cdots & \lambda_2^{*4} \\ \lambda_2^{*4} & 0 & \lambda_2^{*4} & \cdots & \lambda_2^{*4} \\ \lambda_2^{*4} & \lambda_2^{*4} & 0 & \cdots & \cdots \\ \cdots & \cdots & \cdots & \ddots & \lambda_2^{*4} \\ \lambda_2^{*4} & \lambda_2^{*4} & \cdots & \lambda_2^{*4} & 0 \end{bmatrix}_{MN\times MN} \quad (1.A.16)$$

Using the definition of the sub-matrices presented above, one obtains

$$[T_{11,kl}]_{MN\times MN} = 4D_1\pi^4 ab\left(\Delta_1^{*1} + \Delta_1^{*2} + \Delta_1^{*3}\right)$$
$$- \left(m_1\omega^2 + j\omega\rho_0 \frac{2\omega e^{2jk_z H}}{k_z\left(1 - e^{2jk_z H}\right)}\right)\cdot\left(\Delta_2^{*1} + \Delta_2^{*2} + \Delta_2^{*3} + \Delta_2^{*4}\right)$$
$$(1.A.17)$$

$$[T_{12,kl}]_{MN\times MN} = j\omega\rho_0 \frac{2\omega e^{jk_z H}}{k_z\left(1 - e^{2jk_z H}\right)}\left(\Delta_2^{*1} + \Delta_2^{*2} + \Delta_2^{*3} + \Delta_2^{*4}\right) \quad (1.A.18)$$

$$[T_{21,kl}]_{MN\times MN} = j\omega\rho_0 \frac{2\omega e^{jk_z H}}{k_z\left(1 - e^{2jk_z H}\right)}\left(\Delta_2^{*1} + \Delta_2^{*2} + \Delta_2^{*3} + \Delta_2^{*4}\right) \quad (1.A.19)$$

$$[T_{22,kl}]_{MN \times MN} = 4D_2\pi^4 ab \left(\Delta_1^{*1} + \Delta_1^{*2} + \Delta_1^{*3}\right)$$
$$- \left(m_2\omega^2 + j\omega\rho_0 \frac{2\omega e^{2jk_zH}}{k_z\left(1 - e^{2jk_zH}\right)}\right) \cdot \left(\Delta_2^{*1} + \Delta_2^{*2} + \Delta_2^{*3} + \Delta_2^{*4}\right)$$
(1.A.20)

Appendix B

The modal coefficients of the three panels are

$$\{\boldsymbol{\alpha}_1\} = \begin{bmatrix} \alpha_{1,11} & \alpha_{1,21} & \cdots & \alpha_{1,M1} & \alpha_{1,12} & \alpha_{1,22} & \cdots & \alpha_{1,M2} & \cdots & \alpha_{1,MN} \end{bmatrix}^T_{MN \times 1} \quad (1.B.1)$$

$$\{\boldsymbol{\alpha}_2\} = \begin{bmatrix} \alpha_{2,11} & \alpha_{2,21} & \cdots & \alpha_{2,M1} & \alpha_{2,12} & \alpha_{2,22} & \cdots & \alpha_{2,M2} & \cdots & \alpha_{2,MN} \end{bmatrix}^T_{MN \times 1} \quad (1.B.2)$$

$$\{\boldsymbol{\alpha}_3\} = \begin{bmatrix} \alpha_{3,11} & \alpha_{3,21} & \cdots & \alpha_{3,M1} & \alpha_{3,12} & \alpha_{3,22} & \cdots & \alpha_{3,M2} & \cdots & \alpha_{3,MN} \end{bmatrix}^T_{MN \times 1} \quad (1.B.3)$$

The generalized forces can be written in vector form as

$$\{\mathbf{F}\} = 2j\omega\rho_0 I \begin{bmatrix} f_{11} & f_{21} & \cdots & f_{M1} & f_{12} & f_{22} & \cdots & f_{M2} & \cdots & f_{MN} \end{bmatrix}^T_{MN \times 1} \quad (1.B.4)$$

$$\lambda^{*1}_{1,mn} = 3\left(\frac{m}{a}\right)^4 + 3\left(\frac{n}{b}\right)^4 + 2\left(\frac{m}{a}\right)^2\left(\frac{n}{b}\right)^2 \quad (1.B.5)$$

$$\Delta_1^{*1} = \begin{bmatrix} \lambda^{*1}_{1,11} & & & & & & & & \\ & \lambda^{*1}_{1,21} & & & & & & & \\ & & \ddots & & & & & & \\ & & & \lambda^{*1}_{1,M1} & & & & & \\ & & & & \lambda^{*1}_{1,12} & & & & \\ & & & & & \lambda^{*1}_{1,22} & & & \\ & & & & & & \ddots & & \\ & & & & & & & \lambda^{*1}_{1,M2} & \\ & & & & & & & & \ddots & \\ & & & & & & & & & \lambda^{*1}_{1,MN} \end{bmatrix}_{MN \times MN} \quad (1.B.6)$$

$$\lambda_{1,n}^{*2} = \frac{2n^4}{b^4} \begin{bmatrix} 0 & 1 & 1 & \cdots & 1 \\ 1 & 0 & 1 & \cdots & 1 \\ 1 & 1 & 0 & \cdots & \cdots \\ \cdots & \cdots & \cdots & \ddots & 1 \\ 1 & 1 & \cdots & 1 & 0 \end{bmatrix}_{M \times N} \quad (1.B.7)$$

$$\Delta_1^{*2} = \begin{bmatrix} \lambda_{1,1}^{*2} & & & \\ & \lambda_{1,2}^{*2} & & \\ & & \ddots & \\ & & & \lambda_{1,N}^{*2} \end{bmatrix}_{MN \times MN} \quad (1.B.8)$$

$$\lambda_1^{*3} = \frac{2}{a^4} \begin{bmatrix} 1^4 & & & \\ & 2^4 & & \\ & & \ddots & \\ & & & M^4 \end{bmatrix}_{M \times N} \quad (1.B.9)$$

$$\Delta_1^{*3} = \begin{bmatrix} 0 & \lambda_1^{*3} & \lambda_1^{*3} & \cdots & \lambda_1^{*3} \\ \lambda_1^{*3} & 0 & \lambda_1^{*3} & \cdots & \lambda_1^{*3} \\ \lambda_1^{*3} & \lambda_1^{*3} & 0 & \cdots & \cdots \\ \cdots & \cdots & \cdots & \ddots & \lambda_1^{*3} \\ \lambda_1^{*3} & \lambda_1^{*3} & \cdots & \lambda_1^{*3} & 0 \end{bmatrix}_{MN \times MN} \quad (1.B.10)$$

$$\Delta_2^{*1} = \frac{9ab}{4} \begin{bmatrix} 1 & & & \\ & 1 & & \\ & & \ddots & \\ & & & 1 \end{bmatrix}_{MN \times MN} \quad (1.B.11)$$

$$\lambda_2^{*2} = \frac{3ab}{2} \begin{bmatrix} 0 & 1 & 1 & \cdots & 1 \\ 1 & 0 & 1 & \cdots & 1 \\ 1 & 1 & 0 & \cdots & \cdots \\ \cdots & \cdots & \cdots & \ddots & 1 \\ 1 & 1 & \cdots & 1 & 0 \end{bmatrix}_{M \times N} \quad (1.B.12)$$

$$\Delta_2^{*2} = \begin{bmatrix} \lambda_2^{*2} & & & \\ & \lambda_2^{*2} & & \\ & & \ddots & \\ & & & \lambda_2^{*2} \end{bmatrix}_{MN \times MN} \quad (1.B.13)$$

$$\lambda_2^{*3} = \frac{3ab}{2} \begin{bmatrix} 1 & & & & \\ & 1 & & & \\ & & \ddots & & \\ & & & & 1 \end{bmatrix}_{M \times N} \tag{1.B.14}$$

$$\Delta_2^{*3} = \begin{bmatrix} 0 & \lambda_2^{*3} & \lambda_2^{*3} & \cdots & \lambda_2^{*3} \\ \lambda_2^{*3} & 0 & \lambda_2^{*3} & \cdots & \lambda_2^{*3} \\ \lambda_2^{*3} & \lambda_2^{*3} & 0 & \cdots & \cdots \\ \cdots & \cdots & \cdots & \ddots & \lambda_2^{*3} \\ \lambda_2^{*3} & \lambda_2^{*3} & \cdots & \lambda_2^{*3} & 0 \end{bmatrix}_{MN \times MN} \tag{1.B.15}$$

$$\lambda_2^{*4} = ab \begin{bmatrix} 0 & 1 & 1 & \cdots & 1 \\ 1 & 0 & 1 & \cdots & 1 \\ 1 & 1 & 0 & \cdots & \cdots \\ \cdots & \cdots & \cdots & \ddots & 1 \\ 1 & 1 & \cdots & 1 & 0 \end{bmatrix}_{M \times N} \tag{1.B.16}$$

$$\Delta_2^{*4} = \begin{bmatrix} 0 & \lambda_2^{*4} & \lambda_2^{*4} & \cdots & \lambda_2^{*4} \\ \lambda_2^{*4} & 0 & \lambda_2^{*4} & \cdots & \lambda_2^{*4} \\ \lambda_2^{*4} & \lambda_2^{*4} & 0 & \cdots & \cdots \\ \cdots & \cdots & \cdots & \ddots & \lambda_2^{*4} \\ \lambda_2^{*4} & \lambda_2^{*4} & \cdots & \lambda_2^{*4} & 0 \end{bmatrix}_{MN \times MN} \tag{1.B.17}$$

In the context of the above sub-matrices, the elemental matrices can be derived as

$$[\mathbf{T}_{11}]_{MN \times MN} = 4D_1 \pi^4 ab \left(\Delta_1^{*1} + \Delta_1^{*2} + \Delta_1^{*3} \right) \\ - \left(m_1 \omega^2 + j\omega \rho_0 \frac{2\omega e^{2jk_z H_1}}{k_z \left(1 - e^{2jk_z H_1}\right)} \right) \cdot \left(\Delta_2^{*1} + \Delta_2^{*2} + \Delta_2^{*3} + \Delta_2^{*4} \right) \tag{1.B.18}$$

$$[\mathbf{T}_{12}]_{MN \times MN} = j\omega \rho_0 \frac{2\omega e^{jk_z H_1}}{k_z \left(1 - e^{2jk_z H_1}\right)} \left(\Delta_2^{*1} + \Delta_2^{*2} + \Delta_2^{*3} + \Delta_2^{*4} \right) \tag{1.B.19}$$

$$[\mathbf{T}_{21}]_{MN \times MN} = j\omega \rho_0 \frac{2\omega e^{jk_z H_1}}{k_z \left(1 - e^{2jk_z H_1}\right)} \left(\Delta_2^{*1} + \Delta_2^{*2} + \Delta_2^{*3} + \Delta_2^{*4} \right) \tag{1.B.20}$$

$$[\mathbf{T}_{22}]_{MN\times MN} = 4D_2\pi^4 ab \left(\Delta_1^{*1} + \Delta_1^{*2} + \Delta_1^{*3}\right)$$
$$+ \left(-m_2\omega^2 + \frac{\omega^2\rho_0\left(e^{jk_z(H_1-H_2)} - e^{jk_z(H_1+H_2)}\right)}{k_z(-1+e^{2jk_zH_1})\sin(k_z(H_1-H_2))}\right)\left(\Delta_2^{*1}+\Delta_2^{*2}+\Delta_2^{*3}+\Delta_2^{*4}\right)$$
(1.B.21)

$$[\mathbf{T}_{23}]_{MN\times MN} = \frac{\omega^2\rho_0}{k_z\sin(k_z(H_1-H_2))}\left(\Delta_2^{*1} + \Delta_2^{*2} + \Delta_2^{*3} + \Delta_2^{*4}\right) \quad (1.B.22)$$

$$[\mathbf{T}_{32}]_{MN\times MN} = \frac{\omega^2\rho_0}{k_z\sin(k_z(H_1-H_2))}\left(\Delta_2^{*1} + \Delta_2^{*2} + \Delta_2^{*3} + \Delta_2^{*4}\right) \quad (1.B.23)$$

$$[\mathbf{T}_{33}]_{MN\times MN} = 4D_3\pi^4 ab \left(\Delta_1^{*1} + \Delta_1^{*2} + \Delta_1^{*3}\right)$$
$$- \left(m_3\omega^2 + \frac{\omega^2\rho_0 e^{-jk_z(H_1-H_2)}}{k_z\sin(k_z(H_1-H_2))}\right)\left(\Delta_2^{*1} + \Delta_2^{*2} + \Delta_2^{*3} + \Delta_2^{*4}\right)$$
(1.B.24)

References

1. Grosveld F (1992) Plate acceleration and sound transmission due to random acoustic and boundary-layer excitation. AIAA J 30(3):601–607
2. Maury C, Gardonio P, Elliott SJ (2002) Model for active control of flow-induced noise transmitted through double partitions. AIAA J 40(6):1113–1121
3. Pietrzko SJ, Mao Q (2008) New results in active and passive control of sound transmission through double wall structures. Aerosp Sci Technol 12(1):42–53
4. Wu SF, Wu G, Puskarz MM et al (1997) Noise transmission through a vehicle side window due to turbulent boundary layer excitation. J Vib Acoust 119(4):557–562
5. Quirt JD (1982) Sound transmission through windows I. Single and double glazing. J Acoust Soc Am 72(3):834–844
6. Quirt JD (1983) Sound transmission through windows II. Double and triple glazing. J Acoust Soc Am 74(2):534–542
7. Lyle K, Mixson J (1987) Laboratory study of sidewall noise transmission and treatment for a light aircraft fuselage. J Aircr 24(9):660–665
8. Xin FX, Lu TJ (2009) Analytical and experimental investigation on transmission loss of clamped double panels: implication of boundary effects. J Acoust Soc Am 125(3): 1506–1517
9. Xin FX, Lu TJ, Chen CQ (2008) Vibroacoustic behavior of clamp mounted double-panel partition with enclosure air cavity. J Acoust Soc Am 124(6):3604–3612
10. Franco F, Cunefare KA, Ruzzene M (2007) Structural-acoustic optimization of sandwich panels. J Vib Acoust 129(3):330–340
11. Xin FX, Lu TJ, Chen CQ (2009) External mean flow influence on noise transmission through double-leaf aeroelastic plates. AIAA J 47(8):1939–1951
12. Xin FX, Lu TJ, Chen CQ (2009) Dynamic response and acoustic radiation of double-leaf metallic panel partition under sound excitation. Comput Mater Sci 46(3):728–732
13. Laulagnet B (1998) Sound radiation by a simply supported unbaffled plate. J Acoust Soc Am 103(5):2451–2462

14. Lomas NS, Hayek SI (1977) Vibration and acoustic radiation of elastically supported rectangular plates. J Sound Vib 52(1):1–25
15. Sewell EC (1970) Transmission of reverberant sound through a single-leaf partition surrounded by an infinite rigid baffle. J Sound Vib 12(1):21–32
16. London A (1950) Transmission of reverberant sound through double walls. J Acoust Soc Am 22(2):270–279
17. Antonio JMP, Tadeu A, Godinho L (2003) Analytical evaluation of the acoustic insulation provided by double infinite walls. J Sound Vib 263(1):113–129
18. Wang J, Lu TJ, Woodhouse J et al (2005) Sound transmission through lightweight double-leaf partitions: theoretical modelling. J Sound Vib 286(4–5):817–847
19. Kropp W, Rebillard E (1999) On the air-borne sound insulation of double wall constructions. Acta Acust Acust 85:707–720
20. Tadeu A, Pereira A, Godinho L et al (2007) Prediction of airborne sound and impact sound insulation provided by single and multilayer systems using analytical expressions. Appl Acoust 68(1):17–42
21. White PH, Powell A (1966) Transmission of random sound and vibration through a rectangular double wall. J Acoust Soc Am 40(4):821–832
22. Cummings A, Mulholland KA (1968) The transmission loss of finite sized double panels in a random incidence sound field. J Sound Vib 8(1):126–133
23. Villot M, Guigou C, Gagliardini L (2001) Predicting the acoustical radiation of finite size multi-layered structures by applying spatial windowing on infinite structures. J Sound Vib 245(3):433–455
24. Chazot JD, Guyader JL (2007) Prediction of transmission loss of double panels with a patch-mobility method. J Acoust Soc Am 121(1):267–278
25. Yairi M, Sakagami K, Sakagami E et al (2002) Sound radiation from a double-leaf elastic plate with a point force excitation: effect of an interior panel on the structure-borne sound radiation. Appl Acoust 63(7):737–757
26. Carneal JP, Fuller CR (2004) An analytical and experimental investigation of active structural acoustic control of noise transmission through double panel systems. J Sound Vib 272(3–5):749–771
27. Brunskog J (2005) The influence of finite cavities on the sound insulation of double-plate structures. J Acoust Soc Am 117(6):3727–3739
28. Beranek LL, Work GA (1949) Sound transmission through multiple structures containing flexible blankets. J Acoust Soc Am 21(4):419–428
29. Mulholland KA, Parbrook HD, Cummings A (1967) The transmission loss of double panels. J Sound Vib 6(3):324–334
30. Fahy F (1985) Sound and structural vibration: radiation, transmission and response. Academic, London
31. Bao C, Pan J (1997) Experimental study of different approaches for active control of sound transmission through double walls. J Acoust Soc Am 102(3):1664–1670
32. Carneal JP, Fuller CR (1995) Active structural acoustic control of noise transmission through double panel systems. AIAA J 33(4):618–623
33. Hongisto V (2000) Sound insulation of doors – Part 1: Prediction models for structural and leak transmission. J Sound Vib 230(1):133–148
34. Tadeu AJB, Mateus DMR (2001) Sound transmission through single, double and triple glazing. Experimental evaluation. Appl Acoust 62(3):307–325
35. Leppington FG, Broadbent EG, Butler GF (2006) Transmission of sound through a pair of rectangular elastic plates. Ima J Appl Math 71(6):940–955
36. Panneton R, Atalla N (1996) Numerical prediction of sound transmission through finite multilayer systems with poroelastic materials. J Acoust Soc Am 100(1):346–354
37. Mahjoob MJ, Mohammadi N, Malakooti S (2009) An investigation into the acoustic insulation of triple-layered panels containing Newtonian fluids: theory and experiment. Appl Acoust 70(1):165–171
38. Jones RE (1979) Intercomparisons of laboratory determinations of airborne sound transmission loss. J Acoust Soc Am 66(1):148–164

References

39. Toyoda M, Kugo H, Shimizu T et al (2008) Effects of an air-layer-subdivision technique on the sound transmission through a single plate. J Acoust Soc Am 123(2):825–831
40. Liu BL, Feng LP, Nilsson A (2007) Influence of overpressure on sound transmission through curved panels. J Sound Vib 302(4–5):760–776
41. Pellicier A, Trompette N (2007) A review of analytical methods, based on the wave approach, to compute partitions transmission loss. Appl Acoust 68(10):1192–1212
42. Cheng L, Li YY, Gao JX (2005) Energy transmission in a mechanically-linked double-wall structure coupled to an acoustic enclosure. J Acoust Soc Am 117(5):2742–2751
43. Lee JH, Kim J (2002) Analysis of sound transmission through periodically stiffened panels by space-harmonic expansion method. J Sound Vib 251(2):349–366
44. Maidanik G (1962) Response of ribbed panels to reverberant acoustic fields. J Acoust Soc Am 34(6):809–826
45. Ruzzene M (2004) Vibration and sound radiation of sandwich beams with honeycomb truss core. J Sound Vib 277(4–5):741–763
46. Liu B, Feng L, Nilsson A (2007) Sound transmission through curved aircraft panels with stringer and ring frame attachments. J Sound Vib 300(3–5):949–973
47. Takahashi D (1995) Effects of panel boundedness on sound transmission problems. J Acoust Soc Am 98(5):2598–2606
48. Lin G-F, Garrelick JM (1977) Sound transmission through periodically framed parallel plates. J Acoust Soc Am 61(4):1014–1018
49. Takahashi D (1983) Sound radiation from periodically connected double-plate structures. J Sound Vib 90(4):541–557
50. Xin FX, Lu TJ (2009) Effects of core topology on sound insulation performance of lightweight all-metallic sandwich panels. Mater Manuf Process 26(9):1213–1221
51. Brunskog J (2005) The influence of finite cavities on the sound insulation of double-plate structures. Acta Astronaut 117(6):3727–3739
52. Pan J, Bao C (1998) Analytical study of different approaches for active control of sound transmission through double walls. J Acoust Soc Am 103(4):1916–1922
53. Kim BK, Kang HJ, Kim JS et al (2004) Tunneling effect in sound transmission loss determination: theoretical approach. J Acoust Soc Am 115(5):2100–2109
54. Farag NH, Pan J (1998) Free and forced in-plane vibration of rectangular plates. J Acoust Soc Am 103(1):408–413
55. Bosmans I, Mees P, Vermeir G (1996) Structure-borne sound transmission between thin orthotropic plates: analytical solutions. J Sound Vib 191(1):75–90
56. Graham WR (2007) Analytical approximations for the modal acoustic impedances of simply supported, rectangular plates. J Acoust Soc Am 122(2):719–730
57. Utley WA, Fletcher BL (1973) The effect of edge conditions on the sound insulation of double windows. J Sound Vib 26(1):63–72
58. Hongisto V, Lindgren M, Keranen J (2001) Enhancing maximum measurable sound reduction index using sound intensity method and strong receiving room absorption. J Acoust Soc Am 109(1):254–265
59. ASTM E 90-04 (2004) Standard test method for laboratory measurement of airborne sound transmission loss of building partitions and elements
60. Guigou C, Li Z, Fuller CR (1996) The relationship between volume velocity and far-field radiated pressure of a planar structure. J Sound Vib 197(2):252–254
61. Leissa AW (1993) Vibrations of plates. Acoustical Society of America, New York
62. Leissa AW (1973) The free vibration of rectangular plates. J Sound Vib 31:257–293
63. Laura PAA, Grossi RO (1981) Transverse vibrations of rectangular plates with edges elastically restrained against translation and rotation. J Sound Vib 75(1):101–107
64. Xin FX, Lu TJ, Chen CQ (2010) Sound transmission through simply supported finite double-panel partitions with enclosed air cavity. J Vib Acoust 132(1):011008: 011001–011011
65. Xin FX, Lu TJ (2010) Analytical modeling of sound transmission across finite aeroelastic panels in convected fluids. J Acoust Soc Am 128(3):1097–1107

66. Xin FX, Lu TJ (2011) Transmission loss of orthogonally rib-stiffened double-panel structures with cavity absorption. J Acoust Soc Am 129(4):1919–1934
67. Xin FX, Lu TJ (2011) Analytical modeling of wave propagation in orthogonally rib-stiffened sandwich structures: sound radiation. Comput Struct 89(5–6):507–516
68. Langley RS, Smith JRD, Fahy FJ (1997) Statistical energy analysis of periodically stiffened damped plate structures. J Sound Vib 208(3):407–426
69. Xin FX, Lu TJ (2010) Sound radiation of orthogonally rib-stiffened sandwich structures with cavity absorption. Compos Sci Technol 70(15):2198–2206
70. Xin FX, Lu TJ (2010) Analytical modeling of fluid loaded orthogonally rib-stiffened sandwich structures: sound transmission. J Mech Phys Solids 58(9):1374–1396
71. Gardonio P, Elliott SJ (1999) Active control of structure-borne and airborne sound transmission through double panel. J Aircr 36(6):1023–1032
72. Kaiser OE, Pietrzko SJ, Morari M (2003) Feedback control of sound transmission through a double glazed window. J Sound Vib 263(4):775–795
73. Gardonio P (2002) Review of active techniques for aerospace vibro-acoustic control. J Aircr 39(2):206–214
74. Sas P, Bao C, Augusztinovicz F et al (1995) Active control of sound transmission through a double panel partition. J Sound Vib 180(4):609–625
75. Mulholland KA, Price AJ, Parbrook HD (1968) Transmission loss of multiple panels in a random incidence field. J Acoust Soc Am 43(6):1432–1435
76. Yairi M, Sakagami K, Morimoto M et al (2003) Effect of acoustical damping with a porous absorptive layer in the cavity to reduce the structure-borne sound radiation from a double-leaf structure. Appl Acoust 64(4):365–384
77. Del Coz Diaz JJ, Alvarez Rabanal FP, Garcia Nieto PJ et al (2010) Sound transmission loss analysis through a multilayer lightweight concrete hollow brick wall by FEM and experimental validation. Build Environ 45(11):2373–2386
78. Sgard FC, Atalla N, Nicolas J (2000) A numerical model for the low frequency diffuse field sound transmission loss of double-wall sound barriers with elastic porous linings. J Acoust Soc Am 108(6):2865–2872
79. Price AJ, Crocker MJ (1970) Sound transmission through double panels using statistical energy analysis. J Acoust Soc Am 47(3A):683–693
80. Craik RJM (2003) Non-resonant sound transmission through double walls using statistical energy analysis. Appl Acoust 64(3):325–341
81. Fahy FJ (1994) Statistical energy analysis: a critical overview. Philos Trans Phys Sci Eng 346(1681):431–447
82. Vinokur RY (1990) Transmission loss of triple partitions at low frequencies. Appl Acoust 29(1):15–24
83. Brekke A (1981) Calculation methods for the transmission loss of single, double and triple partitions. Appl Acoust 14(3):225–240
84. Taylor RL, Govindjee S (2004) Solution of clamped rectangular plate problems. Commun Numer Methods Eng 20(10):757–765
85. Ingard U (1959) Influence of fluid motion past a plane boundary on sound reflection, absorption, and transmission. J Acoust Soc Am 31(7):1035–1036

Chapter 2
Vibroacoustics of Uniform Structures in Mean Flow

Abstract This chapter is organized as three parts: in the first part, an analytic approach is formulated to account for the effects of mean flow on sound transmission across a simply supported rectangular aeroelastic panel. The application of the convected wave equation and the displacement continuity condition at the fluid-panel interfaces ensures the exact handling of the complex aeroelastic coupling between panel vibration and fluid disturbances. To explore the mean flow effects on sound transmission, three different cases (i.e., mean flow on incident side only, on radiating side only, and on both sides) are separately considered in terms of refraction angular relations and sound transmission loss (STL) plots. Obtained results show that the influence of the incident side mean flow upon sound penetration is significantly different from that of the transmitted side mean flow. The contour plot of refraction angle versus incident angle for the case when the mean flow is on the transmitted side is just a reverse of that when the mean flow is on the incident side. The aerodynamic damping effects on the transmission of sound are well captured by plotting the STL as a function of frequency for varying Mach numbers. However, as the Mach number is increased, the coincidence dip frequency increases when the flow is on the incident side but remains unchanged when in the flow is on the radiating side. In the most general case when the fluids on both sides of the panel are convecting, the refraction angular relations are significantly different from those when the fluid on one side of the panel is moving and that on the other side is at rest.

In the second part, the transmission of external jet noise through a double-leaf skin plate of aircraft cabin fuselage in the presence of external mean flow is analytically studied. An aero-acoustic-elastic theoretical model is developed and applied to calculate the sound transmission loss (STL) versus frequency curves. Four different types of acoustic phenomenon (i.e., the mass-air-mass resonance, the standing-wave attenuation, the standing-wave resonance, and the coincidence resonance) for a flat double-leaf plate as well as the ring frequency resonance for a curved double-leaf plate are identified. Independent of the proposed theoretical model, simple closed-form formulae for the natural frequencies associated with the above acoustic phenomena are derived using physical principles. Excellent agreement between the model predictions and the closed-form formulae is achieved. Systematic parametric investigation with the model demonstrates that the presence of the mean flow as well as the sound incidence angles affects substantially the sound transmission behavior of the double-leaf structure. The influences of panel curvature together with cabin internal pressure on jet-noise transmission are also significant and should be taken into account when designing aircraft cabin fuselages.

In the third part, a theoretical model is developed to investigate the influence of external mean flow on the transmission of sound through an infinite double-leaf panel filled with porous sound absorptive materials. The sound transmission process in the porous material is modeled using the method of equivalent fluid-structure coupling conditions that are accounted for to ensure displacement continuity at fluid-structure interfaces. Analytic solutions for the sound transmission loss of the whole structure are obtained. For validation, the model predictions are compared with existing experimental results. Numerical investigations with the model are subsequently performed to quantify how a set of systematic parameters affect the sound transmission loss. It is demonstrated that the porous material affects the STL curve in terms of both the absorption effect and the damping effect. Besides, the material loss factor and the thickness of the faceplates also have an influence on the coincidence dip of the STL curve. At frequencies below the coincidence frequency, the external mean flow increases the STL values due to the added damping effect of the mean flow while shifting the coincidence frequency upward because of the refraction effect of the mean flow. In addition, the coincidence frequency decreases with increasing azimuth angle between the sound incident direction and mean flow direction.

2.1 Finite Single-Leaf Aeroelastic Plate

2.1.1 Introduction

The vibroacoustic behavior of a flexible panel immersed in convected fluid flow is of practical importance for high-speed transportation vehicles (e.g., aircrafts, trains, and automobiles), and its prediction requires the combined knowledge of aeroelastics and structural acoustics [1–21]. For example, the interior noise of an aircraft stems mainly from the noise induced by external turbulent boundary layer (TBL) and the engine exhaust noise [6, 17, 22–25]. Therefore, the reduction of noise transmission into aircraft cabin interior has been a long-lasting issue for the airplane industry.

Earlier theoretical studies [26, 27] on the somewhat idealized but significant problem of reflection, transmission, and amplification of sound propagation into a moving medium from a static fluid in the absence of a panel have helped to understand the physical mechanisms associated with the transmission of sound through panels immersed in convected fluids. For instance, by studying the total reflection arising from the refraction effect of mean flow and the sound amplification due to self-excited disturbance feeding, it has been found that there exist two critical angles for the occurrence of total reflection when the flow speed is more than twice that of sound [27].

Subsequently, with the focus placed upon the transmission of noise into aircraft cabin interior, Koval [1] studied theoretically the effects of airflow, panel curvature, and internal pressurization on the transmission loss of field incidence through an infinite single panel. It is found that both the mean flow and the panel curvature can enhance the sound transmission loss (STL), whereas the internal pressurization can lead to a slight decrease of the STL. Built upon Koval's work, Xin et al. [23] proposed recently a theoretical model to quantify the influence of external mean flow upon noise transmission through double-leaf aircraft fuselage into cabin interior. A series of closed-form analytic formulae were also derived from physical principles governing mass-air-mass resonance, standing-wave attenuation and resonance, and coincidence resonance, which compare favorably with model predictions.

Sgard et al. [15] developed a coupled FEM-BEM (finite element method-boundary element method) approach to study the vibroacoustic behavior of planar structures in the presence of mean flow, with formulations explicitly accounting for the effects of mean flow on the STL in terms of added mass, panel stiffness, and radiating damping. With due considerations of the effects of structural nonlinearities induced by in-plane forces and shearing forces due to plate bending, Wu and Maestrello [16] studied theoretically the vibroacoustic response of a finite plate supported on a rigid baffle and subjected to TBL excitations. By assuming that the aircraft fuselage is locally flat, Howe and Shah [22] solved analytically the problem of noise generation by subsonic, high-speed turbulent flow, and the reverse flow reciprocal theorem was employed to determine Green's functions for treating the scattering of boundary layer wall pressures at panel edges. Graham [24, 25, 28] proposed a theoretical model as a compromise between the requirement of simplicity and retaining the physical features of the aircraft interior noise problem and used asymptotic expressions to study the effects of mean flow on the radiation efficiency of a rectangular plate.

Recently, Frampton and Clark [2, 4, 6–8, 10, 11, 17, 19] studied systematically the influence of convected fluid loading coupling on the vibroacoustic behavior of finite-sized panels. A theoretical model combining the aerodynamic loading of panels and linearized potential flow aerodynamics was firstly developed in Ref. [4] by adopting singular value decomposition [2], and detailed analyses with the TBL-induced noise disturbance taken into account were further proposed in Refs. [6–8, 10, 17]. In particular, the effect of in-plane forces on sound radiation of convected fluid-loaded plates was theoretically analyzed [19], and it is found that the state of stress in the plate exerts a significant effect on the radiation efficiency of the plate.

There also exist typical investigations regarding that active control of vibration and noise transmission. For instance, Clark and Frampton [11] designed a static, constant-gain, output-feedback control compensators to increase the *STL* across a panel subjected to mean flow, and Maury et al. [12, 14, 29] investigated the active control of flow-induced noise transmission across single and double panels.

Although there exist numerous studies on the coupling of panel vibration with fluid flow, such as aeroelastic panel flutter and sound radiation due to flow-induced panel vibration, a few issues regarding its underlying physical mechanisms remain unclear, such as the mechanism underlying sound power flux penetration and the effect of aeroelastic coupling feedback. Furthermore, while a few studies [1, 7, 17, 23] addressed specifically the practically significant problem of sound transmission across a finite aeroelastic panel immersed in convected fluid on one side (either the incident side or the transmitted side), the more general case when the aeroelastic panel is loaded by convected fluids on both sides has not been solved. The solution of this general case not only is complementary to the existing research so that an overlook of this issue can be acquired but also has practical implications particularly in soundproofing machines that require effective cooling by convected flow. For typical instance, the nozzle of the jet engine of an aircraft works in high-temperature environment. It is well known that increasing the temperature of the jet can increase the efficiency and speed of aircraft, but few materials are able to work in such high-temperature environment particularly in hypersonic vehicles. In such cases, active cooling with convective fluid flow in between the sandwich structure (with open-celled cellular core) of the jet nozzle has emerged as a promising approach. Consequently, how the convected flow on both sides of the panel affects the sound transmission should be of concern. This chapter is mainly aimed to remedy this deficiency and to provide more details in our understanding of the interdisciplinary topic of acoustics and aeroelastics/dynamics. After presenting the modeling part in Sect. 2.1.2, in Sects. 2.1.3–2.1.5 of numerical result discussion, the refraction angular relations between the incident and transmitted sound waves are presented in terms of contour plots for three different cases (i.e., panel immersed in convected fluid on the incident side only, on the transmitted side only, and on both sides). Detailed analyses of sound power flux penetration through the convected fluid-loaded panel are also performed in terms of *STL* versus frequency characteristic curves for the three cases.

2.1.2 Modeling of Aeroelastic Coupled System

Consider a finite, rectangular aeroelastic panel simply supported in an infinite acoustic rigid baffle (as shown in Fig. 2.1a, b). The panel has length a along x-direction, width b along y-direction, and thickness h along z-direction, with $h \ll a$ and $h \ll b$ assumed. The panel divides the spatial region into two regimes, i.e., the incident field ($z < 0$) and the transmitted field ($z > 0$) which, for convenience, are numbered below by 1 and 2, respectively (Fig. 2.1b). An oblique plane sound

2.1 Finite Single-Leaf Aeroelastic Plate

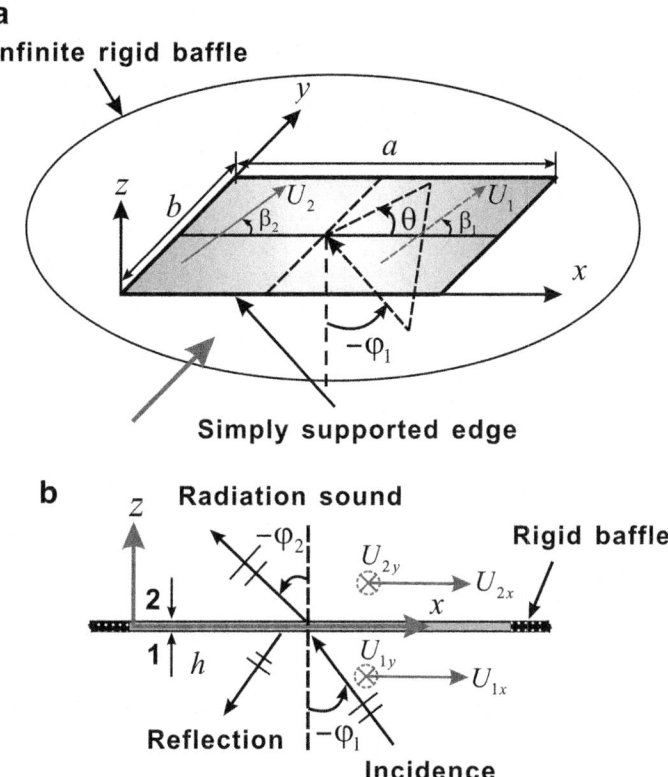

Fig. 2.1 Schematic of sound transmission through a simply supported aeroelastic panel immersed in convected fluid flow on both sides: (**a**) overall view; (**b**) view from the direction of *arrow* in (**a**) (With permission from Acoustical Society of America)

wave varying harmonically in time is incident on the bottom side of the panel, with elevation angle φ_1 and azimuth angle θ (Fig. 2.1a).

The panel is immersed in convected fluid flow on both sides which are parallel to the panel but may flow along different directions (Fig. 2.1) and have different physical properties (including density and sound speed). Without loss of generality, let the flow on the incident side ($z < 0$) have azimuth angle β_1 and speed U_1 and that on the transmitted side ($z > 0$) have azimuth angle β_2 and speed U_2; see Fig. 2.1b.

The panel vibration induced by the incident sound and fluid flow together also creates a pressure disturbance in the surrounding fluid media, including the reflected pressure wave $p^i_{\text{reflected}}$ and the radiating pressure wave $p^i_{\text{radiating}}$ in the incident field and the radiating pressure wave $p^t_{\text{radiating}}$ in the transmitted field. The pressure changes caused by this disturbance will in turn significantly influence the panel vibration, resulting in the so-called aeroelastic coupling [4, 6, 10, 11, 17]. In the present study, it is assumed that the panel deforms out of plane (in the z-direction), positive upward.

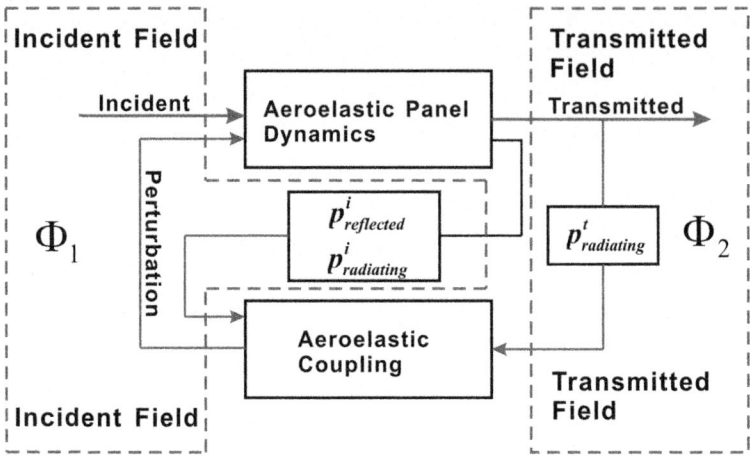

Fig. 2.2 Block diagram of sound transmission across a finite aeroelastic panel immersed in convected fluids on both sides. The *dashed frame* on the *left* indicates the incident field, including the incident sound pressure wave as well as the reflected and radiated pressure waves. The *dashed frame* on the *right* stands for the transmitted field, containing the resultant radiating pressure wave (With permission from Acoustical Society of America)

As an illustration of the sound transmission process, a block diagram (Fig. 2.2) is drawn to highlight the interaction of panel motion and fluid dynamics. Notice first that the left dashed frame represents the incident field, including the incident sound pressure wave as well as the reflected and radiating pressure waves stemming from panel vibration, and this field is characterized by the perturbation acoustic velocity potential Φ_1. The right dashed frame for the transmitted field contains the resultant transmitted pressure wave and is characterized by the perturbation acoustic velocity potential Φ_2. The connections between the two central solid frames in Fig. 2.2 signify essentially the aeroelastic coupling between the panel vibration and the surrounding convected fluids. Subsequent theoretical modeling will consider this aeroelastic coupling by employing the convected wave equation, fluid momentum equation, and displacement continuity condition between the proximal fluid particle and the panel particle.

The dynamic displacement of an aeroelastic panel immersed in convected fluids on both sides and subjected to a uniform, plane sound wave varying harmonically can be described by [3, 5]

$$D\nabla^4 w(x,y;t) + m\frac{\partial^2 w(x,y;t)}{\partial t^2} - j\omega\left[\rho_1\Phi_1(x,y,0;t) - \rho_2\Phi_2(x,y,0;t)\right] = 0 \tag{2.1}$$

where D and m are the flexural rigidity and surface density of the panel and ω is the angular frequency of the incident sound. With c_t denoting the speed of the

2.1 Finite Single-Leaf Aeroelastic Plate

trace wave in the aeroelastic panel, the wavenumber of the trace wave is $k_t = \omega/c_t$. Hence, the displacement of the aeroelastic panel induced by the incident sound and the convected fluid flow can be expressed as:

$$w(x,y;t) = w_0 e^{-j[(k_t \cos\theta)x + (k_t \sin\theta)y - \omega t]} \tag{2.2}$$

With the assumption of idealized fluid (i.e., inviscid, irrotational, and incompressible), the convected fluid field can be regarded as full potential flow. Let Φ_i ($i = 1, 2$) denote the velocity potentials for the acoustic fields in the proximity of the aeroelastic panel, corresponding to the sound incidence and the transmitted field (fields 1 and 2 in Fig. 2.1b), respectively:

$$\Phi_1(x,y,z;t) = I e^{-j(k_{1x}x + k_{1y}y + k_{1z}z - \omega t)} + \beta e^{-j(k_{1x}x + k_{1y}y - k_{1z}z - \omega t)} \tag{2.3}$$

$$\Phi_2(x,y,z;t) = \varepsilon e^{-j(k_{2x}x + k_{2y}y + k_{2z}z - \omega t)} \tag{2.4}$$

where I and β are separately the amplitude of the positive-going wave (incident sound) and the negative-going wave (including the reflected wave and the radiating wave) in the incident acoustic field and ε is the amplitude of the positive-going wave in the transmitted field. The sound wavenumber components in (3) and (4) depend upon the incident sound elevation angle φ_1, the azimuth angle θ and the refraction angle φ_2 according to

$$k_{1x} = k_1 \sin\varphi_1 \cos\theta, \quad k_{1y} = k_1 \sin\varphi_1 \sin\theta, \quad k_{1z} = k_1 \cos\varphi_1 \tag{2.5}$$

$$k_{2x} = k_2 \sin\varphi_2 \cos\theta, \quad k_{2y} = k_2 \sin\varphi_2 \sin\theta, \quad k_{2z} = k_2 \cos\varphi_2 \tag{2.6}$$

For the convenience of describing the modal response of the aeroelastic panel, its dynamic displacement can be rewritten by using the in vacuo orthogonal panel eigenfunctions and the generalized coordinates as

$$w(x,y;t) = \sum_{m=1}^{\infty}\sum_{n=1}^{\infty} \varphi_{mn}(x,y) q_{mn}(t) \tag{2.7}$$

where the modal functions of a simply supported rectangular panel and the generalized coordinates are taken as

$$\varphi_{mn}(x,y) = \sin\frac{m\pi x}{a} \sin\frac{n\pi y}{b} \tag{2.8}$$

$$q_{mn}(t) = \alpha_{mn} e^{j\omega t} \tag{2.9}$$

Similarly, the acoustic velocity potentials of Eqs. (2.3) and (2.4) are expressed as

$$\Phi_1(x,y,z;t) = \sum_{m=1}^{\infty}\sum_{n=1}^{\infty} I_{mn}\varphi_{mn} e^{-j(k_{1z}z-\omega t)} + \sum_{m=1}^{\infty}\sum_{n=1}^{\infty} \beta_{mn}\varphi_{mn} e^{-j(-k_{1z}z-\omega t)} \tag{2.10}$$

$$\Phi_2(x,y,z;t) = \sum_{m=1}^{\infty}\sum_{n=1}^{\infty} \varepsilon_{mn}\varphi_{mn} e^{-j(k_{2z}z-\omega t)} \tag{2.11}$$

The conversion relation between the general forms of Eqs. (2.2), (2.3), and (2.4) and the generalized forms (with modal functions) of Eqs. (2.7), (2.10), and (2.11) can be obtained by utilizing the sine Fourier transform as

$$\tilde{\lambda}_{mn} = \frac{4}{ab}\int_0^b\int_0^a \tilde{\lambda} e^{-j(k_{ix}x+k_{iy}y)} \sin\frac{m\pi x}{a}\sin\frac{n\pi y}{b} dxdy \quad (i=1,2) \tag{2.12}$$

where $\tilde{\lambda}$ refers to any of the symbols I, β, ε, and a (with w_0). Note that the expressions in terms of either traveling wave or panel modal functions are completely equivalent in physical nature when they are both subjected to the same boundary conditions.

The acoustic velocity potentials of (2.3) and (2.4) for an inviscid, irrotational, and incompressible fluid moving in a plane parallel to the aeroelastic panel should satisfy the convected wave equation [1, 30–33], given by

$$\frac{D^2\Phi_1}{Dt^2} = \left(\frac{\partial}{\partial t} + \mathbf{U}_1\cdot\nabla\right)^2 \Phi_1 = c_1^2\nabla^2\Phi_1 \tag{2.13}$$

$$\frac{D^2\Phi_2}{Dt^2} = \left(\frac{\partial}{\partial t} + \mathbf{U}_2\cdot\nabla\right)^2 \Phi_2 = c_2^2\nabla^2\Phi_2 \tag{2.14}$$

which, with fluid velocities expressed as $\mathbf{U}_i = U_{ix}\hat{e}_x + U_{iy}\hat{e}_y$ ($i=1,2$), can be expanded as follows:

$$\left[\frac{\partial^2}{\partial t^2} + U_{1x}^2\frac{\partial^2}{\partial x^2} + U_{1y}^2\frac{\partial^2}{\partial y^2} + 2U_{1x}\frac{\partial^2}{\partial x\partial t} + 2U_{1y}\frac{\partial^2}{\partial y\partial t} + 2U_{1x}U_{1y}\frac{\partial^2}{\partial x\partial y}\right]\Phi_1 = c_1^2\nabla^2\Phi_1 \tag{2.15}$$

$$\left[\frac{\partial^2}{\partial t^2} + U_{2x}^2\frac{\partial^2}{\partial x^2} + U_{2y}^2\frac{\partial^2}{\partial y^2} + 2U_{2x}\frac{\partial^2}{\partial x\partial t} + 2U_{2y}\frac{\partial^2}{\partial y\partial t} + 2U_{2x}U_{2y}\frac{\partial^2}{\partial x\partial y}\right]\Phi_2 = c_2^2\nabla^2\Phi_2 \tag{2.16}$$

Substitution of (2.3) and (2.4) into (2.15) and (2.16) leads to

2.1 Finite Single-Leaf Aeroelastic Plate

$$k_1 = \frac{k_1^*}{1 + M_{1x} \sin\varphi_1 \cos\theta + M_{1y} \sin\varphi_1 \sin\theta} \quad (2.17)$$

$$k_2 = \frac{k_2^*}{1 + M_{2x} \sin\varphi_2 \cos\theta + M_{2y} \sin\varphi_2 \sin\theta} \quad (2.18)$$

where $k_1^* = \omega/c_1$ and $k_2^* = \omega/c_2$ are the wavenumbers in the absence of airflow, while $M_1 = U_1/c_1$ and $M_2 = U_2/c_2$ are the Mach number of mean flow in the incident field and that in the transmitted field, respectively.

The sound waves traveling in fluid media proximal to the panel and the bending wave propagating in the flexural aeroelastic panel should be in coherence with each other in their wavelengths [1], that is,

$$k_{1x} = k_t \cos\theta = k_{2x}, \quad k_{1y} = k_t \sin\theta = k_{2y} \quad (2.19)$$

so that

$$\varphi_2 = \arcsin\left(\frac{c_2 \sin\varphi_1}{c_1 + \left[(c_1 M_{1x} - c_2 M_{2x})\cos\theta + (c_1 M_{1y} - c_2 M_{2y})\sin\theta\right]\sin\varphi_1}\right) \quad (2.20)$$

Note that the corresponding transmitted waves become evanescent waves, when

$$\left|\frac{c_2 \sin\varphi_1}{c_1 + \left[(c_1 M_{1x} - c_2 M_{2x})\cos\theta + (c_1 M_{1y} - c_2 M_{2y})\sin\theta\right]\sin\varphi_1}\right| > 1. \quad (2.21)$$

In such cases, total reflection of the incident sound appears, and the value of φ_2 should be taken as a complex one with real and imaginary parts (either both positive or both negative), such that the value of k_{2z} takes the form of $-j\hbar$ (where $\hbar$ is a positive real number).

2.1.2.1 Displacement Continuity Condition at Fluid-Panel Interfaces

Let ξ_1 and ξ_2 represent the acoustic particle displacement in the incident and transmitted fluid medium, respectively. The fluid particle displacement and the acoustic pressure are related by the fluid momentum equation for inviscid, irrotational, and incompressible fluid as [22]

$$\frac{D^2\xi_1}{Dt^2} = \left(\frac{\partial}{\partial t} + \mathbf{U}_1 \cdot \nabla\right)^2 \xi_1 = -\frac{1}{\rho_1}\frac{\partial p_1}{\partial z}\bigg|_{z=0} \quad (2.22)$$

$$\frac{D^2\xi_2}{Dt^2} = \left(\frac{\partial}{\partial t} + \mathbf{U}_2 \cdot \nabla\right)^2 \xi_2 = -\frac{1}{\rho_2}\frac{\partial p_2}{\partial z}\bigg|_{z=0} \quad (2.23)$$

where the acoustic pressure can be expressed by the acoustic velocity potentials through Bernoulli's equation as [10, 30]

$$p_i = \rho_i \left[\frac{\partial \Phi_i}{\partial t} + U_{ix} \frac{\partial \Phi_i}{\partial x} + U_{iy} \frac{\partial \Phi_i}{\partial y} \right] \quad (i = 1, 2) \tag{2.24}$$

The displacements of the fluid particle adjacent to the panel can be expressed as

$$\xi_1 = \xi_{10} e^{-j(k_{1x}x + k_{1y}y - \omega t)} \tag{2.25}$$

$$\xi_2 = \xi_{20} e^{-j(k_{2x}x + k_{2y}y - \omega t)} \tag{2.26}$$

Substituting (2.24), (2.25), and (2.26) into (2.22) and (2.23) and applying the acoustic velocity potentials of (2.10) and (2.11), one can obtain

$$\xi_{10} = \sum_{m=1}^{\infty} \sum_{n=1}^{\infty} (I_{mn}\varphi_{mn} - \beta_{mn}\varphi_{mn}) \frac{\omega k_{1z}}{(\omega - U_{1x}k_{1x} - U_{1y}k_{1y})^2} e^{j(k_{1x}x + k_{1y}y)} \tag{2.27}$$

$$\xi_{20} = \sum_{m=1}^{\infty} \sum_{n=1}^{\infty} \varepsilon_{mn}\varphi_{mn} \frac{\omega k_{2z}}{(\omega - U_{2x}k_{2x} - U_{2y}k_{2y})^2} e^{j(k_{2x}x + k_{2y}y)} \tag{2.28}$$

The factual case that the aeroelastic panel immersed in a convected fluid medium requires that the displacements of the fluid particles adjacent to the panel should be the same as those of the attached panel particles. Accordingly, the displacement continuity condition can be written as [1, 31]

$$\xi_{10} = w_0, \quad \xi_{20} = w_0 \tag{2.29}$$

Together with (2.2) and (2.27) and (2.28), meanwhile utilizing the following relation between coefficients α_{mn} and w_0,

$$\alpha_{mn} = \frac{4mn\pi^2 w_0 \left[1 - (-1)^m e^{-jk_x a} - (-1)^n e^{-jk_y b} + (-1)^{m+n} e^{-j(k_x a + k_y b)} \right]}{(k_x^2 a^2 - m^2\pi^2)(k_y^2 b^2 - n^2\pi^2)} \tag{2.30}$$

One can express the coefficients in the acoustic velocity potentials by the panel displacement coefficients as

$$\beta_{mn} = I_{mn} - \frac{(\omega - U_{1x}k_{1x} - U_{2x}k_{2x})^2}{\omega k_{1z}} \alpha_{mn} \tag{2.31}$$

$$\varepsilon_{mn} = \frac{(\omega - U_{2x}k_{2x} - U_{2y}k_{2y})^2}{\omega k_{2z}} \alpha_{mn} \tag{2.32}$$

2.1 Finite Single-Leaf Aeroelastic Plate

Substituting (2.7) into (2.1) and applying the orthogonality of the modal functions, one gets

$$\ddot{q}_{mn}(t) + \omega_{mn}^2 q_{mn}(t)$$
$$- \frac{j\omega}{m}\left[\rho_1 I_{mn} e^{-j(k_{1z}z-\omega t)} + \rho_1 \beta_{mn} e^{-j(-k_{1z}z-\omega t)} - \rho_2 \varepsilon_{mn} e^{-j(k_{2z}z-\omega t)}\right] = 0 \tag{2.33}$$

where the natural frequencies of the aeroelastic panel are determined by the panel properties as

$$\omega_{mn}^2 = \frac{D \iint_A \nabla^4 \varphi_{mn} \cdot \varphi_{mn} dA}{m \iint_A \varphi_{mn} \cdot \varphi_{mn} dA} \tag{2.34}$$

Equation (2.33) can be readily converted to the following form:

$$\alpha_{mn} = \frac{2j\omega\rho_1 I_{mn}}{m} \cdot$$
$$\left[\omega_{mn}^2 - \omega^2 + \frac{j\omega\rho_1 \left(\omega - U_{1x}k_{1x} - U_{1y}k_{1y}\right)^2}{m\,\omega k_{1z}} + \frac{j\omega\rho_2 \left(\omega - U_{2x}k_{2x} - U_{2y}k_{2y}\right)^2}{m\,\omega k_{2z}}\right]^{-1}$$
$$\tag{2.35}$$

Once the panel displacement coefficients α_{mn} are known, the acoustic velocity potentials will be known, given by

$$\Phi_1(x,y,0) = 2I e^{-j(k_{1x}x+k_{1y}y)} - \sum_{m=1}^{\infty}\sum_{n=1}^{\infty} \frac{\left(\omega - U_{1x}k_{1x} - U_{1y}k_{1y}\right)^2}{\omega k_{1z}} \alpha_{mn}\varphi_{mn}(x,y) \tag{2.36}$$

$$\Phi_2(x,y,0) = \frac{\left(\omega - U_{2x}k_{2x} - U_{2y}k_{2y}\right)^2}{\omega k_{2z}} \sum_{m=1}^{\infty}\sum_{n=1}^{\infty} \alpha_{mn}\varphi_{mn}(x,y) \tag{2.37}$$

2.1.2.2 Definition of Sound Transmission Loss

Different from previous researches [3, 5, 34, 35] that consider only the incident and reflected sound pressures, the proposed theoretical formulations are capable of accurately modeling all the pressure components in the incident field, and hence all the pressure components are taken into account in the present STL calculations. Actually, this is closer to the physical nature of the factual experimental measurements [36]; thus, the power of incident sound is defined as

$$\Pi_1 = \frac{1}{2}\text{Re}\iint_A p_1 \cdot v_1^* dA \qquad (2.38)$$

where the asterisk symbol denotes complex conjugate, $v_1 = p_1/(\rho_1 c_1)$ is the local acoustic velocity, and

$$p_1 = j\rho_1 \left(\omega - U_{1x}k_{1x} - U_{1y}k_{1y}\right) \Phi_1(x, y, 0)$$
$$= j\rho_1 \left(\omega - U_{1x}k_{1x} - U_{1y}k_{1y}\right)$$
$$\times \left[2Ie^{-j(k_{1x}x+k_{1y}y)} - \sum_{m=1}^{\infty}\sum_{n=1}^{\infty} \frac{\left(\omega - U_{1x}k_{1x} - U_{1y}k_{1y}\right)^2}{\omega k_{1z}} \alpha_{mn}\varphi_{mn}(x, y)\right] \qquad (2.39)$$

is the sound pressure in the incident field. Substituting p_1 and v_1 into (2.38) yields

$$\Pi_1 = \frac{\rho_1 (\omega - U_{1x}k_{1x} - U_{1y}k_{1y})^2}{2c_1} \iint_A |\Phi_1(x, y, 0)|^2 dA$$

$$= \frac{\rho_1 (\omega - U_{1x}k_{1x} - U_{1y}k_{1y})^2}{2c_1}$$

$$\times \iint_A \left|2Ie^{-j(k_{1x}x+k_{1y}y)} - \sum_{m=1}^{\infty}\sum_{n=1}^{\infty} \frac{(\omega - U_{1x}k_{1x} - U_{1y}k_{1y})^2}{\omega k_{1z}} \alpha_{mn}\varphi_{mn}(x, y)\right|^2 dA$$

$$= \frac{\rho_1 (\omega - U_{1x}k_{1x} - U_{1y}k_{1y})^2}{2c_1}$$

$$\times \left| 4I^2 \iint_A e^{-2j(k_{1x}x+k_{1y}y)} dA - 4I \frac{(\omega - U_{1x}k_{1x} - U_{1y}k_{1y})^2}{\omega k_{1z}} \right.$$

$$\times \sum_{m=1}^{\infty}\sum_{n=1}^{\infty} \alpha_{mn} \iint_A e^{-j(k_{1x}x+k_{1y}y)} \varphi_{mn}(x, y) dA$$

$$\left. + \frac{(\omega - U_{1x}k_{1x} - U_{1y}k_{1y})^4}{\omega^2 k_{1z}^2} \sum_{m=1}^{\infty}\sum_{n=1}^{\infty}\sum_{k=1}^{\infty}\sum_{l=1}^{\infty} \alpha_{mn}\alpha_{kl} \iint_A \varphi_{mn}(x, y)\varphi_{kl}(x, y) dA \right| \qquad (2.40)$$

In a similar manner, the radiated sound power can be defined as

$$\Pi_2 = \frac{1}{2}\text{Re}\iint_A p_2 \cdot v_2^* dA \qquad (2.41)$$

where $v_2 = p_2/(\rho_2 c_2)$ is the local acoustic velocity and

$$p_2 = j\rho_2 \left(\omega - U_{2x}k_{2x} - U_{2y}k_{2y}\right) \Phi_2(x, y, 0)$$
$$= j\rho_2 \frac{(\omega - U_{2x}k_{2x} - U_{2y}k_{2y})^3}{\omega k_{2z}} \sum_{m=1}^{\infty}\sum_{n=1}^{\infty} \alpha_{mn}\varphi_{mn}(x, y) \qquad (2.42)$$

2.1 Finite Single-Leaf Aeroelastic Plate

is the sound pressure in the transmitted field. Combination of Eqs. (2.41) and (2.42) and the expression of v_2 results in

$$\Pi_2 = \frac{\rho_2 \omega^2}{2c_2} \iint_A |\Phi_2(x,y,0)|^2 dA = \frac{\rho_2(\omega - U_{2x}k_{2x} - U_{2y}k_{2y})^2}{2c_2}$$

$$\times \iint_A \left| \frac{(\omega - U_{2x}k_{2x} - U_{2y}k_{2y})^2}{\omega k_{2z}} \sum_{m=1}^{\infty} \sum_{n=1}^{\infty} \alpha_{mn} \varphi_{mn}(x,y) \right|^2 dA$$

$$= \frac{\rho_2(\omega - U_{2x}k_{2x} - U_{2y}k_{2y})^2}{2c_2}$$

$$\times \left| \frac{(\omega - U_{2x}k_{2x} - U_{2y}k_{2y})^4}{\omega^2 k_{2z}^2} \sum_{m=1}^{\infty} \sum_{n=1}^{\infty} \sum_{k=1}^{\infty} \sum_{l=1}^{\infty} \alpha_{mn} \alpha_{kl} \iint_A \varphi_{mn}(x,y) \varphi_{kl}(x,y) dA \right|$$

(2.43)

The power transmission coefficient can be obtained as

$$\tau(\varphi_1, \theta, U_1, U_2) = \frac{\Pi_2}{\Pi_1} \tag{2.44}$$

which is dependent upon the sound incident angles (φ_1 and θ) and the mean flow velocities U_1 and U_2. Finally, the sound transmission loss across the panel, defined as the inverse of the power transmission coefficient in decibel scale, is given by

$$STL = 10 \log_{10} \left(\frac{1}{\tau} \right) \tag{2.45}$$

2.1.3 Effects of Mean Flow in Incident Field

To gain fundamental insight into the effects of mean flow on the incident side, the fluid medium on the transmitted side is taken to be at rest. The corresponding contour plot of refraction angle φ_2 versus incident angle φ_1 as well as the *STL* versus frequency plots for a series of different Mach numbers (M_1, incident side) is presented below.

The refraction angle φ_2 is plotted in Fig. 2.3 as a function of incident angle φ_1 for selected Mach numbers. When $M_1 = 0$, the refraction angle is equal to the incident angle φ_1 (in other words, no refraction occurs), which is attributed to the fact that the fluid media considered in the present study are identical (i.e., air) on both sides of the panel. Apart from the case of $M_1 = 0$, the refraction angle is related to the incident angle in a nonlinear fashion due to the refraction effect of the convected flow stream [1, 37]. Another noteworthy point is that when $M_1 > 2$, the φ_2–φ_1 curves are

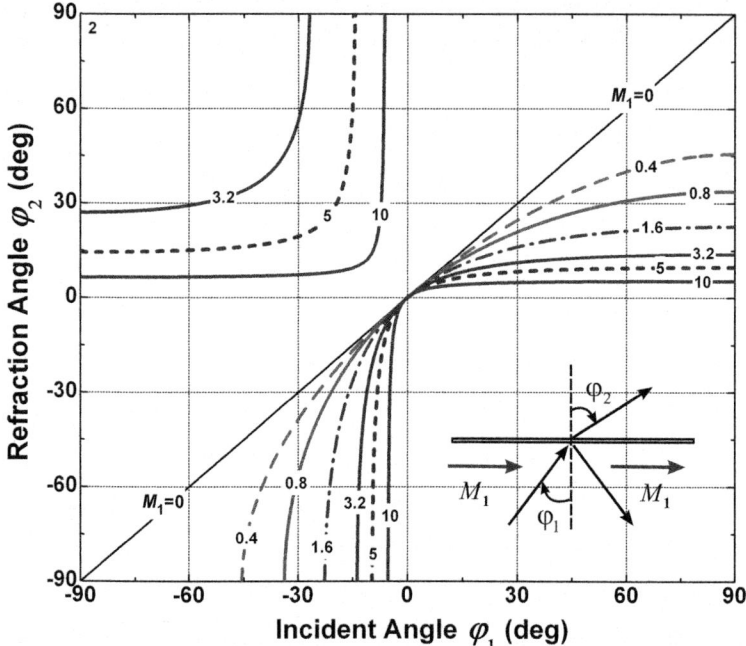

Fig. 2.3 Refraction angle φ_2 plotted as a function of incident angle φ_1 for selected values of Mach number M_1 associated with fluid medium on incident side moving parallel to panel; fluid medium on transmitted side is at rest (i.e., $M_2 = 0$) (With permission from Acoustical Society of America)

divided into two branches: the positive refraction branch and the negative refraction branch, the gap between them along the φ_1-axis corresponding to the total reflection. In the bigger branch (positive refraction), the refraction angle has the same sign as the incident angle, whereas the reverse holds in the smaller branch (negative refraction).

It is seen from Fig. 2.3 that total reflection occurs only when the sound is incident along the upstream direction of the (incident side) mean flow. As previously mentioned by Ribner [26] and Xin et al. [23], when total reflection occurs, while a disturbance on the incident side does penetrate through the panel into the transmitted side, its amplitude attenuates exponentially over a distance on the order of $\hbar$. Furthermore, the transmitted sound carries no energy due to the fact that the pressure and the particle velocity are fully out of phase (i.e., 90°).

To demonstrate the effects of mean flow, the *STL* of an infinite aeroelastic panel is plotted in Fig. 2.4 as a function of incident frequency with incident angle $\varphi_1 = 60°$ and $\theta = 0°$ for different Mach numbers ($M_1 = 0$, 0.4, and 0.8). Similarly, Fig. 2.5 presents the *STL* versus frequency plots of a finite aeroelastic panel for $M_1 = 0$ and 0.8, with their infinite counterparts included as reference. It can be observed from these results that the dip related to coincidence resonance shifts noticeably to higher frequencies as the Mach number (i.e., M_1) is increased. This shift is caused mainly

2.1 Finite Single-Leaf Aeroelastic Plate

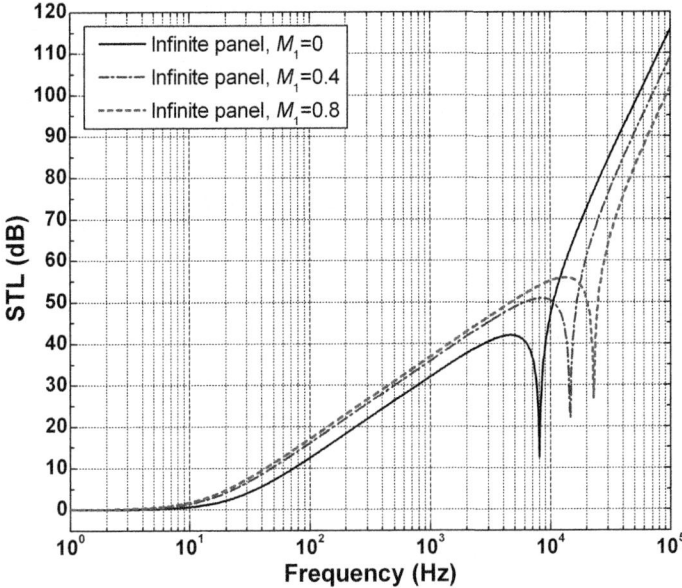

Fig. 2.4 STL plotted as a function of frequency of an infinite panel immersed in convected fluid for selected Mach numbers (*solid line* $M_1 = 0$, *dash-dot line* $M_1 = 0.4$, and *short dash line* $M_1 = 0.8$), when sound is incident with elevation angle $\varphi_1 = 60°$ and azimuth angle $\theta = 0°$. The mean flow on the incident side is completely aligned with the x-direction (Fig. 2.1), while the fluid on the transmitted side is at rest (i.e., $M_2 = 0$) (With permission from Acoustical Society of America)

by the refraction effect of the mean flow [1, 26, 37], i.e., the factual incident angle $\tilde{\varphi}_2$ in the proximal field of the panel has been significantly altered from its original angle of φ_1 by the moving stream on the incident side.

Note also that the factual incident angle $\tilde{\varphi}_2$ in the proximal field of the panel actually is equal to the refraction angle φ_2 when the fluid medium (i.e., air) on the incident side is the same as that on the transmitted side. This effect can be explained by considering the physical origin of the coincidence resonance phenomenon, i.e., the resonance effect between the incident sound wave and the bending wave in the panel when their phases are matched. The coincidence resonance frequency can be predicted as [23]

$$f_c = \frac{c_1^2}{2\pi h \sin \tilde{\varphi}_2} \sqrt{\frac{12\rho(1-v^2)}{E}} \qquad (2.46)$$

In the case when the original incident angle $\varphi_1 = 60°$, one sees from Fig. 2.3 that the factual incident angle $\tilde{\varphi}_2 \ (=\varphi_2)$ decreases as M_1 is increased; as a result, the coincidence resonance frequency should also increase, as predicted by Eq. (2.46).

Another observation from Fig. 2.4 is that the coincidence resonance dip divides the *STL* plot into two regimes within which the mean flow has opposite effects on

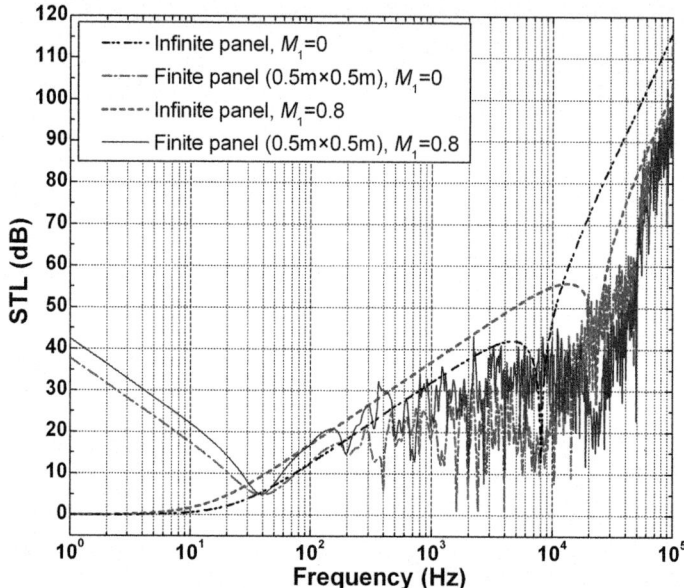

Fig. 2.5 STL plotted as a function of frequency of a finite panel immersed in convected fluid for selected Mach numbers ($M_1 = 0$ and $M_1 = 0.8$), when sound is incident with elevation angle $\varphi_1 = 60°$ and azimuth angle $\theta = 0°$. The mean flow on the incident side is completely aligned with the x-direction (Fig. 2.1), while the fluid on the transmitted side is at rest (i.e., $M_2 = 0$). The corresponding STL versus frequency plots of an infinite panel are included as reference (With permission from Acoustical Society of America)

the *STL*. At frequencies below the coincidence dip, a higher velocity of the mean flow enhances more significantly the *STL* value of the aeroelastic panel; in contrast, at frequencies beyond the coincidence dip, the *STL* value decreases with increasing mean low velocity.

In comparison with an infinite panel, the *STL* versus frequency plot of a finite panel immersed in convected fluid medium should reflect its modal behavior due to the boundary condition, which is affirmed by the appearance of dense peaks and dips in Fig. 2.5. With reference to the counterpart of the infinite panel, the *STL* plot of the finite panel exhibits similar tendencies beyond the (1, 1) modal frequency at approximately 40Hz, with the former setting an asymptotic maximum of the latter. However, at frequencies below the (1, 1) modal frequency, the tendency is remarkably different between the two panels as a result of the boundary effects [3, 5].

The dip associated with the coincidence resonance can also be distinguished from the *STL* plot of the finite panel, which agrees well with the dip in the *STL* plot of the infinite panel (Fig. 2.5). Moreover, similar to the infinite panel, the presence of mean flow on the incident side leads to enhanced *STL* below the coincidence resonance frequency and diminished *STL* beyond it. The present result that the *STL* increases significantly with increasing Mach number over a broad frequency range (approximately 40–10,000 Hz) is consistent with previously published results [6, 7,

2.1 Finite Single-Leaf Aeroelastic Plate

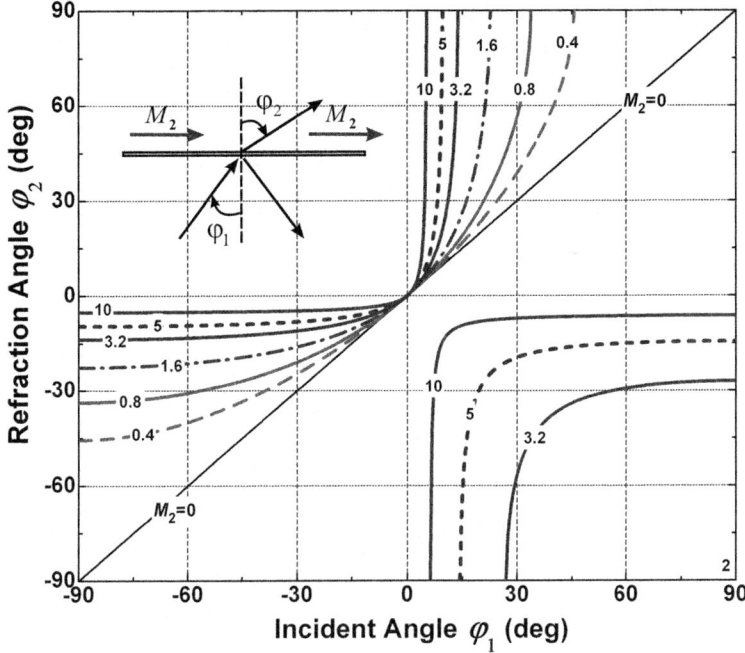

Fig. 2.6 Refraction angle φ_2 plotted as a function of incident angle φ_1 for selected Mach numbers (M_2) of moving fluid (aligned with x-direction) on transmitted side; fluid medium on incident side is at rest (i.e., $M_1 = 0$) (With permission from Acoustical Society of America)

17, 18]. The influence of mean flow on the transmission of sound across a panel can be attributed to the so-called aerodynamic damping effect [38]. Here, the role played by aerodynamic damping is twofold: as the Mach number is increased, it not only increases the sound power radiated to the convected fluid but also modifies the panel stiffness (via convected fluid loading), leading to an overall alteration of the panel impedance [7, 17].

2.1.4 Effects of Mean Flow in Transmitted Field

To evaluate the effects of mean flow on the transmitted side of the aeroelastic panel, the fluid on the incident side is set at rest (i.e., $M_1 = 0$), while that on the transmitted side travels along the x-direction with Mach number M_2 (Fig. 2.1). In such cases, the predicted dependence of the refraction angle φ_2 upon M_2 and the incident angle φ_1 is presented in Fig. 2.6 in the form of contour plots. The presence of mean flow on the transmitted side has an influence completely opposite to that shown in Fig. 2.3 when $M_1 \neq 0$ but $M_2 = 0$, as if the two angles φ_1 and φ_2 are reversed. Similar to Fig. 2.3, the contour plots of Fig. 2.6 are decomposed into two branches (i.e., the

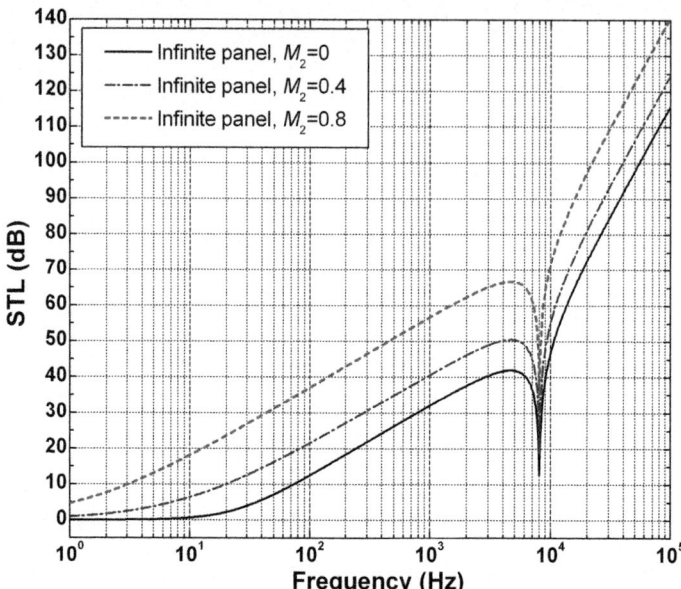

Fig. 2.7 STL versus frequency plots of an infinite panel for sound incident with elevation angle $\varphi_1 = 60°$ and azimuth angle $\theta = 0°$. The mean flow on the transmitted side is completely aligned with the x-direction (three cases: solid line $M_2 = 0$, dash-dot line $M_2 = 0.4$, and short dash line $M_2 = 0.8$), while the fluid on the incident side is at rest ($M_1 = 0$) (With permission from Acoustical Society of America)

positive refraction branch and the negative refraction branch) with respect to the critical Mach number value of $M_2 = 2$. Again, the gap along the φ_1-axis is the region of total reflection, where the wavenumber has an imaginary component in the z-direction [27]. As schematically demonstrated in the inset of Fig. 2.6, the moving stream on the transmitted side exerts a significant refraction effect on the penetration of sound, just as the incident side moving stream does, which is attributable to the transfer effect of adjacent fluid particles in the moving stream. Note also that a normal incident sound penetrates straight through the panel into the other side. In other words, the refraction angle becomes zero when the incident angle is zero, whether the fluid on the transmitted (or incident) side is stationary or moving.

For a given incident angle, the refraction angle φ_2 increases with increasing Mach number M_2 until total reflection occurs. When total reflection occurs (corresponding to the gap along the φ_1-axis between the two branches in Fig. 2.6), there is no energy flux along with the penetrated disturbance as stated previously. Interestingly, a sufficiently large incident angle together with a large Mach number generates a negative refraction angle disturbance, causing the negative refraction effect (corresponding to the smaller branch in Fig. 2.6).

Figure 2.7 plots the *STL* as a function of frequency for an infinite aeroelastic panel immersed in static fluid on the incident side and moving fluid on the transmitted side, with elevation angle $\varphi_1 = 60°$ and azimuth angle $\theta = 0°$ selected

2.1 Finite Single-Leaf Aeroelastic Plate

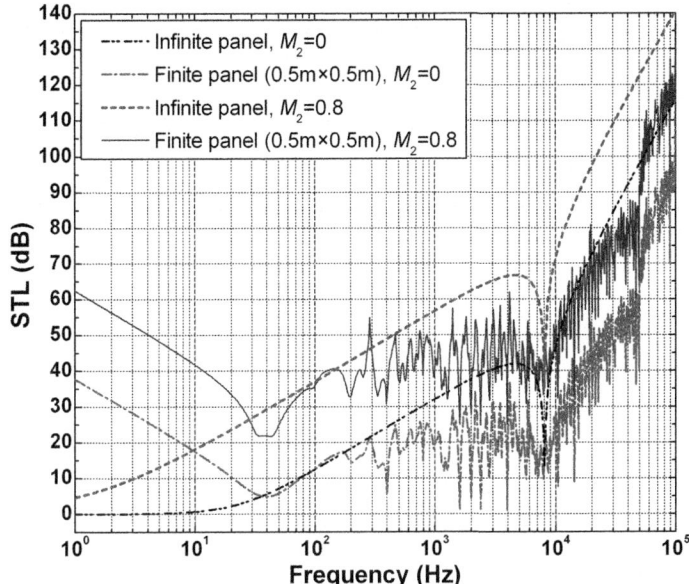

Fig. 2.8 STL versus frequency plots of a simply supported finite panel for sound incident with elevation angle $\varphi_1 = 60°$ and azimuth angle $\theta = 0°$. The mean flow on the transmitted side is completely aligned with the x-direction (two cases: $M_2 = 0$ and $M_2 = 0.8$), while the fluid on the incident side is at rest ($M_1 = 0$). For comparison, the corresponding plots for an infinite panel are included (With permission from Acoustical Society of America)

for the incident sound and three different Mach numbers ($M_2 = 0, 0.4$, and 0.8). The corresponding STL versus frequency plots of a finite aeroelastic panel are presented in Fig. 2.8.

Completely different from the tendencies shown in Fig. 2.4, there is nearly no change of the coincidence dip as Mach number M_2 is varied, while the STL increases dramatically in value over the whole frequency regime considered with increasing M_2. The remarkable discrepancy between Figs. 2.4 and 2.7 suggests that the effects of incident side fluid flow on sound transmission are significantly different from those of fluid flow on the transmitted side.

As aforementioned, the dense peaks and dips appearing in the STL versus frequency plots of Fig. 2.8 reflect the modal behavior of a simply supported finite panel; for comparison, the corresponding STL plots of an infinite panel are included in Fig. 2.8. Analogous to Fig. 2.5, the STL curves for the two cases (i.e., infinite panel and finite panel) are in a global agreement above the (1,1) panel resonance frequency, with the infinite panel setting upper bounds for the finite bounded panel. The boundary constraint effect of the finite panel causes the discrepancy between the two cases below the (1,1) panel resonance frequency. Note also that, despite the presence of dense peaks and dips, the coincidence dip associated with the finite panel coincides with its counterpart of the infinite one. The pronounced increase of STL with increasing M_2 in the finite case is in accordance with the conclusion drawn previously for the infinite case.

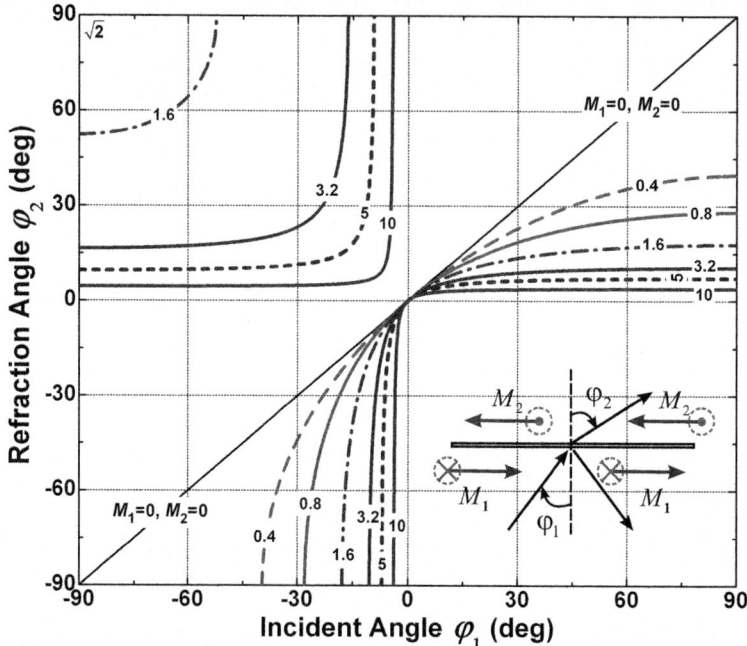

Fig. 2.9 Refraction angle φ_2 plotted as a function of incident angle φ_1 for the case when the moving fluid medium on the incident side is aligned with $\beta_1 = 45°$ from the x-direction and that on the transmitted side has $\beta_2 = 45°$. The two streams have the same Mach numbers, i.e., $M_1 = M_2$, but move in opposite directions. The labeled numbers denote the values of M_1 and absolute values of M_2: for example, the curve labeled by 1.6 represents the case of $M_1 = 1.6$ and $M_2 = -1.6$ (With permission from Acoustical Society of America)

2.1.5 Effects of Incident Elevation Angle in the Presence of Mean Flow on Both Incident Side and Transmitted Side

In terms of angular contour plots as well as *STL* versus frequency plots, the above two sections have addressed the influence of mean flow residing in either the incident or transmitted side on the process of sound tunneling through an aeroelastic panel. To explore the physical nature of mean flow effects in more general cases, we consider next the case when the fluid media on both sides of an infinite panel are moving. The moving stream on the incident side is aligned with the direction of $\beta_1 = 45°$ from the x-direction, while that on the transmitted side has $\beta_2 = 45°$ (Fig. 2.1a). The predicted contour plots of refraction angle are shown in Fig. 2.9, where the labeled numbers denote the absolute Mach number values of the two flows; for simplicity, it is assumed that the two streams have the same Mach number but flow in opposite directions.

2.1 Finite Single-Leaf Aeroelastic Plate

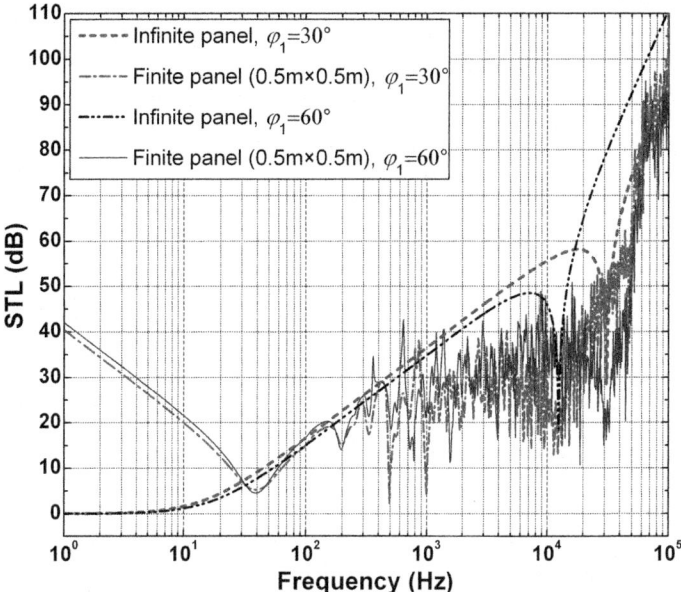

Fig. 2.10 STL plotted as a function of frequency for infinite and finite panels immersed in convected fluids, when the sound is incident with different elevation angles ($\varphi_1 = 30°$ and $60°$) at the same azimuth angle $\theta = 0°$. The mean flow on the incident side is specified by $M_1 = 0.4$ and $\beta_1 = 45°$, while that on the transmitted side by $M_2 = -0.4$ and $\beta_2 = 45°$ (With permission from Acoustical Society of America)

It can be observed that the general trend exhibited by the curves of Fig. 2.9 is similar to that shown in Fig. 2.3 for the case when the fluid on the incident side is moving ($M_1 \neq 0$) and that on the transmitted side is at rest ($M_2 = 0$). However, the critical Mach number for the occurrence of negative refraction (beyond which the contour curves break into two branches) changes from 2 in Fig. 2.3 to $\sqrt{2}$ in Fig. 2.9. This is attributed to the presence of fluid flow on both sides of the infinite panel, which also leads to the significant alterations of the contour plots in Figs. 2.3 and 2.9. Moreover, when the directions of the two flows are reversed with respect to the case shown in Fig. 2.9, it has been established that the resulting contour plots are analogous to those of Fig. 2.6.

With $M_1 = 0.4$ and $\beta_1 = 45°$ for flow on the incident side and $M_2 = -0.4$ and $\beta_2 = 45°$ for flow on the transmitted side, Fig. 2.10 plots the *STL* as a function of frequency for both the infinite and finite bounded panels when the sound is incident in the downstream direction ($\varphi_1 = 30°$ and $60°$). The corresponding plots when the sound is incident in the upstream direction ($\varphi_1 = -30°$ and $-60°$) are presented in Fig. 2.11.

The overall trends of the curves in Fig. 2.11 are similar to those of Fig. 2.10, implying that the intrinsic physical nature is not changed when sound is incident upstream. For example, the coincidence dip is shifted to a significantly lower

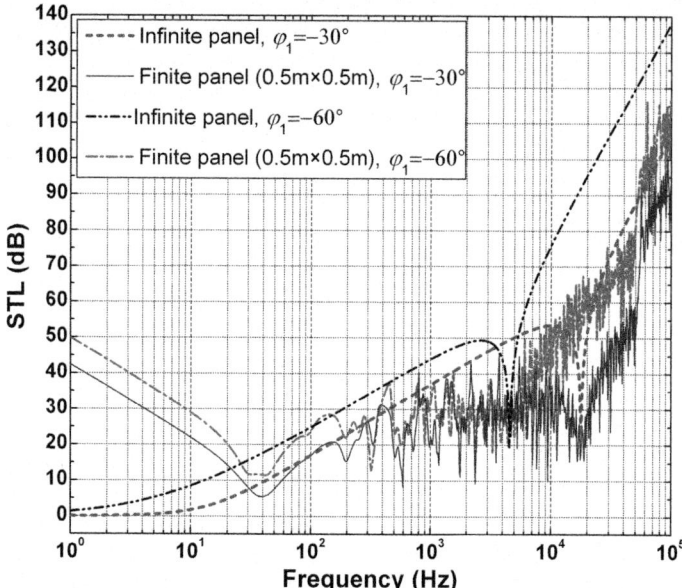

Fig. 2.11 STL plotted as a function of frequency for infinite and finite panels immersed in convected fluids, when the sound is incident with different elevation angles ($\varphi_1 = -30°$ and $-60°$) at the same azimuth angle $\theta = 0°$. The mean flow on the incident side is specified by $M_1 = 0.4$ and $\beta_1 = 45°$, while that on the transmitted side by $M_2 = -0.4$ and $\beta_2 = 45°$ (With permission from Acoustical Society of America)

frequency when the incident angle is increased from $\varphi_1 = 30°$ to $60°$ in Fig. 2.10, and the same thing occurs in Fig. 2.11 upon changing φ_1 from $-30°$ to $-60°$. However, the coincidence frequencies in Fig. 2.11 are considerably lower than their counterparts in Fig. 2.10, reflecting the difference between the upstream incident sound and the downstream incident sound. Moreover, the difference in *STL* values between the two cases of $\varphi_1 = 30°$ and $60°$ in the lower frequency range below the coincidence dip is enlarged when the incident sound is changed from downstream in Fig. 2.10 to upstream in Fig. 2.11.

2.1.6 Conclusions

A theoretical model is proposed for the vibroacoustic behavior of a finite aeroelastic rectangular panel, which is immersed in convected fluids on both sides and simply embedded in an infinite acoustic baffle. The model accounts for the aeroelastic coupling of the convected fluid disturbance induced by panel motion by employing the convected wave equation for inviscid irrotational fluid flow and the displacement continuity condition at fluid-panel interfaces. Different from previous studies, the general case of sound transmission across a finite panel enveloped by two-dimensional fluid flows (aligned along an arbitrary direction parallel to the panel surface) on both sides is theoretically formulated.

The influence of the incident side mean flow upon sound penetration is significantly different from that of the radiating (transmitted) side mean flow. Two branches are found in the contour plots of the refraction angle versus incident angle, corresponding to positive refraction and negative refraction, respectively. The contour plot for the case when the mean flow is on the transmitted side is just a reverse of that when the mean flow is on the incident side. The aerodynamic damping effects on the transmission of sound for both cases are well captured by plotting the *STL* as a function of frequency for varying Mach numbers. However, the frequency of coincidence dip differs significantly between the two cases: as the Mach number is increased, the coincidence dip frequency increases when the flow is on the incident side but remains unchanged when in the flow is on the radiating side.

In the most general case when the fluids on both sides of the panel are moving with mean flow, the predicted contour plots of the refraction angle versus incident angle are significantly different from those when the fluid on one side of the panel is moving and that on the other side is at rest. Furthermore, pronounced discrepancies in panel vibroacoustic behavior are found when the incident sound is changed from upstream to downstream.

2.2 Infinite Double-Leaf Aeroelastic Plates

2.2.1 *Introduction*

The reduction of sound transmission into aircraft interiors is a classical structural acoustic topic of paramount importance for the successful development of supersonic (or high subsonic) civil and military aircrafts [1–7, 12, 18, 39–47]. In general, the construction of such aircrafts is made by thin-walled structural elements. For example, double-leaf aeroelastic plates with a dissipative layer placed in between are commonly used for constructing the aircraft cabin fuselage [12, 24, 25, 39, 41–43, 46, 48, 49]. From the vibroacoustic point of view, the use of double-leaf partitions provides much more effective noise insulation over a wide frequency range than single-leaf plates do. The air cavity formed in between the outer panel (i.e., the source panel) and the trim panel (i.e., the radiating panel) is usually filled with high-density fiberglass blankets to improve thermal insulation and noise attenuation. However, the use of fiberglass blankets leads to increase of weight and thus offsets the above benefits to some extent.

Turbulent boundary layer (TBL)-induced noise and engine exhaust noise have been recognized as the primary sources of the interior noise of aircraft cabins [6, 7, 12, 14, 22, 25, 39, 41, 42, 45, 50–52]. A great deal of work is available concerning the influences of convected fluid loaded on aircraft skin plates. For example, the acoustic power radiated by thin flexible panels subjected to TBL wall-pressure fluctuations was estimated by Davies [53] using a modal analysis method, in which the light fluid loading effects were considered. Also utilizing the method of modal expansion, Dowell [40] theoretically analyzed the transmission of TBL-induced noise through a flexible plate into a closed cavity by accounting for the effects of nonlinear plate stiffness and interaction between the plate and the external airflow. On the basis of the variational method for the vibration of a plate, a set of formulae for sound radiation from rectangular baffled plates having arbitrary boundary conditions were developed by Berry et al. [54]. This was later extended by Atalla and Nicolas [18] to study inviscid uniform subsonic flow, where the effects of the fluid flow were explicitly shown in terms of added mass and radiation resistance to avoid integration in the complex domain. Subsequently, by employing a suitable polynomial function to describe the displacement of a fluid-loaded plate having elastic boundary conditions, Berry [55] developed a new formulation for the vibration and sound radiation of the plate. Based upon the radiation of sound from a single, flat, elastic plate under TBL excitation, Graham [24] proposed a model to address the design problems associated with aircraft cabins, although the effect of mean flow was not taken in account. Graham [25] then developed an extended model consisting of a boundary layer-excited flat plate with its interior covered by two dissipative layers to simulate a factual aircraft cabin plate and found that the presence of the dissipative layers greatly reduces the radiation efficiency compared to a bare plate. A coupled FEM-BEM (finite element method-boundary element method) approach was adopted by Sgard et al. [15] to investigate the effects of the mean flow upon the vibroacoustic behavior of a flat plate subjected to a point force, where it was assumed that the mean flow remains undisturbed by the vibrating plate. The formulation [15] can explicitly show the effects of the mean flow in terms of added mass, damping, and stiffness, as done by Atalla and Nicolas [18]. The dynamic and acoustic responses of a finite baffled plate excited by TBL were investigated by Wu and Maestrello [56], in which the effect of structural nonlinearities induced by in-plane forces was considered. Recently, Clark and Frampton performed systematic studies on the aeroelastic structural response of a single-leaf panel excited by TBL noise and coupled with full potential flow aerodynamics [2, 4, 6–8, 10, 11]. The model accounting for the aerodynamic loading of panels and linearized potential flow aerodynamics was firstly developed in Ref. [4], and further analyses with the TBL-induced noise disturbance taken into account were presented in Refs. [6, 8, 10, 17]. In addition, numerous numerical, theoretical, and experimental investigations have been devoted to studying the transmission of airborne sound across double-panel partitions immersed in static fluid [5, 13, 21, 34, 35, 41, 48, 49, 57–68].

Most of the aforementioned investigations, however, focus either on the TBL-induced noise transmission or the transmission of sound from static fluid (irrespective of mean flow) or on the effects of the mean flow on structural stability (i.e., panel flutter, self-excited vibrations; see, e.g., Crighton [69] for a list of references).

2.2 Infinite Double-Leaf Aeroelastic Plates

Only a few studies [1, 7] have specifically considered the influence of the mean flow on external noise transmission, although this is of particular importance for studying jet-noise transmission into an aircraft. To squarely address this issue, we develop in the present study an aero-acoustic-elastic theoretical model to quantify the influence of external mean flow on sound transmission through a double-leaf aeroelastic plate. This chapter is organized as follows. First, an aero-acoustic-elastic theoretical model to emulate the transmission of jet-power or propeller-induced noise into cabin interior is presented for the sound transmission loss (*STL*), in which the plate dynamics and fluid-structure coupling are accounted for. Second, the physical mechanisms associated with the sound transmission process are discussed, and a set of simple closed-form formulae for the associated natural frequencies are derived from physical principles independent of the theoretical model. These formulae are then used to validate the model predictions, as no suitable experimental or numerical or theoretical results exist in the open literature that can be used to check the validity of the present model. The effects of a few relevant parameters (e.g., Mach number, direction of mean flow, sound incidence angle, panel curvature, and cabin internal pressure) on the *STL* are systematically explored. This chapter is finished with concluding remarks drawn from the obtained results.

2.2.2 Statement of the Problem

To mimic the transmission of engine exhaust noise into the interior of an airplane cabin under typical cruise conditions, a uniform plane sound wave varying harmonically in time is assumed to transmit through a double-leaf aeroelastic plate from the external mean flow side to the interior static fluid side. As shown in Fig. 2.12, the considered system consists of two infinite parallel flexural plates made of homogenous and isotropic materials and is immersed in inviscid, irrotational fluid media. The upper, middle, and bottom fluid media separated by the two plates occupy the spaces of $z<0$, $h_1<z<H+h_1$, and $z>H+h_1+h_2$, respectively, and are characterized by (ρ_1, c_1), (ρ_2, c_2), and (ρ_3, c_3) in terms of mass density and sound speed, respectively. Here, H is the depth of the air gap, and h_1 and h_2 are the thicknesses of the two panels. The mean fluid flow with uniform speed v is assumed to move along the x-axis direction. The incident sound wave transmitting from the external mean flow side is characterized by elevation angle φ_1 and azimuth angle β with respect to the defined coordinate system (see Fig. 2.12). The incident sound is partially reflected and partially transmitted through the structure via the upper plate, middle fluid medium, and bottom plate into the static fluid medium side. The double-layer plate is modeled initially as a flat double-leaf aeroelastic partition, with both the external mean flow and the aeroelastic coupling accounted for. Subsequently, to better emulate the curved skin of an aircraft fuselage and the real process of jet-noise penetration into aircraft interior, the effects of the panel curvature and cabin internal pressurization on sound transmission are quantified.

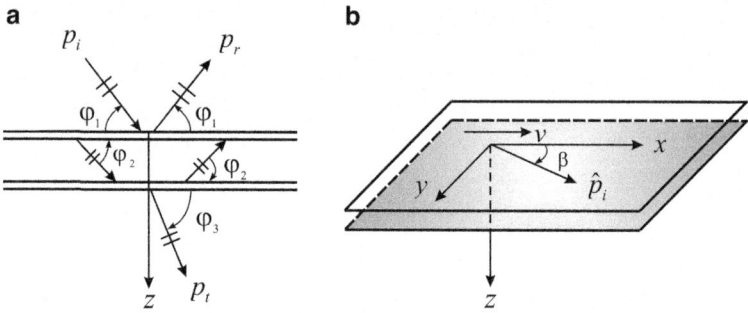

Fig. 2.12 Schematic illustration of sound transmission through a double-leaf aeroelastic plate in the presence of external mean flow: (**a**) side view and (**b**) global view

Several simplifying assumptions are adopted in the present analysis of the system shown in Fig. 2.12: (1) The two plates are modeled by the classical Kirchhoff thin plate theory. (2) The fluid media are inviscid and irrotational [2, 4, 6–8, 10, 11, 15, 18, 56]. (3) The plate surface adjacent to the mean flow is sufficiently smooth so that it is appropriate to represent the fluid-plate interface by the streamline of the fluid flow, i.e., the total flow is tangential to the acoustically deformed boundary [1, 31]. Note that several previous studies [26, 32, 70] also considered the problem of sound reflection and transmission associated with a moving fluid medium but without accounting for the interaction with a thin plate.

2.2.3 Formulation of Plate Dynamics

As shown in Fig. 2.12, the acoustic field is divided into three regimes by the two parallel flat plates which, without loss of generality, are assumed below to have the same thickness. The corresponding sound pressures for the incident field, the middle field, and the transmission field (denoted by indices 1, 2, and 3, respectively) can be expressed as

$$p_1 = P_i e^{i\omega t - i(k_{1x}x + k_{1y}y + k_{1z}z)} + \sum_n \beta_n e^{i\omega t - i(k_{1x}x + k_{1y}y - k_{1z}z)} \quad (2.47)$$

$$p_2 = \sum_n \varepsilon_n e^{i\omega t - i(k_{2x}x + k_{2y}y + k_{2z}z)} + \sum_n \zeta_n e^{i\omega t - i(k_{2x}x + k_{2y}y - k_{2z}z)} \quad (2.48)$$

$$p_3 = \sum_n \xi_n e^{i\omega t - i(k_{3x}x + k_{3y}y + k_{3z}z)} \quad (2.49)$$

The summation index n is introduced in Eqs. (2.47), (2.48), and (2.49) as a consequence of the modal decomposition of the pressure p_2 onto the standing-wave

2.2 Infinite Double-Leaf Aeroelastic Plates

modes of the cavity, which justifies through the continuity equations [see Eqs. (2.63) and (2.64)] the present modal formulation for the reflected and the transmitted pressure amplitudes. Note however that this index does not appear explicitly through the wavenumber components in the argument of the exponential factors. The wavenumber components in Eqs. (2.47)–(2.49) can be written as

$$k_{1x} = k_1 \cos\varphi_1 \cos\beta, \quad k_{1y} = k_1 \cos\varphi_1 \sin\beta, \quad k_{1z} = k_1 \sin\varphi_1 \tag{2.50}$$

$$k_{2x} = k_2 \cos\varphi_2 \cos\beta, \quad k_{2y} = k_2 \cos\varphi_2 \sin\beta, \quad k_{2z} = k_2 \sin\varphi_2 \tag{2.51}$$

$$k_{3x} = k_3 \cos\varphi_3 \cos\beta, \quad k_{3y} = k_3 \cos\varphi_3 \sin\beta, \quad k_{3z} = k_3 \sin\varphi_3 \tag{2.52}$$

Let c_t be the trace wave speed in the plate and the trace wavenumber be given by $k_t = \omega/c_t = 2\pi/\lambda_t$, where λ_t is the wavelength of the panel trace wave. Accordingly, the transverse deflections of the two plates induced by the incident sound can be expressed as

$$w_1(x, y; t) = w_{10} e^{i\omega t - i(k_t \cos\beta)x - i(k_t \sin\beta)y} \tag{2.53}$$

$$w_2(x, y; t) = w_{20} e^{i\omega t - i(k_t \cos\beta)x - i(k_t \sin\beta)y} \tag{2.54}$$

In the incident acoustic field, there exists a uniform flow of velocity V tangential to the acoustically deformed boundary (i.e., the fluid-plate interface). The convected wave equation for the pressure in the fluid is thence given by [1]

$$\frac{D^2 p_1}{Dt^2} = \left(\frac{\partial}{\partial t} + \mathbf{V} \cdot \nabla\right)^2 p_1 = c_1^2 \nabla^2 p_1 \tag{2.55}$$

As stated above, for simplicity, the mean flow is aligned along the x-axis on the fluid-plate interface. Consequently, Eq. (2.55) can be simplified as

$$\left(\frac{\partial}{\partial t} + v \cdot \frac{\partial}{\partial x}\right)^2 p = c_1^2 \nabla^2 p_1 \tag{2.56}$$

Substituting Eq. (2.47) into Eq. (2.56), one obtains the wavenumber in the flowing fluid as

$$k_1 = \frac{k_1^*}{(1 + M \cos\varphi_1 \cos\beta)} \tag{2.57}$$

where $k_1^* = \omega/c_1$ is the acoustic wavenumber in the fluid at rest and $M = v/c_1$ is the Mach number of the mean flow.

Different from the incident field coupled with a mean flow, the fluid medium in between the two plates and that in the transmitted field are both static. In such cases, the propagation of sound obeys the classical wave equation, so that the wavenumbers in the two fluid media are given by

$$k_2 = \frac{\omega}{c_2}, \quad k_3 = \frac{\omega}{c_3} \tag{2.58}$$

In order for the sound waves to "fit" at the boundary (i.e., the plate), the trace wavelengths must match [1], namely,

$$k_{1x} = k_t \cos \beta = k_{2x} = k_t \cos \beta = k_{3x} \tag{2.59}$$

$$k_{1y} = k_t \sin \beta = k_{2y} = k_t \sin \beta = k_{3y} \tag{2.60}$$

Incorporating Eqs. (2.50)–(2.52) into Eqs. (2.59) and (2.60), one obtains the directions of sound propagation in the middle and the transmitted fluid media as

$$\varphi_2 = \arccos \left(\frac{c_2}{c_1} \frac{\cos \varphi_1}{1 + M \cos \varphi_1 \cos \beta} \right) \tag{2.61}$$

$$\varphi_3 = \arccos \left(\frac{c_3}{c_1} \frac{\cos \varphi_1}{1 + M \cos \varphi_1 \cos \beta} \right) \tag{2.62}$$

which also describe the refraction laws for sound transmission from one medium to another. Note that $\varphi_1 = \varphi_2 = \varphi_3$ and $c_1 = c_2 = c_3$ in the absence of the mean flow ($M = 0$). Thus, to refract the wave at the plate is one noticeable effect of the mean flow [1]. In fact, in the presence of the mean flow, the wave would be refracted (and partially reflected) even if the plates were not present.

2.2.4 Consideration of Fluid-Structure Coupling

To determine the unknown parameters appearing in the above equations, supplementary boundary conditions are needed, i.e., the displacement continuity condition between the plate particle and the adjacent fluid particle and the driving relation between the incident sound and the plate dynamic response. In general, the continuity condition for the fluid-structure coupling is described through the velocity of the particles pertaining separately to the fluid medium and the solid medium when the fluid is at rest. In the case of a moving flow, however, the transfer effect of the fluid motion needs to be considered [31], and hence the primary particle displacement continuity should be applied, as described below.

2.2.4.1 Displacement Continuity Condition

Let $\tilde{\lambda}_1$ and $\tilde{\lambda}_2$ denote separately the displacements of the fluid particles in the incident field and the middle field, both adjacent to the upper panel, and let $\tilde{\lambda}_3$ and $\tilde{\lambda}_4$ denote separately the displacements of the fluid particles in the middle field and the transmitted field, both adjacent to the bottom panel. These displacements should satisfy the Navier-Stokes equation for an inviscid and irrotational fluid, namely,

$$\frac{D^2\tilde{\lambda}_1}{Dt^2} = -\frac{1}{\rho_1}\frac{\partial p_1}{\partial z}\bigg|_{z=0}, \quad \frac{D^2\tilde{\lambda}_2}{Dt^2} = -\frac{1}{\rho_2}\frac{\partial p_2}{\partial z}\bigg|_{z=h_1} \quad (2.63)$$

$$\frac{D^2\tilde{\lambda}_3}{Dt^2} = -\frac{1}{\rho_2}\frac{\partial p_2}{\partial z}\bigg|_{z=H+h_1}, \quad \frac{D^2\tilde{\lambda}_4}{Dt^2} = -\frac{1}{\rho_3}\frac{\partial p_3}{\partial z}\bigg|_{z=H+h_1+h_2} \quad (2.64)$$

For harmonic sound wave excitation, the fluid particle displacements take the form of

$$\tilde{\lambda}_j = \tilde{\lambda}_{j0} e^{i\omega t - i(k_{1x}x + k_{1y}y)} \quad (j = 1, 2, 3, 4) \quad (2.65)$$

Substitution of Eqs. (2.47)–(2.49) and (2.65) into Eqs. (2.63) and (2.64) gives the amplitudes of the fluid particle displacements as

$$\tilde{\lambda}_{10} = -\frac{ik_{1z}}{\rho_1} \frac{\left(P_i - \sum_n \beta_n\right)}{(\omega - vk_{1x})^2} \quad (2.66)$$

$$\tilde{\lambda}_{20} = -\frac{ik_{2z} \left(\sum_n \varepsilon_n - \sum_n \zeta_n\right)}{\rho_2 \omega^2} \quad (2.67)$$

$$\tilde{\lambda}_{30} = -\frac{ik_{2z}}{\rho_2 \omega^2} \left(\sum_n \varepsilon_n e^{-ik_{2z}H} - \sum_n \zeta_n e^{ik_{2z}H}\right) \quad (2.68)$$

$$\tilde{\lambda}_{40} = -\frac{ik_{3z}}{\rho_3 \omega^2} \sum_n \xi_n e^{-ik_{3z}H} \quad (2.69)$$

In view of the wavenumber relationships, Eqs. (2.59) and (2.60), and the dynamic deflections of the two plates, Eqs. (2.53) and (2.54), the continuity condition of the particle displacements is given by

$$\tilde{\lambda}_{10} = w_{10} = \tilde{\lambda}_{20}, \quad \text{and} \quad \tilde{\lambda}_{30} = w_{20} = \tilde{\lambda}_{40} \quad (2.70)$$

Substitution of Eqs. (2.53) and (2.54) and (2.66)–(2.69) into (2.70) leads to

$$P_i - \sum_n \beta_n = \frac{\rho_1 c_1 \sin \varphi_2}{\rho_2 c_2 \sin \varphi_1} \frac{1}{1 + M \cos \varphi_1 \cos \beta} \left(\sum_n \varepsilon_n - \sum_n \zeta_n \right) \quad (2.71)$$

$$\sum_n \varepsilon_n - \sum_n \zeta_n = \frac{\rho_2 c_2}{\sin \varphi_2} \dot{w}_{10} \quad (2.72)$$

$$\sum_n \varepsilon_n e^{-ik_{2z}H} - \sum_n \zeta_n e^{ik_{2z}H} = \frac{\rho_2 c_2 \sin \varphi_3}{\rho_3 c_3 \sin \varphi_2} \sum_n \xi_n e^{-ik_{3z}H} \quad (2.73)$$

$$\sum_n \xi_n e^{-ik_{3z}H} = \frac{\rho_3 c_3}{\sin \varphi_3} \dot{w}_{20} \quad (2.74)$$

where $\dot{w}_{10} = i\omega w_{10}$ and $\dot{w}_{20} = i\omega w_{20}$.

2.2.4.2 Driving Relations

The driving relations for the pressure difference across the panel and the panel vibration response can be described as

$$P_i + \sum_n \beta_n - \sum_n \varepsilon_n - \sum_n \zeta_n = Z_{p1} \dot{w}_{10} \quad (2.75)$$

$$\sum_n \varepsilon_n + \sum_n \zeta_n - \sum_n \xi_n = Z_{p2} \dot{w}_{20} \quad (2.76)$$

Detailed derivations of Eqs. (2.75)–(2.76) are given in Sect. 2.2.6 for both the flat and curved aeroelastic plates.

For simplicity, the equivalent characteristic impedances for the three separate fluid media associated with the transmission of sound across the double-leaf plate of Fig. 2.12 are defined as

$$Z_1 = \frac{\rho_1 c_1}{\sin \varphi_1 (1 + M \cos \varphi_1 \cos \beta)} \quad (2.77)$$

$$Z_2 = \frac{\rho_2 c_2}{\sin \varphi_2} \quad (2.78)$$

$$Z_3 = \frac{\rho_3 c_3}{\sin \varphi_3} \quad (2.79)$$

It is readily seen that in addition to being strongly dependent on the intrinsic property of the fluid (i.e., density and sound speed), the equivalent characteristic impedances of the fluid media are also determined by the sound incident angle and flow velocity. Actually, these characteristic impedances reflect the close relationship between the sound pressure and the fluid particle velocity.

2.2.5 Definition of Sound Transmission Loss

The transmissivity $\tau(\varphi_1, \beta)$ is defined to quantify the sound transmission through the double-leaf plate [1] as

$$\tau(\varphi_1, \beta) = \frac{\rho_1 c_1}{\rho_2 c_2} \left| \frac{\sum_n \xi_n}{P_i} \right|^2 \tag{2.80}$$

where

$$\frac{P_i}{\sum_n \xi_n} = \frac{1}{4Z_2 Z_3 \cos(k_{2z} H)}$$

$$\begin{bmatrix} (Z_1 Z_{p2} + Z_{p1} Z_{p2}) \left(e^{-i(-k_{2z}+k_{3z})H} - e^{-i(k_{2z}+k_{3z})H} \right) \\ + Z_2 Z_{p2} \left(e^{-i(-k_{2z}+k_{3z})H} + e^{-i(k_{2z}+k_{3z})H} \right) + 2Z_2 Z_3 \cos(k_{2z} H) \\ + 2i Z_3 Z_{p1} \sin(k_{2z} H) + 2i Z_1 Z_3 \sin(k_{2z} H) + 2Z_1 Z_2 e^{-ik_{3z}H} \end{bmatrix} \tag{2.81}$$

Accordingly, the sound transmission loss is defined as a decibel scale of the transmissivity:

$$\text{STL} = -10 \log_{10} \tau(\varphi_1, \beta) \tag{2.82}$$

2.2.6 Characteristic Impedance of an Infinite Plate

2.2.6.1 Infinite Flat Plate

Consider an infinite plate immersed in a fluid medium and subjected to a harmonic incident sound excitation on one side. The governing equation for the deflection of the plate is given by

$$D \nabla^4 w + m \frac{\partial^2 w}{\partial t^2} = p e^{i\omega t - i(k_{2x} x + k_{2y} y)} \tag{2.83}$$

Note that, in the present study, the structural loss factor η is accounted for by introducing the complex bending stiffness D [3, 5] as

$$D = \frac{E h^3 (1 + j\eta)}{12 (1 - \nu^2)} \tag{2.84}$$

Fig. 2.13 Shallow cylindrical panel under biaxial membrane stresses

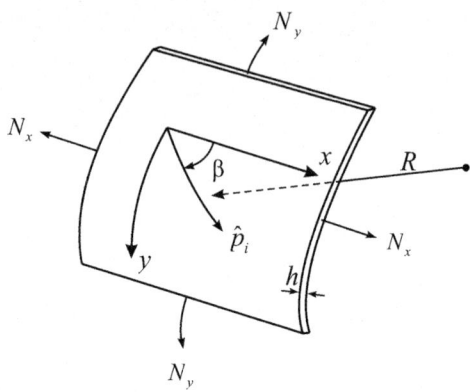

Since the incident sound is harmonic, the deflection of the plate is assumed to take the form of

$$w = w_0 e^{i\omega t - i[(k_t \cos\beta)x + (k_t \sin\beta)y]} \tag{2.85}$$

where w_0 is determined by substituting (2.85) into (2.83) as

$$w_0 = \frac{p}{D\left(k_{2x}^2 + k_{2y}^2\right)^2 - m\omega^2} \tag{2.86}$$

Accordingly, the panel impedance that determines the relation between the sound pressure and the particle velocity is given by

$$Z_p = \frac{p}{\dot{w}} = \frac{p}{i\omega w_0} = im\omega \left(1 - \frac{D\omega^2}{mc_2^4}\cos^4\varphi_2\right) \tag{2.87}$$

2.2.6.2 Infinite Curved Plate with Biaxial Membrane Stresses

In reality, the aircraft fuselage skin is a typical shallow cylindrical panel structure with internal pressurization during cruise condition. To better emulate the actual fuselage skin and the real process of jet-noise penetration into the aircraft interior, the Donnell-Mushtari-Vlasov shallow cylindrical shell theory [1, 71] is adopted. The geometry and coordinates of a shallow cylindrical panel of thickness h under biaxial membrane stresses (i.e., N_x in the x-direction and N_y in the y-direction) are schematically illustrated in Fig. 2.13. The governing equation for its lateral deformation w can be expressed as

2.2 Infinite Double-Leaf Aeroelastic Plates

$$D\nabla^8 w + \frac{Eh}{R^2}\frac{\partial^4 w}{\partial x^4} - \nabla^4\left(N_x\frac{\partial^2 w}{\partial x^2} + N_y\frac{\partial^2 w}{\partial y^2}\right) + m\nabla^4\left(\frac{\partial^2 w}{\partial t^2}\right)$$
$$= \nabla^4\left(pe^{i\omega t - i(k_{2x}x + k_{2y}y)}\right) \qquad (2.88)$$

After a few algebraic manipulations similar to those leading to Eq. (2.86), one obtains

$$w_0 = p \cdot \left[D\left(k_{2x}^2 + k_{2y}^2\right)^2 + \frac{Eh}{R^2}\frac{k_{2x}^4}{\left(k_{2x}^2 + k_{2y}^2\right)^2} + \left(N_x k_{2x}^2 + N_y k_{2y}^2\right) - m\omega^2\right]^{-1} \qquad (2.89)$$

which, together with the definition of the panel impedance, leads to

$$Z_p = \frac{p}{\dot{w}} = \frac{p}{i\omega w_0} = \frac{1}{i\omega}\left[D\left(k_{2x}^2 + k_{2y}^2\right)^2 + \frac{Eh}{R^2}\frac{k_{2x}^4}{\left(k_{2x}^2 + k_{2y}^2\right)^2}\right.$$
$$\left. + \left(N_x k_{2x}^2 + N_y k_{2y}^2\right) - m\omega^2\right]$$
$$= im\omega\left[1 - \frac{D\omega^2}{mc_2^4}\cos^4\varphi_2 - \frac{Eh}{R^2}\frac{\cos^4\beta}{m\omega^2} - \frac{\cos^2\varphi_2}{mc_2^2}\left(N_x\cos^2\beta + N_y\sin^2\beta\right)\right] \qquad (2.90)$$

2.2.7 Physical Interpretation for the Appearance of STL Peaks and Dips

Figure 2.14 plots the predicted *STL* as a function of incident sound frequency for Mach number $M = 0.05$, with the sound incidence elevation angle fixed at $\varphi_1 = 30°$ and the azimuth angle at $\beta = 0°$ (i.e., completely aligned with the downstream direction). In Fig. 2.14, four physical phenomena (i.e., mass-air-mass resonance, standing-wave attenuation, standing-wave resonance, and coincidence resonance) associated with the transmission of sound from the external mean flow side across the double-leaf partition can be clearly identified, which are marked with symbols in the frequency range considered (from 0 to 10,000 Hz). To predict the inherent frequencies associated with these phenomena in the presence of mean flow, a set of simple closed-form formulae are derived based purely on physical principles (see Appendix). Since the derivation of these formulae is independent of the aero-acoustic-elastic theoretical model presented in Sects. 2.2.3–2.2.6, they may be used

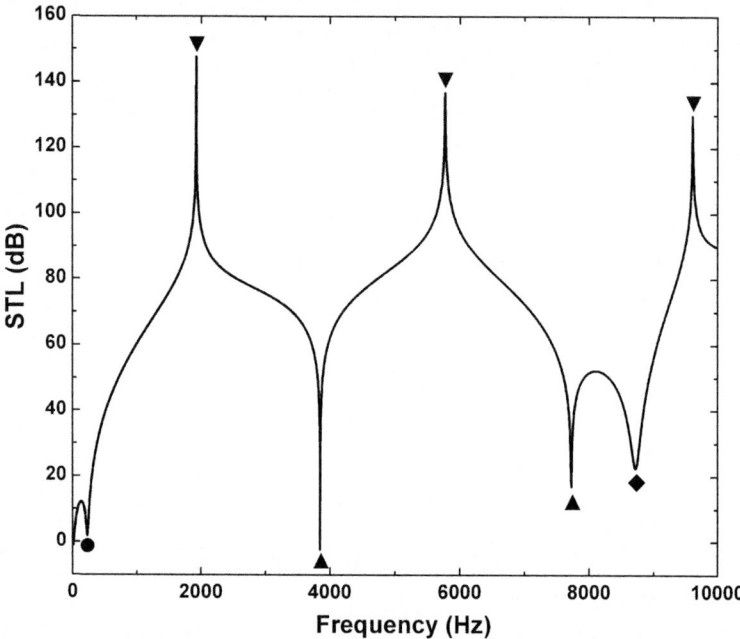

Fig. 2.14 Predicted *STL* as a function of frequency for mean flow speed $M = 0.05$, sound incident elevation angle $\varphi_1 = 30°$, and azimuth angle $\beta = 0°$; ● mass-air-mass resonance, ▼ standing-wave attenuation, ▲ standing-wave resonance, ♦ coincidence resonance

to check the validity of the model predictions as no other suitable experimental or theoretical work exists. In Fig. 2.14, the Mach number is selected as $M = 0.05$ only because, at this small Mach number, all the four physical phenomena can be clearly identified in the considered frequency range of 0–10,000 Hz. Of course, the present model is suitable for handling higher Mach number cases (e.g., $M = 0.4$, 0.8, and 1.2 as shown in Figs. 2.16, 2.17, 2.19, and 2.20), although the coincidence resonance occurs beyond 10,000 Hz at large Mach numbers when the sound is incident downstream. Furthermore, when the sound is incident upstream, the three phenomena (i.e., mass-air-mass resonance, standing-wave attenuation, standing-wave resonance) all disappear, leaving only the coincidence dip in the *STL* curve (see Fig. 2.18).

As shown in Fig. 2.14, the first dip is associated with the mass-air-mass resonance that is particularly marked by the filled circle. The mass-air-mass resonance in the absence of the mean flow usually occurs when the two panels move in opposite phases [3, 5, 34, 59, 65], with the air gap behaving like an elastic spring. With the influence of the mean flow accounted for, the frequency of the mass-air-mass resonance is (see Appendix for detailed derivations)

$$f_\alpha = \frac{1}{2\pi \sin \varphi_2} \sqrt{\frac{\rho_2 c_2^2}{H} \frac{(m_1 + m_2)}{m_1 m_2}} \quad (2.91)$$

2.2 Infinite Double-Leaf Aeroelastic Plates

In view of the expression for φ_2 in Eq. (2.61), the above relation indicates that f_α is dependent of the Mach number M as well as sound incidence angles φ_1 and β, which is different from that of a double-leaf plate immersed in static fluid.

The second phenomenon relates to the three peaks on the *STL* versus frequency curve (marked by the inverted and filled triangle ▼ in Fig. 2.14), corresponding separately to the first-order, second-order, and third-order standing-wave attenuations. Wave attenuation occurs when the distance difference between the routes that the two intervening waves pass through is the odd number of the one-quarter wavelength of the incident sound. Thus, the frequencies for these standing-wave attenuations are given by (Appendix)

$$f_{p,n} = \frac{(2n-1)c_2}{4H \sin \varphi_2}, (n = 1, 2, 3 \ldots) \tag{2.92}$$

which are again dependent upon the Mach number and the sound incidence angles. When standing-wave attenuation occurs, the destructive interference between the positive- and negative-going waves causes the wave amplitude to significantly decrease before the sound is transmitted across the partition, resulting in maximum sound reduction.

In contrast to the standing-wave attenuation, the third phenomenon (i.e., the standing-wave resonance) occurs when the distance difference between the routes that the two intervening waves pass through is the multiple of the half wavelength of the incidence sound (denoted by the filled triangle ▲ in Fig. 2.14). In such circumstances, the constructive interference of the positive- and negative-going waves leads to an enhanced sound transmission through the partition. The standing-wave resonance as a result of the enhanced effect occurs at the following frequencies (Appendix):

$$f_{d,n} = \frac{nc_2}{2H \sin \varphi_2}, (n = 1, 2, 3 \ldots) \tag{2.93}$$

The fourth phenomenon (i.e., the coincidence resonance) occurs when the wavelength of the flexural bending wave in the panel matches the trace wavelength of the incidence sound. The corresponding dip is marked in Fig. 2.14 by the filled diamond ♦. Due to the influence of the mean flow, the resonance frequency differs from that in the static fluid case, given by (Appendix)

$$f_c = \frac{c_2^2}{2\pi h \cos^2 \varphi_2} \sqrt{\frac{12\rho(1-v^2)}{E}} \tag{2.94}$$

In the absence of mean flow ($M = 0$), according to Eq. (2.61), the angle φ_2 is equal to the incident angle φ_1 for the case considered here (i.e., $c_1 = c_2$, both panels immersed in air). Consequently, in this limiting case (static fluid), the frequencies for the above four phenomena are simply obtained by replacing φ_2 with φ_1 in Eqs. (2.91), (2.92), (2.93), and (2.94).

Table 2.1 Comparison between theoretical model predictions and simple closed-form formulae for *STL* peaks and dips ($M = 0.05$, $\varphi_1 = 30°$, $\beta = 0°$)

Mass-air-mass resonance f_α (Hz)		Standing-wave attenuation $f_{p,n}$ (Hz)		Standing-wave resonance $f_{d,n}$ (Hz)		Coincidence resonance f_c (Hz)	
Theory	Eq. (2.91)	Theory	Eq. (2.92)	Theory	Eq. (2.93)	Theory	Eq. (2.94)
227.73	231.69	1922.2	1922.2	3844.3	3844.3	8714.9	8726.2
		5766.5	5766.5	7720.5	7688.6		
		9610.8	9610.8	11533	11533		

The existence of the four distinct acoustic phenomena influences significantly the shape of the *STL* versus frequency curves, as evidenced by the intense peaks and dips appearing in Fig. 2.14. The frequencies of the four phenomena predicted from the theoretical model, i.e., Eqs. (2.47), (2.48), (2.49), (2.50), (2.51), (2.52), (2.53), (2.54), (2.55), (2.56), (2.57), (2.58), (2.59), (2.60), (2.61), (2.62), (2.63), (2.64), (2.65), (2.66), (2.67), (2.68), (2.69), (2.70), (2.71), (2.72), (2.73), (2.74), (2.75), (2.76), (2.77), (2.78), (2.79), (2.80), (2.81), (2.82), (2.83), (2.84), (2.85), (2.86), and (2.87), are compared in Table 2.1 with the closed-form formulae, i.e., Eqs. (2.91), (2.92), (2.93), and (2.94). Excellent agreement is achieved, which in a way validates the theoretical model since the closed-form formulae are derived completely independent of the model. Of course, it would be more desirable to validate the present model predictions with other theories or experimental measurements, but, unfortunately, none exists in the open literature.

2.2.8 Effects of Mach Number

As discussed in the previous section, all the four acoustic phenomena associated with the *STL* peaks and dips depend on the Mach number of the mean fluid flow. It is thus expected that the Mach number plays an important role in the transmission process of sound through a double-leaf partition. Two typical cases for sound incidence in the downstream direction and in the upstream direction, respectively, are studied below to explore further the Mach number influence.

2.2.8.1 Sound Incidence Along the Downstream Direction

Consider first the case of sound incidence having an elevation angle of $\varphi_1 = 30°$ and an azimuth angle of $\beta = 45°$, with its wave vector component in the downstream direction being positive. Figure 2.15 presents the predicted *STL* versus frequency curves for selected Mach numbers, $M = 0, 0.4, 0.8$, and 1.2. It is seen from Fig. 2.15 that changes in the Mach number lead to noticeable shifts of the *STL* curves: as the Mach number is increased, the *STL* peaks and dips are all shifted to lower frequencies, resulting in an increase of the *STL* value over a relatively broad

2.2 Infinite Double-Leaf Aeroelastic Plates

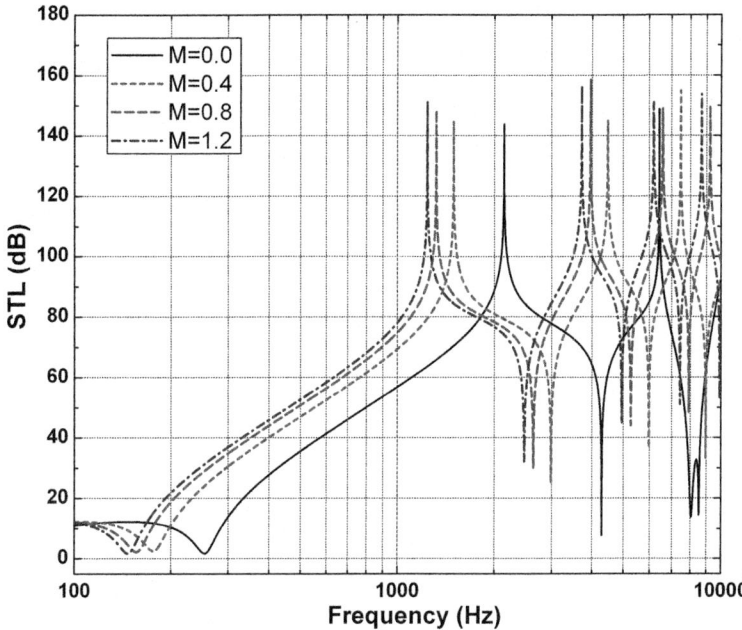

Fig. 2.15 Predicted *STL* plotted as a function of frequency for selected Mach numbers with sound incidence elevation angle $\varphi_1 = 30°$ and azimuth angle $\beta = 45°$

frequency range. While the noticeable decrease of the *STL* peaks and dips can be attributed to the added-mass effect of the convected fluid loading, the increase of the *STL* value in the frequency range considered agrees well with existing results [6, 7]. One may expect that the *STL* value corresponding to the peaks and dips should also increase as the Mach number is increased, because the aerodynamic damping effect increases when the convected flow becomes more turbulent [7]. However, the increase of the *STL* value related to the peaks and dips is not as remarkable as anticipated (Fig. 2.15). This may be attributed to the fact that the present study assumes irrotational, inviscid potential flow, which is much different from the turbulent boundary layer considered by Frampton and Clark [7]. Another possible reason may be that as the Mach number is increased, the complex fluid-structure coupling effects overwhelm the aerodynamic damping effect when the *STL* peaks and dips move considerably away from their original locations.

By extracting the frequencies associated with the *STL* peaks and dips for different Mach numbers, the dependence of these frequencies on the Mach number is obtained, as shown in Fig. 2.16a–d using different symbols (e.g., hollow circles, hollow diamonds, and hollow squares). For comparison, predictions from the closed-form formulae are also included in Fig. 2.16b, c, denoted by different lines (e.g., solid line, dash line, and dash-dot line). Note that the first three orders of the frequencies are plotted for the standing-wave attenuation and standing-wave resonance in Fig. 2.16. Similar to the results of Table 2.1, it is seen from Fig. 2.16

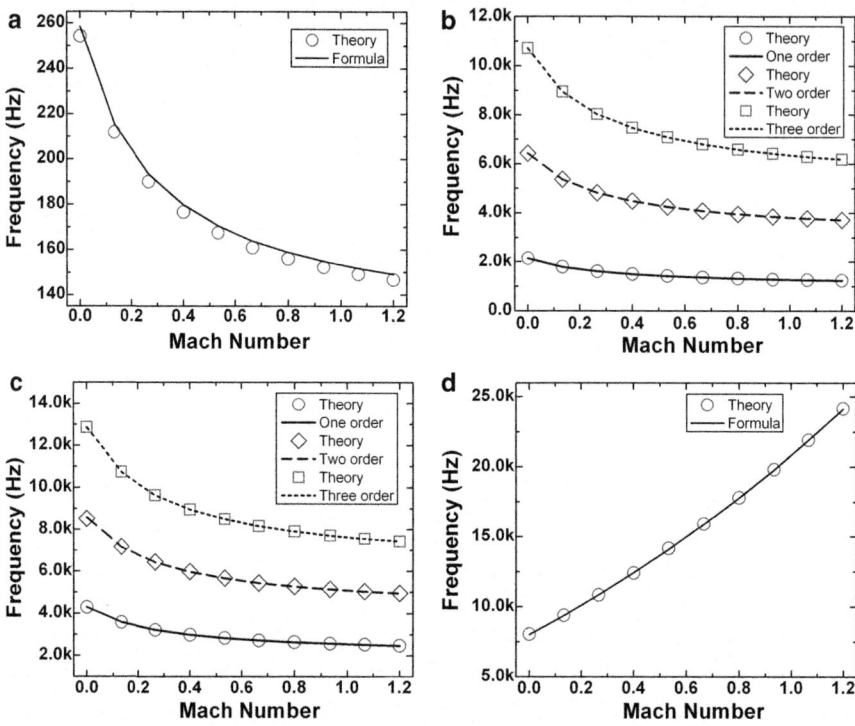

Fig. 2.16 Effects of Mach number on the frequencies of *STL* peaks and dips for sound incidence with elevation angle $\varphi_1 = 30°$ and azimuth angle $\beta = 45°$: (**a**) mass-air-mass resonance, (**b**) standing-wave attenuation, (**c**) standing-wave resonance, and (**d**) coincidence resonance. Symbols (e.g., *hollow circles*, *hollow diamonds*, and *hollow squares*) refer to theoretical predictions. Lines (e.g., *solid line*, *dash line*, and *dash-dot line*) denote the calculated results from Eqs. (2.91), (2.92), (2.93), and (2.94)

that the model predictions agree very well with Eqs. (2.91), (2.92), (2.93), and (2.94), and the same can be said regarding the other cases shown in Figs. 2.17, 2.19, and 2.20. The results of Fig. 2.16 demonstrate that, except for the coincidence resonance frequencies, the frequencies for the mass-air-mass resonance, standing-wave attenuation, and standing-wave resonance decrease as the Mach number increases, due mainly to the added-mass effects of the convected fluid loading. The exception of the coincidence resonance is attributed to the fact that the significant refraction effect of the mean flow has overwhelmed the added-mass effect on the coincidence resonance.

2.2.8.2 Sound Incidence Along the Upstream Direction

Consider next the case of sound incidence in the upstream direction, with elevation angle $\varphi_1 = 30°$ and azimuth angle $\beta = 135°$. The effects of the Mach number

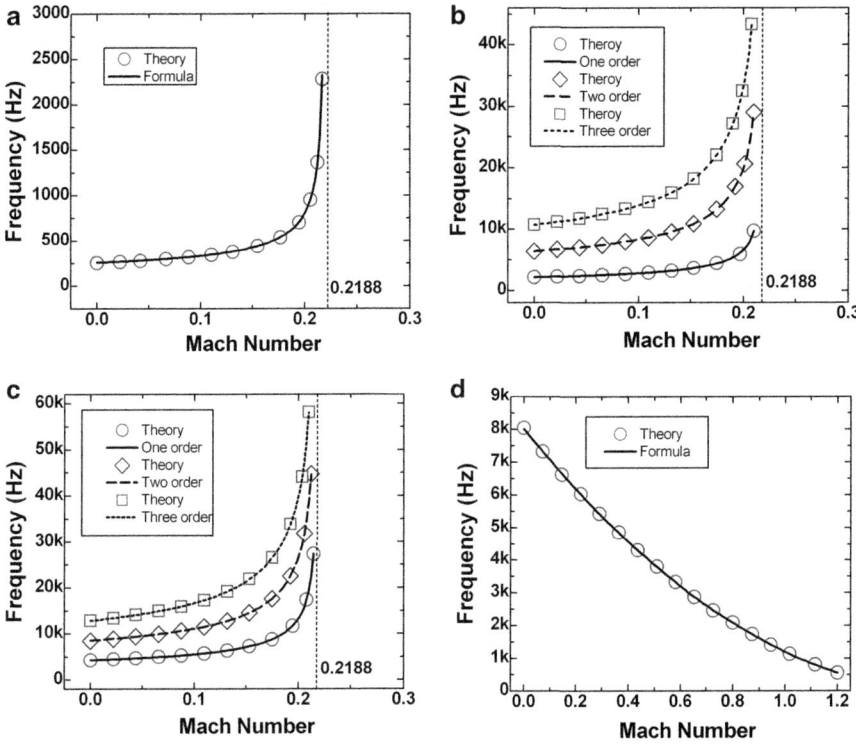

Fig. 2.17 Effects of Mach number on the frequencies of *STL* peaks and dips for sound incidence with elevation angle $\varphi_1 = 30°$ and azimuth angle $\beta = 135°$: (**a**) mass-air-mass resonance, (**b**) standing-wave attenuation, (**c**) standing-wave resonance, and (**d**) coincidence resonance

on the mass-air-mass resonance, the standing-wave attenuation, the standing-wave resonance, and the coincidence resonance frequencies are displayed in Fig. 2.17a–d. The frequencies for the first three acoustic phenomena increase with increasing Mach number until a critical value of $M = 0.2188$ is reached, beyond which these phenomena all disappear. The existence of the critical Mach number can be explained as follows. Note from Eqs. (2.91), (2.92), and (2.93) that $\sin\varphi_2$ is present in each of the denominators and φ_2 depends on the Mach number through Eq. (2.61). It is seen from (2.61) that for sound incidence with $\varphi_1 = 30°$ and $\beta = 135°$, $\sin\varphi_2$ approaches zero as the Mach number approaches 0.2188, thus giving rise to infinitely large values of the frequencies. Consequently, the mass-air-mass resonance, the standing-wave attenuation, and the standing-wave resonance do not exist for $M > 0.2188$. On the other hand, the transmitted evanescent wave for coincidence resonance is enhanced by the coincidence effect of the bottom panel. Accordingly, the corresponding frequency in the present upstream case (see Fig. 2.17d) decreases as the Mach number is increased, a feature that is opposite to that in the downstream case shown in Fig. 2.11d.

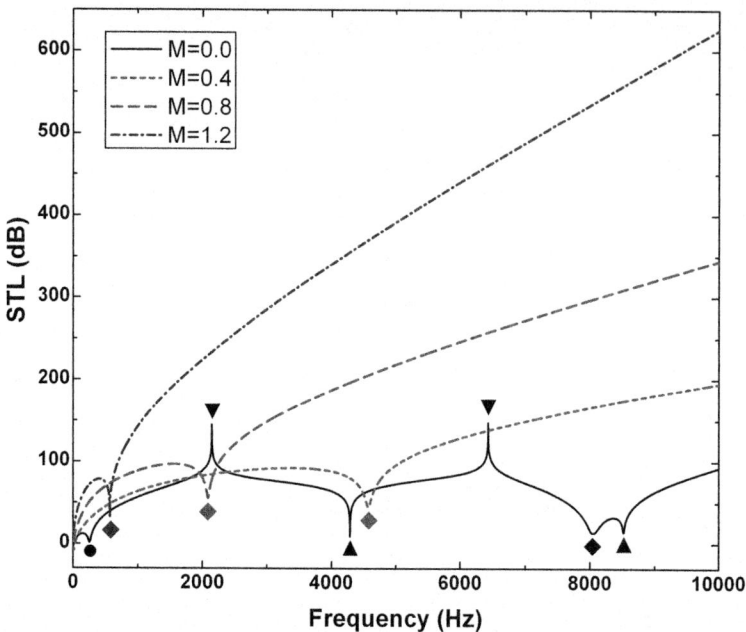

Fig. 2.18 Variations of *STL* with incident frequency for selected Mach numbers, with elevation angle $\varphi_1 = 30°$ and azimuth angle $\beta = 135°$; ● mass-air-mass resonance, ▼ standing-wave attenuation, ▲ standing-wave resonance, ♦ coincidence resonance

To illustrate further the shift of the *STL* peaks and dips with varying flow velocity, Fig. 2.18 plots the predicted *STL* versus frequency curves for selected Mach numbers, with elevation angle $\varphi_1 = 30°$ and azimuth angle $\beta = 135°$. It is seen that all the four acoustic phenomena appear in the case of static fluid ($M = 0$), which are marked by different symbols in Fig. 2.18. As the Mach number increases, the mass-air-mass resonance, the standing-wave attenuation, and the standing-wave resonance gradually disappear, consistent with the results of Fig. 2.17a–c. This also backs the selection of a small Mach number ($M = 0.05$) for plotting the results in Fig. 2.14, so that all the four different acoustic phenomena can be clearly identified within the considered frequency range. In addition to the disappearing of *STL* peaks and dips with increasing Mach number, a dramatic increase of the *STL* value over a broad frequency range is observed in Fig. 2.18. It should be clarified that the significant increase of the *STL* value (up to 600 dB) is caused not only by the structural damping but also by the total reflection phenomenon occurring in the specific case associated with Fig. 2.18. When total reflection occurs, a disturbance penetrates through the double-leaf panel into the transmitted side fluid medium, while the wavenumber component k_z in the z-direction takes the form of $-j\hbar$ ($\hbar$ being a positive real number), resulting in a rapid exponential delay of the wave amplitude in the form of $\exp(-\hbar)$. Although the physical nature of the total reflection has been addressed in detail by Ribner [26], we believe that its influence on *STL* may have been quantified for the first time in Fig. 2.18.

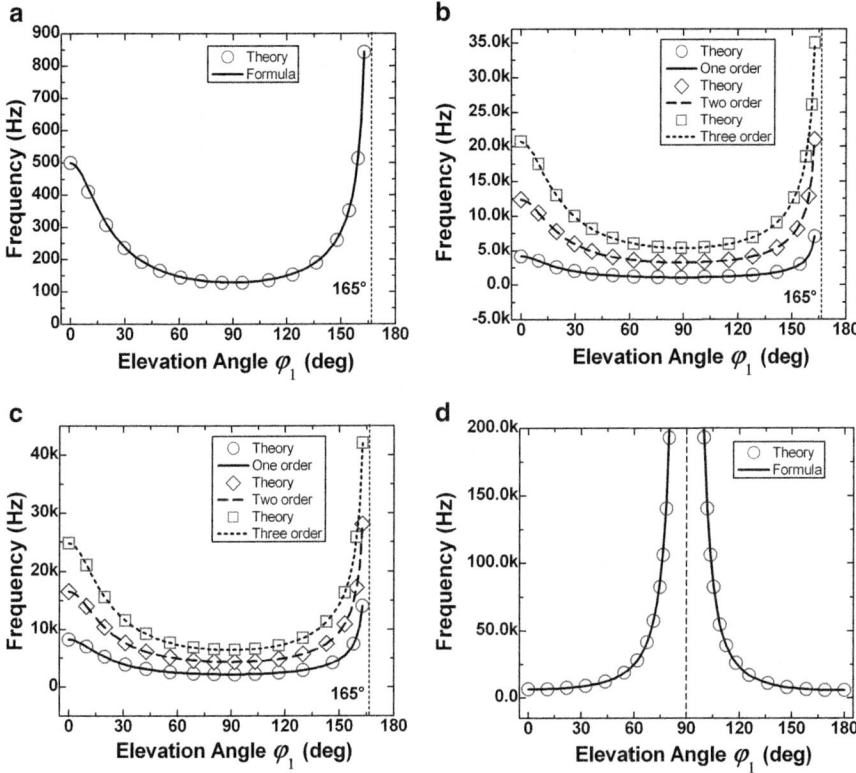

Fig. 2.19 Effects of incident elevation angle on the frequencies of *STL* peaks and dips for sound incidence with azimuth angle $\beta = 45°$ and Mach number $M = 0.05$: (**a**) mass-air-mass resonance, (**b**) standing-wave attenuation, (**c**) standing-wave resonance, and (**d**) coincidence resonance

2.2.9 Effects of Elevation Angle

It has been reported that the sound incidence elevation angle has a noticeable effect on the transmission of sound through a partition immersed in static fluid [65]. For the problem considered here, its influence in the presence of mean flow is quantified. Obtained results for a fixed azimuth angle of $\beta = 45°$ and a fixed Mach number of $M = 0.05$ are presented in Fig. 2.19, where it is seen that the four distinct acoustic phenomena all exhibit significant dependence on the elevation angle. It is interesting to find that for the first three phenomena (i.e., the mass-air-mass resonance, the standing-wave attenuation, and the standing-wave resonance), critical values of the elevation angle exist, beyond which all three phenomena instantaneously vanish. For example, for the specific case of $\beta = 45°$ and $M = 0.05$, the critical value is found to be $\varphi_1 = 165°$, whereas for the case of $\beta = 135°$ and $M = 0.05$ (results not shown here for brevity), the three phenomena are suppressed when the elevation angle lies within the range between 0 and 15°. Moreover, the coincidence resonance frequency becomes infinitely large when the elevation angle approaches $\pi/2$, and the system is

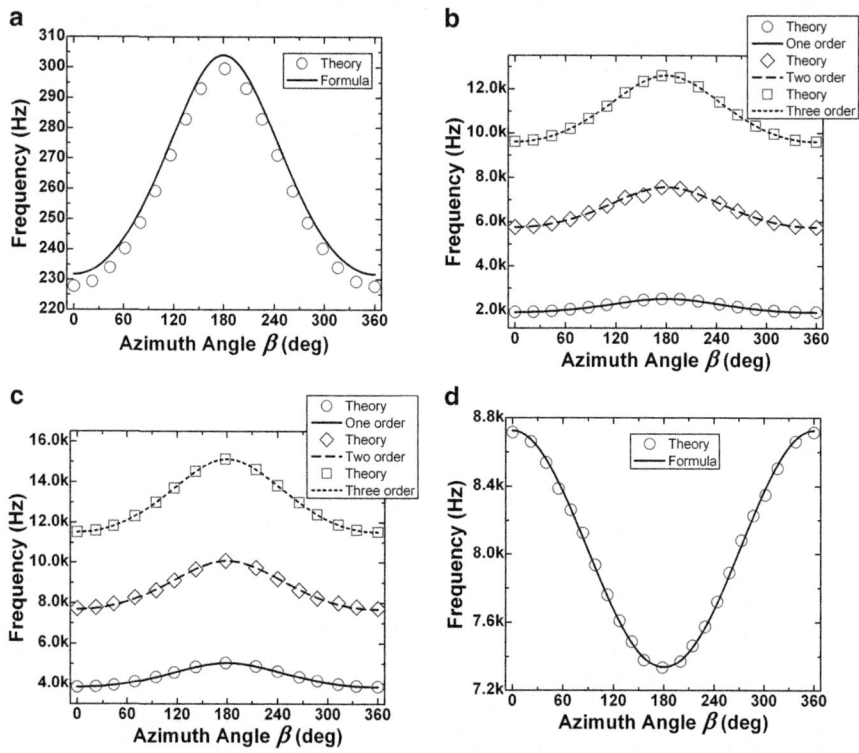

Fig. 2.20 Effects of incident azimuth angle on the frequencies of *STL* peaks and dips for sound incidence with elevation angle $\varphi_1 = 30°$ and Mach number $M = 0.05$: (**a**) mass-air-mass resonance, (**b**) standing-wave attenuation, (**c**) standing-wave resonance, and (**d**) coincidence resonance

symmetrical with respect to the sound incidence angle and the flow direction (e.g., the dependence of the four critical frequencies upon the elevation angle for the case of $\beta = 135°$ is simply obtained by inverting Fig. 2.19 for the case of $\beta = 45°$).

2.2.10 Effects of Azimuth Angle

For sound transmission through sandwich panels with corrugated cores immersed in static fluid, it has been found that the sound incidence azimuth angle β plays a negligible role due to the symmetrical property of the considered system [5, 65]. This, however, is no longer valid if the system is immersed in a flowing fluid. For the present double-leaf plate, Fig. 2.20 plots the predicted critical frequencies as functions of the azimuth angle β in the specific case of $M = 0.05$ and $\varphi_1 = 30°$. The azimuth angle is seen to have a significant effect on all the four acoustic phenomena. Note that varying the azimuth angle from 0 to 2π corresponds to four processes in sequence, i.e., first downstream phase from 0 to $\pi/2$, first upstream phase from $\pi/2$ to π, second upstream phase from π to $3\pi/2$, and second downstream phase from $3\pi/2$ to 2π. The symmetry of the results shown in Fig. 2.20 with respect to $\beta = \pi$ is therefore readily understandable. Actually, alteration of these frequencies with respect to the incident azimuth angle confirms the refraction effect of the mean flow.

2.2.11 Effects of Panel Curvature and Cabin Internal Pressurization

As for the investigation of external jet-noise penetration through fuselage skin into aircraft interior, it would be of great interest to estimate the effects of panel curvature and cabin internal pressurization on the noise transmission. To this end, we extend Koval's work [1] regarding the effects of panel curvature and cabin internal pressurization on sound transmission through a single-leaf aeroelastic plate, and for computational simplicity (assuming that the apparent contradiction of an unlimited curved panel can be overlooked), an infinite slightly curved double-leaf panel is considered. However, we believe that the results obtained with this somewhat idealized model are reasonable because all the physical phenomena have been well captured (as shown below), including the mass-air-mass resonance, the ring frequency resonance, the standing-wave attenuation and resonance, and the coincidence resonance. Particularly, the predicted ring frequency resonance is consistent with existing results [1, 72] concerning single-leaf curved panels.

Again, to clearly demonstrate all the significant acoustic phenomena associated with a curved double-leaf panel for frequencies below 10,000 Hz, a small Mach number of $M = 0.05$ is selected, with the incident elevation angle arbitrarily fixed at $\varphi_1 = 30°$. Under such conditions, the effects of panel curvature and internal pressurization are shown in Fig. 2.21 using a set of combinations, i.e., a flat panel $R = \infty$ with and without internal pressure $P = 0.1$ MPa and a curved panel $R = 6$ m with and without internal pressure $P = 0.1$ MPa. It is seen from Fig. 2.21 that, irrespective of the panel curvature or internal pressurization, all the four acoustic phenomena can be distinctively identified. In the two cases concerning curved double-leaf panels, i.e., ($P = 0$ Pa, $R = 6$ m) and ($P = 0.1$ MPa, $R = 6$ m), a newly added ring frequency resonance dip (i.e., the first dip) occurs, which is absent in flat double-leaf panels. In the absence of internal pressure, the ring frequency of the curved panel can be predicted by [1, 72]

$$f_R = \frac{1}{2\pi R}\sqrt{\frac{Eh}{m}} \qquad (2.95)$$

In addition to generating the ring frequency resonance, the panel curvature also shifts the mass-air-mass resonance dip to a higher frequency, which can be seen by comparing the second dip in the curve of ($P = 0$ Pa, $R = 6$ m) with the first dip

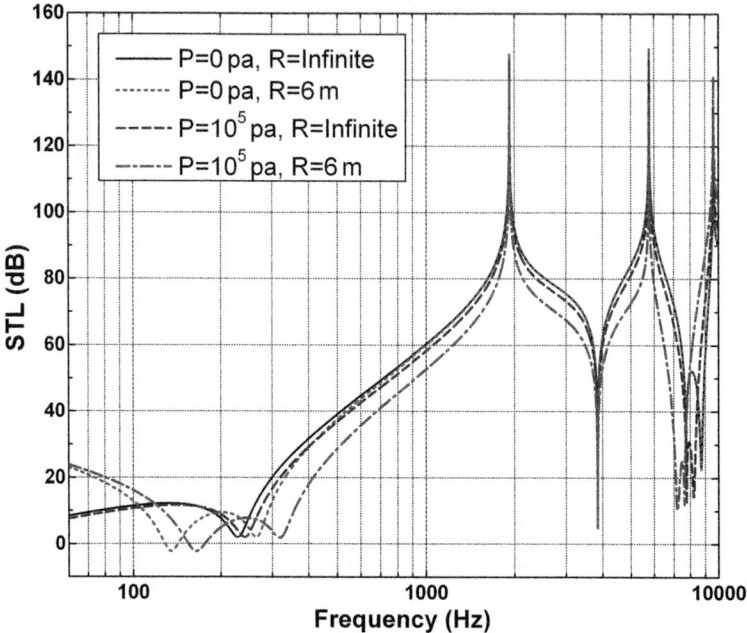

Fig. 2.21 Effects of panel curvature and cabin internal pressurization on *STL* for sound incidence with elevation angle $\varphi_1 = 30°$ in the presence of external mean flow ($M = 0.05$)

in the curve of ($P = 0$ Pa, $R = \infty$). Otherwise, the general trend of the *STL* versus frequency curve of the flat panel agrees well with that of the curved panel (Fig. 2.21).

By comparing the two cases of ($P = 0$ Pa, $R = \infty$) and ($P = 0.1$ MPa, $R = \infty$), it is seen that the internal pressurization shifts the mass-air-mass resonance dip to a higher frequency but only has a small influence on the second-order standing-wave resonance and coincidence dips. In comparison, the panel curvature has a noticeable influence on all of these phenomena. Therefore, when designing a practical aircraft fuselage, the noticeable combination effects of the panel curvature and internal pressurization on noise transmission need to be carefully considered, especially in the relatively low-frequency range where the ring frequency resonance occurs.

2.2.12 Conclusions

The effects of external mean flow on sound transmission through double-leaf aeroelastic plates have been quantified analytically, with the intention to simulate the transmission of engine exhaust noise through typical aircraft fuselage skin panels into cabin interiors. Theoretical formulations have been developed for the analysis of the fluid-plate coupling problem, and the *STL* versus frequency curves for various specific cases (Mach number, direction of mean flow, sound incidence angle, panel curvature, and internal pressurization) are obtained, with the added-mass effects of the convected fluid loading well captured. Four distinct acoustic phenomena (i.e., the mass-air-mass resonance, the standing-wave attenuation, the standing-wave resonance, and the coincidence resonance) for flat double-leaf plates as well as the ring frequency resonance for curved double-leaf plates are clearly identified. Simple closed-form formulae for predicting the natural frequencies associated with these acoustic phenomena in the presence of mean flow are subsequently derived from physical principles, which are completely independent upon the theoretical model. In the absence of other relevant theoretical or experimental work, the excellent agreement between these formulae and the model predictions serve to validate the two theories against each other.

Systematic parametric studies are subsequently conducted to quantify the effects of Mach number, the direction of mean flow, the sound incidence elevation and azimuth angles, the panel curvature, as well as the internal pressurization on the STL. As the Mach number is increased, in the case of sound incidence along the downstream direction, the STL values increase over a broad frequency range, and the natural frequencies for the associated acoustic phenomena (except for the coincidence resonance) are shifted considerably to the lower frequency range due to the added-mass effects of the mean flow. The exception of the coincidence resonance is attributed to its strong dependence on the refraction angle φ_2 but not on the convected fluid loading.

For sound incidence along the upstream direction, the corresponding frequencies increase until the Mach number is increased up to a critical value, except again for the coincidence resonance. Further increase of the Mach number beyond the critical value results in the disappearance of the mass-air-mass resonance, the standing-wave attenuation, and the standing-wave resonance, but the coincidence resonance is always existent. The increase of the Mach number induces a noticeable increment of the STL value over a relatively wide range of frequency due to the total reflection effects.

In the presence of external mean flow, the transmission of sound is significantly influenced by the sound incidence elevation angle and azimuth angle, and the noticeable combination effects of the panel curvature and internal pressurization should be taken into account in the practical design of aircraft fuselages.

2.3 Double-Leaf Panel Filled with Porous Materials

2.3.1 Introduction

Improving the sound transmission insulation performance of aircraft cabin fuselage panels is helpful for reducing aircraft interior noise. As different kinds of sandwich panels have been applied to construct cabin fuselages to provide more effective noise isolation, such as double-leaf thin panels filled with porous absorbent materials [39, 42, 43], it is important to evaluate the sound transmission characteristics of such sandwich constructions. Further, since there exists high-speed airflow outside the cabin in typical cruise condition, it is even more significant to investigate the influence of external airflow upon sound transmission.

For double-leaf panels filled with porous absorbent materials, before exploring the sound insulation capability of the whole structure, it is necessary to accurately model the propagation process of sound in the porous material. Two main approaches have been developed to address the issue. One is the well-known Biot theory [73, 74], which assumes that stress wave propagation in a fluid-saturated porous material can be described by four nondimensional parameters and a characteristic frequency, and there exist two dilatational waves and one rotational wave in the material. In addition, the theory provides a set of functions of characteristic parameters to govern the acoustic medium. However, when the pore diameter is equal to the quarter wavelength of the acoustic wave, the Biot theory breaks down as stated by Lighthill [75]. The Biot model has been extensively applied to investigate the acoustic properties of different materials including saturated sand [76], rectangular and triangular pores [77], and catalytic converters [78]. The other approach is the semiempirical model [25, 79–81]. Delany and Bazley [79] showed empirically that the characteristic impedance and the propagation coefficient are functions of frequency f divided by static flow resistance R, i.e., f/R, and provided a method to estimate R of a material from its bulk density. Bies and Hansen [82] presented static flow resistance information for typical porous materials and suggested the most common applications of the flow resistance. Graham [25] used this approach to examine the characteristic impedance of dissipative materials for cabin inside treatment. Allard and Champoux [80] proposed a set of new semiempirical equations for sound propagation in rigid frame fibrous materials by modeling the porous material as an equivalent fluid with dynamic density and dynamic bulk modulus. They also suggested that the equations could be used when f/R is smaller than 1 kg/m^3 for most cases.

The transmission loss of sound across double-leaf panels separated by porous materials has been extensively investigated. For typical example, Lauriks et al. [83] developed a transfer matrix model to study the transmission loss through panels with solid porous layers, with the Biot model employed to describe sound propagation through the porous material, while Brouard et al. [84] proposed a general method to model sound propagation in layered systems such as fluid-saturated porous layers. Panneton et al. [34] presented a three-dimensional (3D) finite element model to calculate the loss of sound transmission through multilayer structures containing porous absorbent materials. The structures considered vary according to whether the filling porous material is bonded or not to the faceplates. Making use of two-dimensional (2D) elasticity theory, Chonan and Kugo [85] presented a model to examine the sound transmission characteristics of a three-layered panel excited by plane waves. Kang et al. [86] employed the method of Gauss distribution function for incident energy to predict the sound transmission loss (*STL*) of multilayered panels such as double-plate structures embedded with porous materials. Bolton et al. [87] calculated random incidence transmission loss through double-leaf panels lined with elastic porous materials.

2.3 Double-Leaf Panel Filled with Porous Materials

As for the influence of external flow interaction on acoustic characteristics of structures, numerous works exist in the open domain. For example, Clark and Frampton [2, 4, 7] systematically investigated the structural response of panels excited by turbulent boundary layer (TBL)-induced noise, and Graham [24, 25] developed an extended theoretical model to investigate aircraft structure response excited by TBL-induced noise. Davies [53] theoretically estimated acoustic power radiation of TBL excited panels, while Koval et al. [1] examined the mean flow effect on *STL* of a single plate and found that the mean flow increased the transmission loss in the whole frequency range. Xin and Lu [23, 88] studied theoretically the effect of external mean flow on the transmission loss of double-leaf panels and identified four different kinds of acoustic phenomena in the sound transmission process through a double-leaf structure. Accounting for the mean flow effect, Sgard et al. [15] developed a coupled FEM-BEM (finite element method-boundary element method) model to investigate the vibroacoustic behavior of planar plates. However, in the presence of external mean flow, the structures studied in the aforementioned investigations are relatively simple and do not consider the presence of additional porous sound absorptive materials.

To more accurately examine the effects of mean flow on jet-engine noise transmission through aircraft fuselages, we propose a theoretical model for the aero-acoustic problem of jet-engine noise transmission through infinite double-leaf panels filled with fibrous sound absorptive materials in the presence of uniform external mean flow. Although a uniform flow across the panel might be an ideal case that would be not happening in reality, the assumption of a uniform mean flow can be warranted as has been done by many researchers [1, 22]. To describe sound propagation in the fibrous material, the equivalent fluid model is employed, and the fluid momentum equations are applied to ensure displacement continuity at fluid-structure interfaces. Upon validating the model predictions against existing experimental results, systematic parameter investigations are carried out using the proposed model, with guidance conclusions for practical aircraft fuselage designs obtained.

2.3.2 Problem Description

Consider the double-leaf panel structure shown schematically in Fig. 2.21, which contains two parallel thin elastic plates filled with fibrous sound absorptive materials in between. The two plates are made of homogenous and isotropic material. For simplicity, the double-leaf panel is taken as infinitely large in plane. A harmonic plane sound wave penetrates through the sandwich panel, accompanied by a uniform external mean flow that moves along the x-axis with uniform speed V on one side of the panel (Fig. 2.22). The fluid on the other side is motionless. The sound wave is incident upon the panel with elevation angle φ_1 and azimuth angle β relative to the coordinate axes of Fig. 2.22. The thicknesses of the two thin plates are h_1 and h_2, respectively, while H is the thickness of the filled porous material. In general,

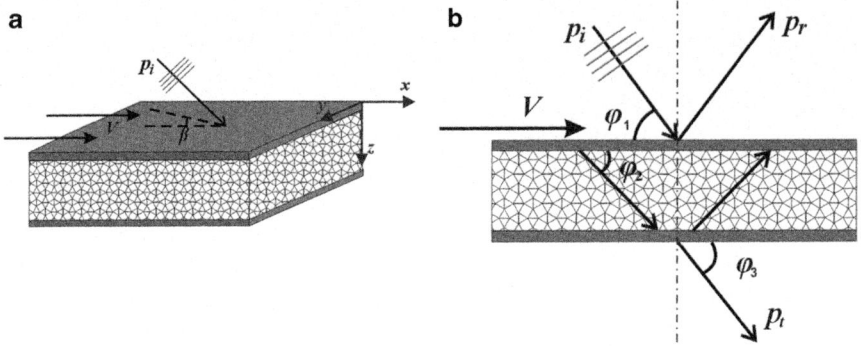

Fig. 2.22 Schematic of sound transmission through a *double-leaf panel* filled with porous material in the presence of uniform external mean flow: (**a**) global view; (**b**) side view

the acoustic field may be divided into three parts: incident field $z<0$, middle field (i.e., sound absorbent field) $h_1 < z < h_1 + H$, and transmitted field $z > h_1 + H + h_2$. The acoustic media in the three fields are defined by mass densities (ρ_1, ρ_2, ρ_3) and sound wavenumbers (k_1, k_2, k_3), respectively.

Two further assumptions are introduced to develop the theoretical model: (1) The porous sound absorptive material is bonded to both faceplates, with no gap between the porous material and either of the faceplates. The equivalent fluid model is employed to describe the acoustic behaviors of the porous material. According to Rebillard et al. [89], when a porous material is bonded to a screen, the equivalent fluid model is applicable only when the skeleton of the porous material is very limp. Consequently, the porous material considered here is assumed to be limp enough to apply the equivalent fluid model, which is usually true for commonly used porous sound absorptive materials. (2) The fluid media are inviscid and irrotational.

2.3.3 Theoretical Model

2.3.3.1 Acoustic Field

The acoustic field is divided into three parts as previously described. Accordingly, sound pressures in the three acoustic fields can be expressed as

Incident field:

$$p_1 = P_{1i} e^{j\omega t - j(k_{1x}x + k_{1y}y + k_{1z}z)} + P_{1r} e^{j\omega t - j(k_{1x}x + k_{1y}y - k_{1z}z)} \qquad (2.96)$$

Middle field:

$$p_2 = P_{2i} e^{j\omega t - j(k_{2x}x + k_{2y}y + k_{2z}z)} + P_{2r} e^{j\omega t - j(k_{2x}x + k_{2y}y - k_{2z}z)} \qquad (2.97)$$

2.3 Double-Leaf Panel Filled with Porous Materials

Transmitted field:

$$p_3 = P_{3t} e^{j\omega t - j(k_{3x}x + k_{3y}y + k_{3z}z)} \tag{2.98}$$

where P_{1i} is the amplitude of the incident wave, P_{1r} is the amplitude of the reflected wave, P_{2i} and P_{2r} are separately the amplitude of the positive and negative waves, P_{3t} is the amplitude of the transmitted wave, (k_1, k_2, k_3) are the wavenumbers of the three acoustic fields, while $(k_{ix}, k_{iy}, k_{iz}, i = 1, 2, 3)$ are the components of the corresponding wavenumbers along $(x$-, y-, z-$)$directions, respectively. Thus, (k_{ix}, k_{iy}, k_{iz}) can be written as

$$k_{ix} = k_i \cos\varphi_i \cos\beta, \quad k_{iy} = k_i \cos\varphi_i \sin\beta, \quad k_{iz} = k_i \sin\varphi_i \tag{2.99}$$

Due to the refraction effect of external mean flow [1], the wavenumber in the incident field is

$$k_1 = \frac{\omega/c_1}{(1 + M\cos\varphi_1 \cos\beta)} \tag{2.100}$$

The equivalent fluid model [80, 90, 91] is employed to calculate the sound wavenumber in the middle field. According to this model, sound propagation in fibrous materials can be described by equivalent dynamic density and dynamic bulk modulus. The dynamic density $\rho(\omega)$ is given by

$$\rho(\omega) = \rho_0 \left[1 + \frac{1}{j2\pi}\left(\frac{R}{\rho_0 f}\right) G_1\left(\frac{\rho_0 f}{R}\right)\right] \tag{2.101}$$

and the dynamic bulk modulus is given by

$$K(\omega) = \gamma_s P_0 \left[\gamma_s - \frac{\gamma_s - 1}{1 + (1/j8\pi N_{\text{Pr}})(\rho_0 f/R)^{-1} G_2(\rho_0 f/R)}\right]^{-1} \tag{2.102}$$

where

$$G_1(\rho_0 f/R) = \sqrt{1 + j\pi(\rho_0 f/R)} \tag{2.103}$$

$$G_2(\rho_0 f/R) = G_1[(\rho_0 f/R) 4 N_{\text{Pr}}] \tag{2.104}$$

Here, R is the static flow resistivity, γ_s is the specific heat ratio of air, ρ_0 is the air density, P_0 is the standard atmospheric pressure, N_{Pr} is the Prandtl number, and f is the frequency of the sound wave. The dynamic density considers inertial and viscous forces per unit volume of air in the material and is related to the averaged molecular displacement of air and the averaged variation of pressure. With k_{cav} denoting the complex wavenumber of the porous material, it can be expressed by the dynamic density and dynamic bulk modulus as

$$k_{cav} = 2\pi f \sqrt{\frac{\rho(\omega)}{K(\omega)}} \tag{2.105}$$

According to Morse and Ingard [92], the complex density of the porous material may be written as

$$\rho_2 = \rho_{cav} = \frac{k_2^2 \rho_0}{k_0^2 \gamma_s \sigma} \tag{2.106}$$

where σ is the porosity of the porous material.

In the transmitted field, the wavenumber is

$$k_3 = \frac{\omega}{c_3} \tag{2.107}$$

where $c_3 = c_0$.

Let the transverse deflection wavenumber in the plate be denoted as k_p. For the sound waves to fit at the boundary, the trace wavelengths must match [1, 23, 88], namely,

$$k_{1x} = k_p \cos\beta = k_{2x} = k_p \cos\beta = k_{3x}$$
$$k_{1y} = k_p \sin\beta = k_{2y} = k_p \sin\beta = k_{3y} \tag{2.108}$$

Substitution of Eqs. (2.100) and (2.105) into Eq. (2.108) yields

$$\cos\varphi_2 = \frac{k_1 \cos\varphi_1}{k_2} = \frac{\cos\varphi_1}{c_1(1 + M\cos\varphi_1 \cos\beta)} \sqrt{\frac{K(\omega)}{\rho(\omega)}} \tag{2.109}$$

$$\cos\varphi_3 = \frac{k_2 \cos\varphi_2}{k_3} = \frac{c_3 \cos\varphi_1}{c_1(1 + M\cos\varphi_1 \cos\beta)} \tag{2.110}$$

2.3.3.2 Fluid-Structure Coupling

To determine the unknown parameters associated with the three acoustic fields, two fluid-structure coupling conditions must be supplemented: (a) displacement continuity between plate particle and adjacent fluid particle and (b) driving relation between incident sound and plate dynamic response.

Let (δ_1, δ_4) represent the fluid particle displacements adjacent to the outer side of the two faceplates and (δ_2, δ_3) represent the porous material particle displacements adjacent to the inner side of the plates, respectively. With the transfer effect of the

2.3 Double-Leaf Panel Filled with Porous Materials

mean flow accounted for, the fluid-structure coupling is described using the particle displacement continuity condition [31]. These particle displacements satisfy the Navier-Stokes equation for inviscid and irrotational fluid as

$$\frac{D^2 \delta_1}{Dt^2} = -\frac{1}{\rho_1} \frac{\partial p_1}{\partial z}\bigg|_{z=0_1} \tag{2.111}$$

$$\frac{D^2 \delta_2}{Dt^2} = -\frac{1}{\rho_2} \frac{\partial p_2}{\partial z}\bigg|_{z=h_1} \tag{2.112}$$

$$\frac{D^2 \delta_3}{Dt^2} = -\frac{1}{\rho_2} \frac{\partial p_2}{\partial z}\bigg|_{z=h_1+H} \tag{2.113}$$

$$\frac{D^2 \delta_4}{Dt^2} = -\frac{1}{\rho_3} \frac{\partial p_3}{\partial z}\bigg|_{z=h_1+H+h_2} \tag{2.114}$$

For an infinite uniform flat plate, the vibration displacement excited by a harmonic wave is given by [1]

$$w_j = w_{j0} e^{j\omega t - j\left[(k_p \cos\beta)x + (k_p \sin\beta)y\right]} \tag{2.115}$$

where $k_p = \omega/c_p$ denotes the transverse deflection wavenumber in the plate and c_p is the trace velocity of the transverse wave in the plate.

Since the porous material and the adjacent fluid are in good contact with the two faceplates, the displacements of the fluid particle and the porous material particle adjacent the two plates take the same form as that of the plates:

$$\delta_i = \delta_{i0} e^{j\omega t - j(k_{ix}x + k_{iy}y)} \tag{2.116}$$

Combining Eqs. (2.111), (2.112), (2.113), (2.114), (2.115), and (2.116) yields

$$\delta_{10} = -\frac{jk_{1z}(P_{1i} - P_{1r})}{\rho_1(\omega - vk_{1x})^2} \tag{2.117}$$

$$\delta_{20} = -\frac{jk_{2z}(P_{2i} - P_{2r})}{\rho_2 \omega^2} \tag{2.118}$$

$$\delta_{30} = -\frac{jk_{2z}}{\rho_2 \omega^2}\left(P_{2i} e^{-jk_{2z}H} - P_{2r} e^{jk_{2z}H}\right) \tag{2.119}$$

$$\delta_{40} = -\frac{jk_{3z}}{\rho_3 \omega^2} P_{3t} e^{-jk_{3z}H} \tag{2.120}$$

Continuity of particle displacements dictates

$$\delta_{10} = w_{10} = \delta_{20}, \delta_{30} = w_{20} = \delta_{40} \tag{2.121}$$

Substitution of (2.115), (2.116), (2.117), (2.118), (2.119), and (2.120) into (2.121) leads to

$$P_{1r} = P_{1i} - \frac{j\rho_0(\omega - vk_{1x})^2 w_{10}}{k_{1z}} \tag{2.122}$$

$$P_{2i} = \frac{\rho_2 \omega^2}{2k_{2z} \sin(k_{2z} H)} \left(w_{10} e^{jk_{2z} H} - w_{20} \right) \tag{2.123}$$

$$P_{2r} = \frac{\rho_2 \omega^2}{2k_{2z} \sin(k_{2z} H)} \left(w_{10} e^{-jk_{2z} H} - w_{20} \right) \tag{2.124}$$

$$P_{3t} = \frac{j\rho_3 \omega^2 w_{20}}{k_{3z}} e^{ik_{3z} H} \tag{2.125}$$

The driving relation between the pressure difference across the plate and the plate response is

$$\begin{aligned} P_{1i} + P_{1r} - P_{2i} - P_{2r} &= Z_{p1} \dot{w}_{10} \\ P_{2i} + P_{2r} - P_{3t} &= Z_{p2} \dot{w}_{20} \end{aligned} \tag{2.126}$$

where Z_{p1} and Z_{p2} represent separately the plate impedances for the upper and lower faceplates (Fig. 2.22) and $\dot{w}_{j0} = i\omega w_{j0}$ ($j = 1, 2$). For infinite plates as considered in the present study, the plate impedance can be calculated as follows.

The governing equation of an infinite plate immersed in a fluid medium excited by a harmonic wave is given by

Upper faceplate:

$$\left(D_1 \nabla^4 - m_1 \omega^2 \right) w_1(x, y) = p_1(x, y, 0) - p_2(x, y, h_1) \tag{2.127}$$

Lower faceplate:

$$\left(D_2 \nabla^4 - m_2 \omega^2 \right) w_2(x, y) = -p_3(x, y, H + h_1 + h_2) + p_2(x, y, H + h_1) \tag{2.128}$$

The complex Young's modulus of the plate material is modified by the structure loss factor η as

$$E'_n = E_n (1 + j\eta_n), \quad n = 1, 2 \tag{2.129}$$

Accordingly, the complex bending stiffness is given by

$$D'_n = D_n (1 + j\eta) = \frac{E_n h^3 (1 + j\eta_n)}{12(1 - \upsilon^2)}, \quad n = 1, 2 \tag{2.130}$$

2.3 Double-Leaf Panel Filled with Porous Materials

Substitution of (2.115) into (2.127) and (2.128) yields

$$w_{10} = \frac{p_1(x,y,0) - p_2(x,y,h_1)}{D_1 k_p^4 - m_1 \omega^2} \frac{1}{e^{j\omega t - j(k_{1x}x + k_{1y})}} \quad (2.131)$$

$$w_{20} = \frac{p_2(x,y,H+h_1) - p_3(x,y,H+h_1+h_2)}{D_2 k_p^4 - m_2 \omega^2} \frac{1}{e^{j\omega t - j(k_{2x}x + k_{2y})}} \quad (2.132)$$

Finally, the impedances of the upper and lower faceplates are obtained as

$$Z_{p1} = \frac{p_1(x,y,0) - p_2(x,y,h_1)}{i\omega w_{10}} = jm_1\omega \left(1 - \frac{D_1 \omega^2 \cos^4 \varphi_2}{m_1} \sqrt{\frac{\rho(\omega)}{K(\omega)}}\right) \quad (2.133)$$

$$Z_{p2} = \frac{-p_3(x,y,H+h_1+h_2) + p_2(x,y,H+h_1)}{j\omega w_{20}}$$

$$= jm_2\omega \left(1 - \frac{D_2 \omega^2 \cos^4 \varphi_2}{m_2} \sqrt{\frac{\rho(\omega)}{K(\omega)}}\right) \quad (2.134)$$

2.3.3.3 Sound Transmission Loss

The transmission loss of sound across a double-leaf panel filled with porous material is defined as

$$STL = -10\log_{10}\tau(f, \varphi_1, \beta) \quad (2.135)$$

where $\tau(\omega, \varphi_1, \beta)$ is the acoustic transmissivity given by

$$\tau(f, \varphi_1, \beta) = \frac{\rho_1 c_1}{\rho_2 c_2} \left|\frac{P_{3t}}{P_{1i}}\right|^2 \quad (2.136)$$

with

$$\frac{P_{1i}}{P_{3t}} = \frac{-j\rho_3 k_{2z} \tan(k_{2z}H)}{2\rho_2 \left(\rho_3 + k_{3z}(Z_{p2}/\omega)e^{-jk_{3z}H}\right) Z_{p1}}$$

$$\left[\begin{pmatrix} \rho_2 \cos(k_{2z}H) + \rho_2 \frac{Z_{p1}}{Z_{p2}} \\ - j(Z_{p2}/\omega) k_{2z} \sin(k_{2z}H) \\ + \frac{\rho_2 \rho_3}{\rho_3 + k_{3z}(Z_{p2}/\omega)e^{-jk_{3z}H}} \end{pmatrix} \begin{pmatrix} \frac{Z_{p1} k_{1z}\omega + \rho_0(\omega - vk_{1x})^2}{jk_{2z}k_{1z}\omega \sin(k_{2z}H)} \end{pmatrix}\right.$$

$$\left. + \frac{\rho_2 \rho_0 \cos(k_{2z}H)(\omega - vk_{1x})^2}{jk_{2z}k_{1z}\omega \sin(k_{2z}H)} \right] \quad (2.137)$$

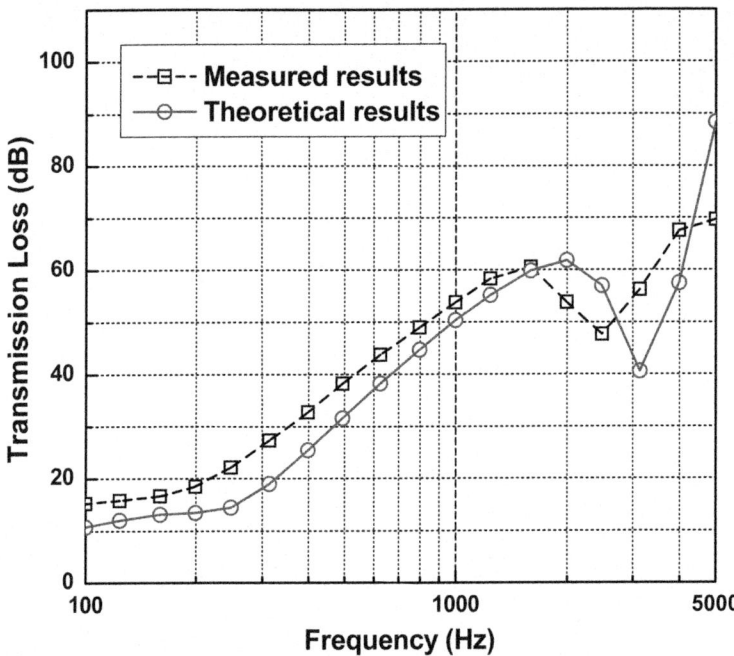

Fig. 2.23 Sound transmission loss of double-leaf panel separated by foam material in the absence of mean flow ($M = 0$): comparison between the present model predictions and experimental results by Pellicier et al. [93]

2.3.4 Validation of Theoretical Model

To validate the theoretical model presented in the previous section, the model predictions are compared with existing experimental results [93] in Fig. 2.23 for the case of no external mean flow. For experimental measurement, the sandwich structure is composed of two plywood plates with thickness of 8.5 mm and separated by an open-celled foam with thickness of 50 mm. The plywood has Young's modulus of 4.25 GPa, Poisson ratio 0.5, density 650 kg/m^3, and loss factor 0.03. The foam material has a porosity of 93 % and a static flow resistivity of 55,000 Ns/m^4. The comparison is conducted by applying these parameters into our theoretical model and setting the Mach number to 0. Since the experimental results were obtained in a diffuse sound field, the model should be adjusted to calculate the *STL* in a diffuse sound field. The azimuth angle makes no sense to the *STL* in the absence of mean flow; in such case, Eq. (2.136) becomes

$$\tau(f, \varphi_1) = \frac{\rho_1 c_1}{\rho_2 c_2} \left| \frac{P_{3t}}{P_{1i}} \right|^2 \quad (2.138)$$

2.3 Double-Leaf Panel Filled with Porous Materials

The diffuse transmissivity is then calculated by

$$\tau_{diff}(f) = \frac{\int_0^{\varphi_{\lim}} \tau(f,\varphi) \sin\varphi \cos\varphi \, d\varphi}{\int_0^{\varphi_{\lim}} \sin\varphi \cos\varphi \, d\varphi} \quad (2.139)$$

Accordingly, the sound transmission loss becomes

$$STL = -10 \log_{10} \tau_{\text{diff}}(f) \quad (2.140)$$

The results of Fig. 2.23 show that the theoretical model predictions for diffuse incidence ($\theta_{\lim} = 78°$) exhibit the same trend as that of the experimental measurements and, overall, the agreement is reasonable. However, there does exist some discrepancy between theory and experiment (e.g., the shift of the coincidence dip location in Fig. 2.23), which is attributed to the fact that finite-sized boundary conditions hold in the experiment measurements [93], while infinite size is assumed in the present model. Moreover, the imperfect connection between the plywood plates and foam material may increase the damping effect of the whole structure, causing the higher measured STL values than those predicted over the frequency range considered. Notice also that as the value of STL at the coincidence dip is also greatly affected by the damping of the whole structure, the measured value that includes the effect of structure material damping and boundary constraints damping is expected to be larger than the prediction which only considers the effect of structure material damping.

2.3.5 Influence of Porous Material and the Faceplates

Figure 2.24 compares the predicted transmission loss of a double-leaf panel filled with porous material with that of a double-leaf panel with air cavity having the same thickness as that of the porous material, for Mach number $M = 0.05$ and sound incident elevation angle $\varphi_1 = 30°$, middle layer thickness $H = 0.08$ m, and azimuth angle $\beta_1 = 0°$ (Fig. 2.22). As previously mentioned in the Introduction, the necessary condition for using the equivalent fluid theory is $f/R < 1$ kg/m^3 [80]. Therefore, the highest frequency of the present analysis is limited to 24,000 Hz. In the following, the model predictions are presented within the frequency range of 0–10,000 Hz, for all the acoustic phenomena of interest can be shown within this range.

The results of Fig. 2.24 show that the presence of the filling porous material significantly weakens the plate-air-plate resonance dip of the system, due to the damping effect (or the absorption effect) of the porous material. Further, in the frequency range of 1,000–8,000 Hz, two kinds of acoustic phenomena exist for double-leaf panel with air cavity: antiresonance and standing-wave resonance. As illustrated schematically in Fig. 2.25, the plane sound wave propagates from point

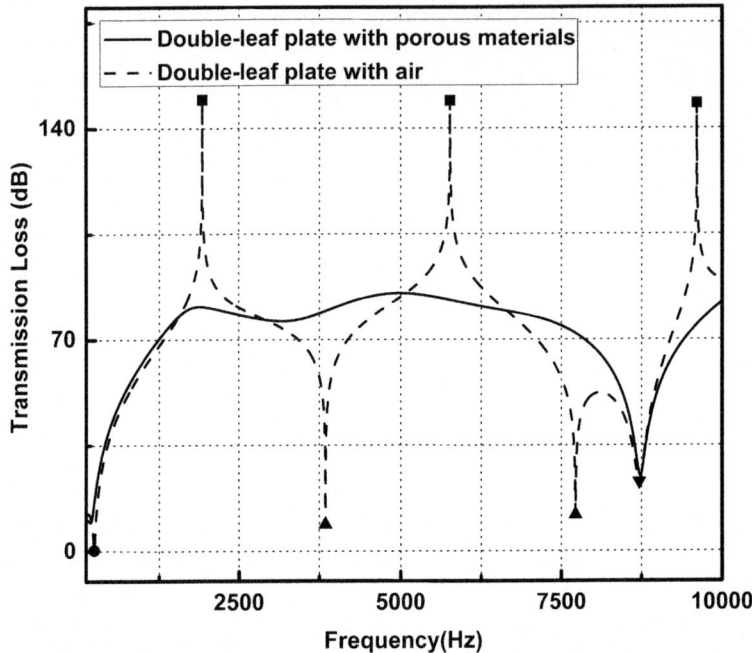

Fig. 2.24 Sound transmission loss of double-leaf panel for Mach number $M = 0.05$, loss factor $\eta = 0.01$, incident elevation angle $\varphi_1 = 30°$, middle layer thickness $H = 0.08$ m, and azimuth angle $\beta_1 = 0°$: ■ antiresonance, • "plate-air-plate" resonance, ▲ standing-wave resonance, ▼ coincidence dip

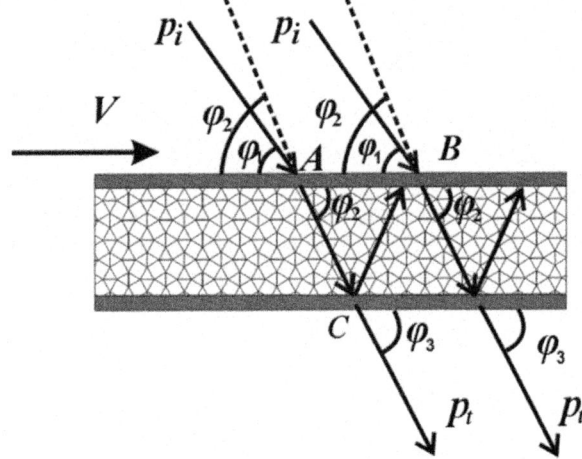

Fig. 2.25 Schematic of standing-wave resonance and antiresonance

2.3 Double-Leaf Panel Filled with Porous Materials

A on the upper plate to point B on the lower plate; it then reflects to point C on the upper plate where it meets another incident sound wave. If the distance between the incident wave and the reflected wave is the odd number of the one-fourth wavelength of the sound wave, the interference between the two waves causes antiresonance. Alternatively, if the distance is the multiple of half the wavelength, the interference becomes standing-wave resonance. In comparison, for a double-leaf panel filled with porous material, since the energy intensity of the reflected wave has been significantly weakened due to the absorption of the porous material, both the standing-wave resonance and antiresonance almost disappear. Accordingly, the curve is relatively smooth in the frequency range of 1,000–8,000 Hz, as shown in Fig. 2.24. The absorption effect of the porous material exists in both the presence and absence of mean flow, which will be discussed in detail in Sect. 2.3.7 of this section.

It can also be observed from Fig. 2.24 that coincidence dip, at which the wavelength of the incident sound wave matches with the wavelength of the flexural bending wave in the plate, exists on both STL and frequency curves at the same frequency. This agrees well with Xin et al. [23, 94] and Wang et al. [59], both pointing out that the coincidence frequency is given by

$$f_c = \frac{c_2^2}{2\pi h \cos^2 \varphi_2} \sqrt{\frac{12\rho(1-v^2)}{E}} \qquad (2.141)$$

Combining Eqs. (2.99) and (2.108) with Eq. (2.141) yields

$$f_c = \frac{c_2^2}{2\pi h \cos^2 \varphi_2} \sqrt{\frac{12\rho(1-v^2)}{E}} = \frac{\omega^2}{2\pi h k_1^2 \cos^2 \varphi_1} \sqrt{\frac{12\rho(1-v^2)}{E}} \qquad (2.142)$$

which means that the frequency of the coincidence dip for the whole structure is independent of the presence of the porous material. However, the value of the STL at the coincidence dip is dependent on the panel damping, as shown in Fig. 2.26, which increases when the loss factor of the faceplate increases. Since the effective value of the loss factor is related with the damping of the attached porous material [25], it deduced that the extent of coincidence dip is significantly affected by the damping of the porous material.

The influence of faceplate thickness on structure transmission loss is shown in Fig. 2.27. As the thickness of the faceplates is increased, the frequency at which coincidence dip occurs decreases significantly, consistent with Eq. (2.141). However, the extent of the coincidence dip does not change significantly with increasing faceplate thickness.

2.3.6 Influence of Porous Material Layer Thickness

Since the presence of a porous material layer affects considerably the transmission of sound across a double-leaf panel, the effect of its thickness is further explored

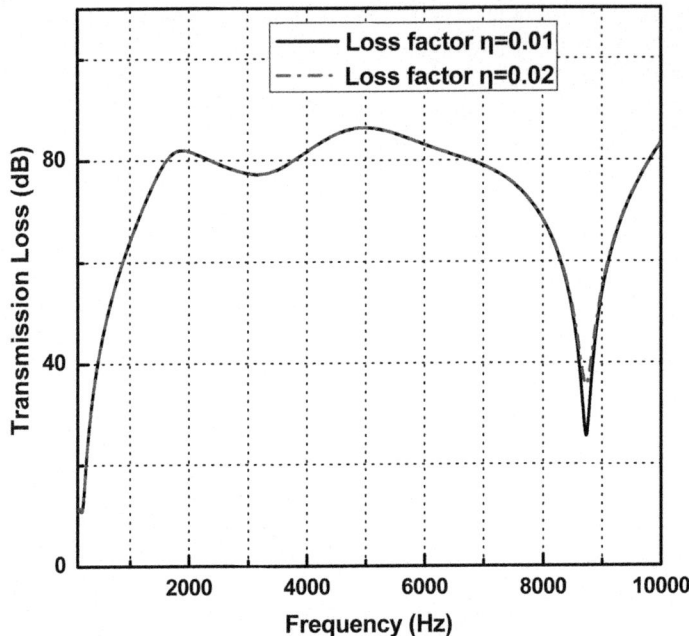

Fig. 2.26 Sound transmission loss of *double-leaf panel* filled with porous material for selected loss factors, with $\varphi_1 = 30°$, $\beta_1 = 0°$, $M = 0.05$, and $H = 0.08$ m

below. Figure 2.28 plots the predicted transmission loss as a function of frequency for selected porous layer thicknesses, with Mach number fixed at $M = 0.05$, sound incident elevation angle at $\varphi_1 = 30°$, and azimuth angle at $\beta_1 = 0°$. It is seen from Fig. 2.28 that at frequencies between the coincidence resonance and the plate-air-plate resonance, both the standing-wave resonance and antiresonance are increasingly weakened and the value of *STL* decreases as the porous layer thickness increases. This should remarkably be affected by the increase of the gap between the panels, which is also affected by the damping loss effect of the porous material within this frequency range [95]. In contrast, at frequencies near or above the coincidence resonance, since the porous material has a strong sound absorption capability at high frequencies, the *STL* may have reached its utmost value with a certain thickness of the porous material. Consequently, at relatively high frequencies, increasing the porous layer thickness further has negligible influence upon the transmission loss, as shown in Fig. 2.28.

2.3.7 Influence of External Mean Flow

Figure 2.29 presents the influence of external mean flow on *STL* across double-leaf panels with filling porous materials for $\varphi_1 = 30°$, $H = 0.08$ m, and $\beta_1 = 0°$. The

2.3 Double-Leaf Panel Filled with Porous Materials

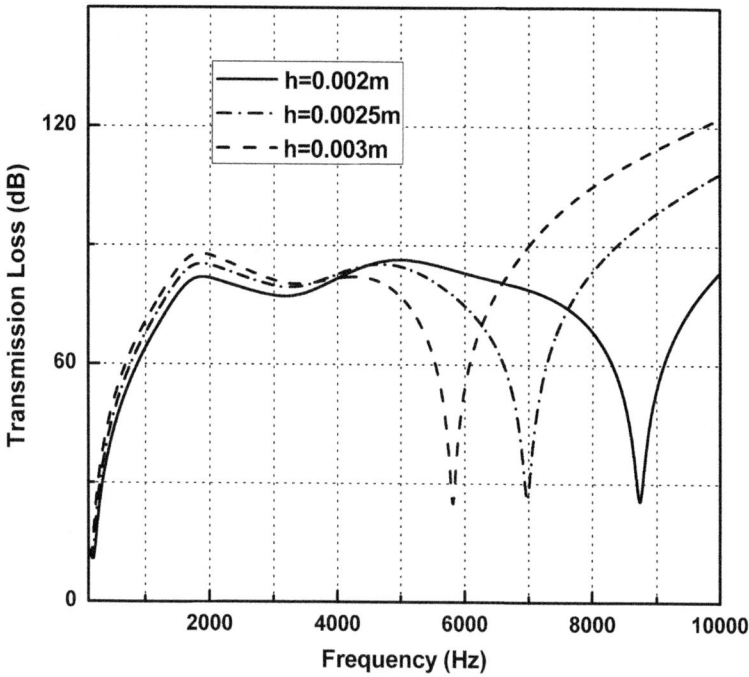

Fig. 2.27 Sound transmission loss of double-leaf panel filled with porous material for selected faceplate thicknesses, with $\varphi_1 = 30°$, $\beta_1 = 0°$, $M = 0.05$, and $H = 0.08$ m

results demonstrate that increasing the Mach number leads to noticeable changes of the *STL*: as the Mach number increases, the *STL* increases over a wide frequency range below the coincidence dip frequency. This is mainly caused by the added damping effect of the mean flow as suggested by Sgard et al. [15]. Therefore, the radiated acoustic power decreases with increasing Mach number of the mean flow, resulting in the increase of the *STL* value.

It is interesting to see from Fig. 2.29 that the amplitudes of the standing-wave resonance and antiresonance for the case of $M = 0.05$ do not change significantly in comparison with those when there is no mean flow. This implies that the absorption effect of the porous material is almost independent of the mean flow.

The effect of Mach number on the coincidence frequency of double-leaf structures is quantified in Fig. 2.30, for $\varphi_1 = 30°$, $H = 0.08$ m, loss factor $\eta = 0.01$, and $\beta_1 = 0°$. With increasing Mach number, the coincidence dip frequency increases almost proportionally. This is attributed to the fact that the refraction effect of external mean flow is noticeably enhanced as the Mach number is increased, which is confirmed by

$$f_c = \frac{\omega^2}{2\pi h k_1^2 \cos^2\varphi_1} \sqrt{\frac{12\rho(1-v^2)}{E}} = \frac{c_1^2(1+M\cos\varphi_1)^2}{2\pi h \cos^2\varphi_1} \sqrt{\frac{12\rho(1-v^2)}{E}} \tag{2.143}$$

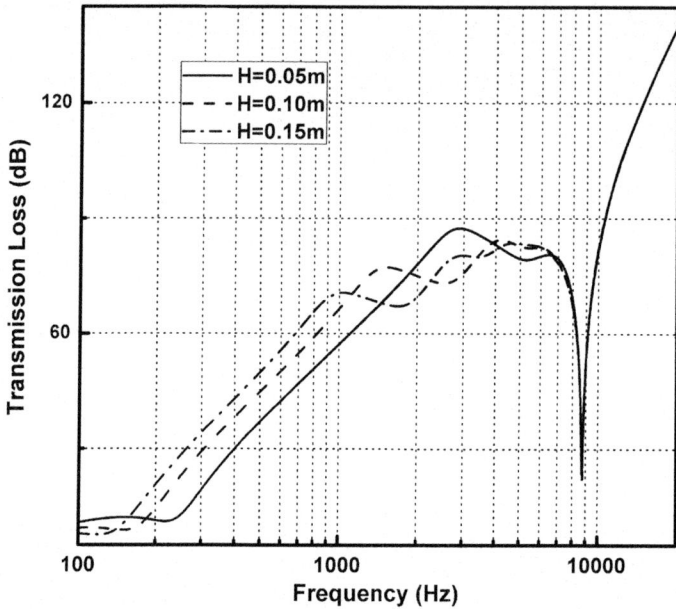

Fig. 2.28 Sound transmission loss of double-leaf panel plotted as a function of frequency for selected porous layer thicknesses, with Mach number $M = 0.05$, loss factor $\eta = 0.01$, incident elevation angle $\varphi_1 = 30°$, and azimuth angle $\beta_1 = 0°$

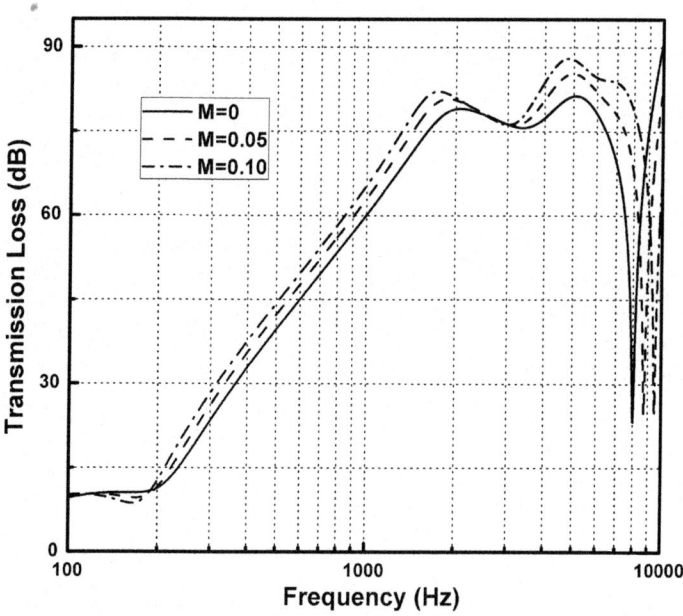

Fig. 2.29 Sound transmission loss of double-leaf panel filled with porous material for selected Mach numbers, with $\varphi_1 = 30°$, loss factor $\eta = 0.01$, $\beta_1 = 0°$, and $H = 0.08$ m

2.3 Double-Leaf Panel Filled with Porous Materials

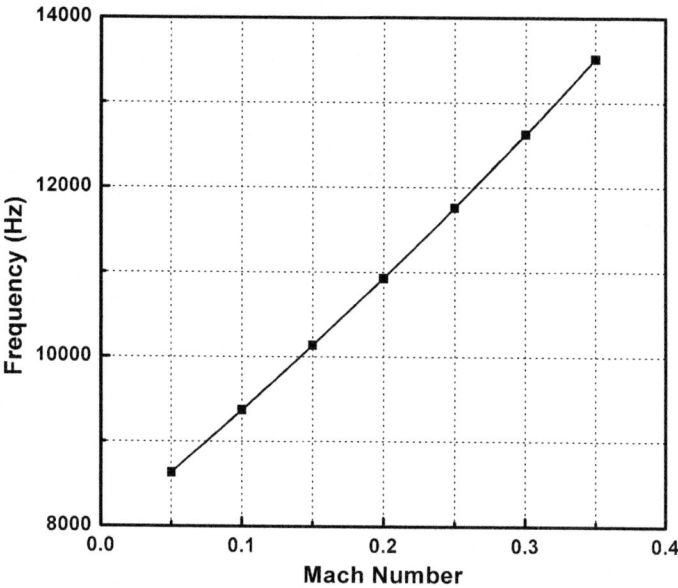

Fig. 2.30 Effect of Mach number on coincidence frequency of double-leaf panel filled with porous material, with $\varphi_1 = 30°$, loss factor $\eta = 0.01$, $\beta_1 = 0°$, and $H = 0.08$ m

Since the value of M considered is small, Eq. (2.143) may be transformed into the form below:

$$f_c = \frac{c_1^2(1 + M\cos\varphi_1)^2}{2\pi h \cos^2\varphi_1}\sqrt{\frac{12\rho(1-v^2)}{E}} \approx \frac{c_1^2(1 + 2M\cos\varphi_1)}{2\pi h \cos^2\varphi_1}\sqrt{\frac{12\rho(1-v^2)}{E}} \tag{2.144}$$

Equation (2.144) exhibits the same trend as that of the present model predictions.

2.3.8 Influence of Incident Sound Elevation Angle

Figure 2.31 plots the *STL* of a double-leaf panel with filling porous material as a function of frequency for selected incident elevation angles, with $M = 0.05$, $\beta_1 = 0°$, and $H = 0.08$ m. The incident angle is seen to have an obvious effect on the transmission loss and the position of coincidence frequency. Between the plate-porous material-plate resonance and the coincidence dip frequency, the *STL* value increases when the angle from the axis normal to the plane of the panel decreases (i.e., the elevation angle increases). However, since the coincidence frequency increases with increasing elevation angle, the *STL* decreases in the frequency range above the coincidence frequency. Besides, when the incident angle is relatively

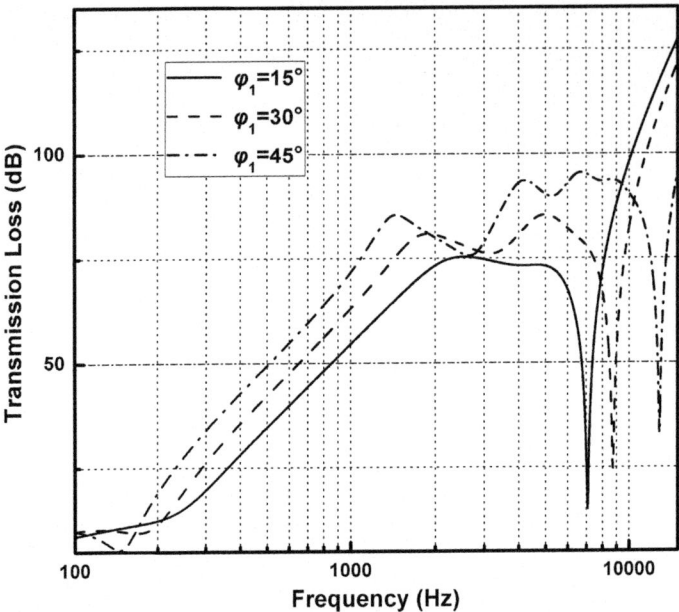

Fig. 2.31 Sound transmission loss of double-leaf panel filled with porous material for selected elevation angles, with $M = 0.05$, $\beta_1 = 0°$, loss factor $\eta = 0.01$, and $H = 0.08$ m

small, the effect of plate-porous material-plate resonance can be ignored since it does not cause an apparent dip in the *STL* versus frequency curve. In sharp contrast, the influence of plate-porous material-plate resonance on transmission loss will be reinforced as the incident angle is increased.

The predicted effect of sound incidence angle on the coincidence frequency is illustrated further in Fig. 2.32. The coincidence frequency increases remarkably with increasing incident elevation angle, agreeing well with the coincidence frequency formula of Eq. (2.142). Also, it can be seen from Fig. 2.32 and Eq. (2.142) that the coincidence dip tends to infinity when the incident angle approaches $\pi/2$, implying that the structure considered here has the best sound insulation ability for normal incident sound.

2.3.9 Influence of Sound Incident Azimuth Angle

It can be seen from Fig. 2.33 that the sound transmission loss of the panel is affected by the azimuth angle of the incident sound, particularly in the high-frequency regime. The coincidence frequency increases as the azimuth angle is increased. The influence of azimuth angle on sound transmission loss is mainly caused by the refraction effect of the mean flow. The results presented in Table 2.2 illustrate

2.3 Double-Leaf Panel Filled with Porous Materials

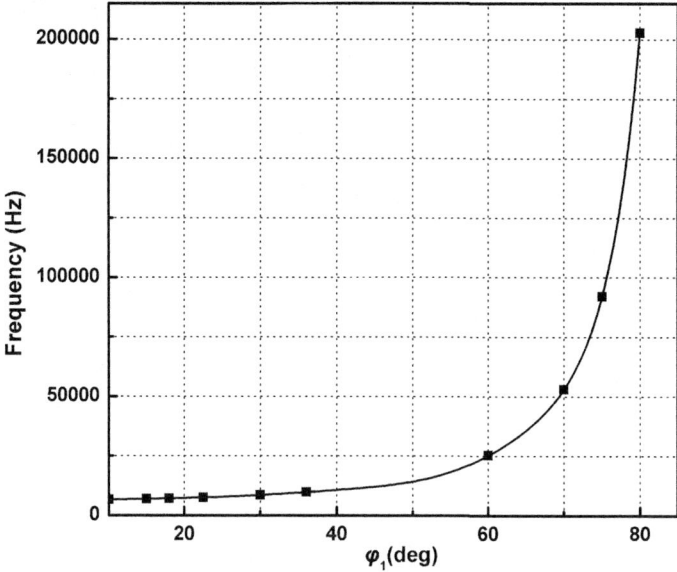

Fig. 2.32 Effect of sound incident angle on coincidence frequency of double-leaf panel filled with porous material, with $M = 0.05$, loss factor $\eta = 0.01$, $\beta_1 = 0°$, and $H = 0.08$ m

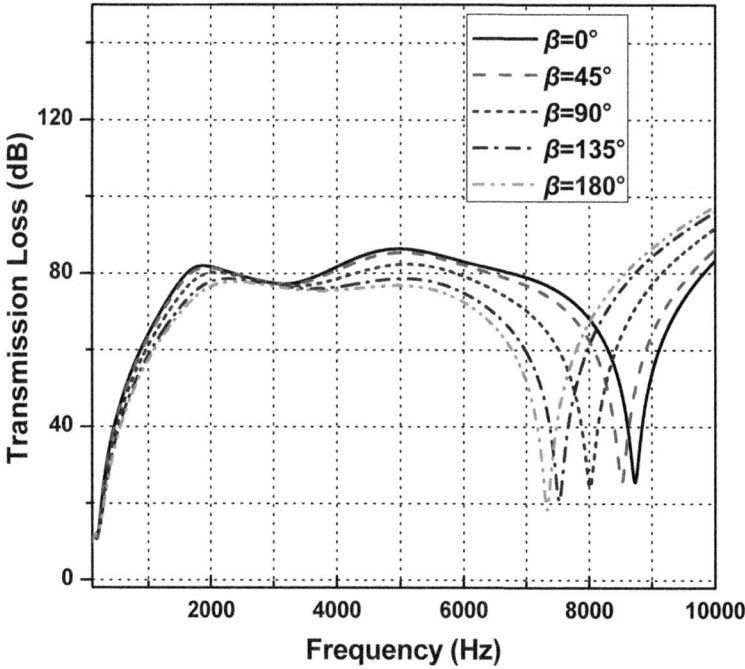

Fig. 2.33 Sound transmission loss of double-leaf panel filled with porous material for selected azimuth angles, with $M = 0.05$, $\varphi_1 = 30°$, loss factor $\eta = 0.01$, and $H = 0.08$ m

Table 2.2 Comparison of coincidence frequency for different azimuth angles and Mach numbers

	$M=0$	$M=0.05$	$M=0.10$	$M=0.15$
f_c (Hz) for $\beta = 0°$	8,020	8,730	9,641	10,230
f_c (Hz) for $\beta = 45°$	8,020	8,517	9,026	9,553
Difference	0	213	615	677

that when the Mach number is increased, the coincidence frequency difference between different azimuth angles is enlarged, which also proves that the refraction effect is enhanced by the increase of Mach number as mentioned in Sect. 2.3.7. As anticipated, since the double-leaf panel is isotropic and homogenous, the sound transmission loss should not change with the azimuth angle when the fluid media are static, as warranted by the results of Table 2.2.

2.3.10 Conclusion

A theoretical model has been developed for sound transmission across infinite double-leaf panels filled with porous absorptive materials in the presence of external mean flow. Based on the model predictions, the following conclusions are drawn:

1. The presence of external mean flow has negligible influence upon the sound absorption ability of the filling porous material. Besides, the damping effect of the porous material affects significantly the magnitude of *STL* at coincidence dip, and the same can be said for the material loss factor of the faceplate material.
2. The added damping effect of the external mean flow prevails at frequencies below the coincidence dip frequency, enhancing the transmission loss as the Mach number is increased. Also, since the radiated acoustic power decreases with increasing Mach number, the peak value of the *STL* increases with increasing Mach number. In addition, the refraction effect of the external mean flow causes an upward shift of the coincidence frequency when the Mach number is increased.
3. The transmission loss of double-leaf panels filled with porous materials is considerably affected by sound incident elevation angle. Increasing the incident elevation angle (i.e., decreasing the angle from the axis normal to the plane of the panel) leads to enhanced transmission loss at frequencies below the coincidence dip as well as reduced transmission loss above the coincidence dip. The azimuth angle of the incident sound also has an influence on the sound transmission loss of the panel. The coincidence frequency decreases with increasing azimuth angle due to the refraction effect of the mean flow.

Appendix

As stated in Sect. 2.2.7, the simple closed-form formulae, i.e., Eqs. (2.91), (2.92), (2.93), and (2.94), are applied to validate the proposed aero-acoustic-elastic theoretical model in Sect. 2.2, because these formulae are developed independent of the theoretical model. The physical nature of the four acoustic phenomena based on which the closed-form formulae are derived is presented below (Fig. 2.34).

Mass-Air-Mass Resonance

In the absence of the external mean flow, the formula for predicting the mass-air-mass resonance frequency of a double-leaf aeroelastic panel has been presented in Refs. [3, 5, 47, 73]. The radially outspreading bending wave in the panel caused by the incident sound leads to the highly directional sound radiation. As a result, the mass-air-mass resonance strongly depends on the incident angle. When the mass-air-mass resonance occurs, the two panels with the air cavity in between behave like a mass-spring-mass system, with the stiffness of the air cavity given by $\rho_2 c_2^2/(H \sin^2\varphi_1)$. However, due to the refraction effect of the mean flow as shown in Fig. 2.34, the incident angle φ_1 in the presence of the mean flow will be changed to φ_2 before the sound penetrates through the incident panel. In other words, the mean flow case is equivalent to the case when the sound is incident on the panel with angle φ_2 in the absence of the mean flow (denoted by the dash lines in Fig. 2.34). Accordingly, the stiffness of the air cavity is changed to $\rho_2 c_2^2/(H \sin^2\varphi_2)$, and the eigenvalue equation of the equivalent mass-spring-mass vibration system becomes

$$\begin{vmatrix} \frac{\rho_2 c_2^2}{H\sin^2\varphi_2} - \omega^2 m_1 & \frac{-\rho_2 c_2^2}{H\sin^2\varphi_2} \\ \frac{-\rho_2 c_2^2}{H\sin^2\varphi_2} & \frac{\rho_2 c_2^2}{H\sin^2\varphi_2} - \omega^2 m_2 \end{vmatrix} = 0 \qquad (2.145)$$

Solving Eq. (2.145) leads to the mass-air-mass resonance given in Eq. (2.91).

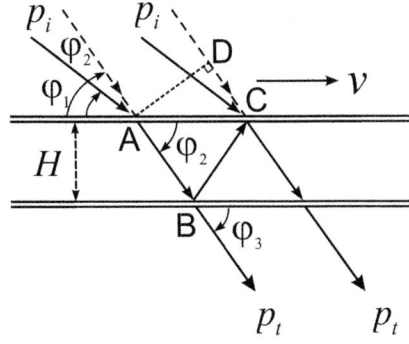

Fig. 2.34 Sketch of sound transmission through double-leaf aeroelastic panel in the presence of external mean flow

Standing-Wave Attenuation

In the presence of external mean flow, the phenomenon of standing-wave attenuation in a double-panel system is caused by the destructive interference between two meeting waves, which occurs only when the distance difference between the routes that the two intervening waves pass through is the odd number of one quarter of the incident sound wavelength. Under such conditions, the interference between the positive-gong wave and the negative-going wave tends to be destructive when the sound is transmitting through the partition, resulting in maximum sound reduction. As shown in Fig. 2.34, the plane sound ray comprised of a set of harmonic sound waves with the same vibration phase obliquely impinges on the incident panel. There indeed exists a case when one sound ray is incident on the upper panel at point A and transmits through the air cavity onto the bottom panel at point B and its reflected portion back to the upper panel at point C meets another sound ray from the external incident side, i.e., the interference between the two sound waves has occurred.

As is well known, the destructive effect between two harmonic waves occurs only when their vibration phases differ by odd numbers of $\pi/2$. In the case considered here (Fig. 2.34), the distance difference between the routes that the two intervening waves pass through is the only cause of the phase difference. For convenience, the equivalent case of sound incident with angle φ_2 in the absence of mean flow (dash lines in Fig. 2.34) is used to represent the case of sound incident with angle φ_1 in the presence of mean flow (solid lines in Fig. 2.34). The distance difference between the routes that the two sound waves pass through is such that

$$\overline{AB} + \overline{BC} - \overline{CD} = \frac{2H}{\sin \varphi_2} - 2H \cot \varphi_2 \cos \varphi_2 = 2H \sin \varphi_2 \qquad (2.146)$$

The occurrence of the destructive effect requires that

$$H \sin \varphi_2 = \frac{(2n-1)\lambda}{4} \quad (n = 1, 2, 3 \ldots) \qquad (2.147)$$

where $\lambda = c/f$ is the wavelength in air. Equation (2.92) for the standing-wave attenuation frequency follows from Eq. (2.147).

Standing-Wave Resonance

As the counterpart of the standing-wave attenuation, the standing-wave resonance is also caused by the interference effect between two meeting waves, which is however constructive rather than destructive. It appears when the distance difference between the routes that the two intervening waves pass through is the multiple of

Appendix

the half wavelength of the incidence sound wave. In such cases, the constructive interference between the positive-going and negative-going waves leads to an enhanced transmission through the double-panel partition. The condition for the appearance of the standing-wave resonance is given by

$$H \sin \varphi_2 = \frac{n\lambda}{2} \quad (n = 1, 2, 3 \ldots) \tag{2.148}$$

With the relation $\lambda = c/f$, Eq. (2.148) can be readily converted to Eq. (2.93).

Coincidence Resonance

Similar to the standing-wave attenuation and resonance, the coincidence resonance is also an interference effect between two intervening waves by physical nature. The difference between the two different types of acoustic phenomenon is that the coincidence resonance is caused by wave interference between the sound wave and the panel flexural bending wave. As shown in Fig. 2.34, there exists a situation when the panel flexural bending wave at point A (excited by the incident sound ray at this point) rapidly outspreads to point C, where it matches another sound ray from the external incident side. The premise for this matching is that the flexural bending wave in the aeroelastic panel propagates faster than sound propagation in air, requiring that the former transmits a larger distance than the latter during the same time period (i.e., $\overline{AC} > \overline{DC}$ in Fig. 2.34), and hence

$$\frac{\overline{AC}}{c_t} = \frac{\overline{DC}}{c} \tag{2.149}$$

This premise is in general satisfied in practice. Consequently, the constructive interference between the two waves results in the coincidence resonance. From the governing equation of panel vibration, we have

$$c_t = \sqrt[4]{\frac{D\omega^2}{m}} \tag{2.150}$$

Substitution of Eq. (55) and the relation $\overline{DC} = \overline{AC} \cdot \cos \varphi_2$ into Eq. (2.149) leads to Eq. (2.94) for the coincidence resonance frequency.

Note that the coincidence resonance will occur in the bottom panel as well when the upper panel generates coincidence resonance. In other words, the incident sound will be enhanced twice, initially by the upper panel and followed by the bottom panel, when the condition for the occurrence of the coincidence resonance is satisfied.

References

1. Koval LR (1976) Effect of air flow, panel curvature, and internal pressurization on field-incidence transmission loss. J Acoust Soc Am 59(6):1379–1385
2. Frampton KD, Clark RL (1996) State-space modeling of aerodynamic forces on plate using singular value decomposition. AIAA J 34(12):2627–2630
3. Xin FX, Lu TJ (2009) Analytical and experimental investigation on transmission loss of clamped double panels: implication of boundary effects. J Acoust Soc Am 125(3):1506–1517
4. Frampton KD, Clark RL, Dowell EH (1996) State-space modeling for aeroelastic panels with linearized potential flow aerodynamic loading. J Aircr 33(4):816–822
5. Xin FX, Lu TJ, Chen CQ (2008) Vibroacoustic behavior of clamp mounted double-panel partition with enclosure air cavity. J Acoust Soc Am 124(6):3604–3612
6. Clark RL, Frampton KD (1997) Aeroelastic structural acoustic coupling: implications on the control of turbulent boundary-layer noise transmission. J Acoust Soc Am 102(3):1639–1647
7. Frampton KD, Clark RL (1997) Power flow in an aeroelastic plate backed by a reverberant cavity. J Acoust Soc Am 102(3):1620–1627
8. Frampton KD (2003) Radiation efficiency of convected fluid-loaded plates. J Acoust Soc Am 113(5):2663–2673
9. Xin FX, Lu TJ, Chen C (2008) Sound transmission through lightweight all-metallic sandwich panels with corrugated cores. Multifunct Mater Struct, Parts 1 and 2 47:57–60
10. Frampton KD (2005) The effect of flow-induced coupling on sound radiation from convected fluid loaded plates. J Acoust Soc Am 117(3):1129–1137
11. Clark RL, Frampton KD (1999) Aeroelastic structural acoustic control. J Acoust Soc Am 105(2):743–754
12. Maury C, Gardonio P, Elliott SJ (2002) Model for active control of flow-induced noise transmitted through double partitions. AIAA J 40(6):1113–1121
13. Xin FX, Lu TJ, Chen CQ (2009) Dynamic response and acoustic radiation of double-leaf metallic panel partition under sound excitation. Comput Mater Sci 46(3):728–732
14. Maury C, Gardonio P, Elliott SJ (2001) Active control of the flow-induced noise transmitted through a panel. AIAA J 39(10):1860–1867
15. Sgard F, Atalla N, Nicolas J (1994) Coupled FEM-BEM approach for mean flow effects on vibro-acoustic behavior of planar structures. AIAA J 32(12):2351–2358
16. Wu SF, Maestrello L (1995) Responses of finite baffled plate to turbulent flow excitations. AIAA J 33(1):13–19
17. Frampton KD, Clark RL (1997) Sound transmission through an aeroelastic plate into a cavity. AIAA J 35(7):1113–1118
18. Atalla N, Nicolas J (1995) A formulation for mean flow effects on sound radiation from rectangular baffled plates with arbitrary boundary conditions. J Vib Acoust 117(1):22–29
19. Schmidt PL, Frampton KD (2009) Effect of in-plane forces on sound radiation from convected, fluid loaded plates. J Vib Acoust 131(2):021001–021008
20. Maury C, Gardonio P, Elliott SJ (2002) A wavenumber approach to modelling the response of a randomly excited panel, Part II: Application to aircraft panels excited by a turbulent boundary layer. J Sound Vib 252(1):115–139
21. Xin FX, Lu TJ, Chen CQ (2010) Sound transmission through simply supported finite double-panel partitions with enclosed air cavity. J Vib Acoust 132(1):011008:011001–011011
22. Howe MS, Shah PL (1996) Influence of mean flow on boundary layer generated interior noise. J Acoust Soc Am 99(6):3401–3411
23. Xin FX, Lu TJ, Chen CQ (2009) External mean flow influence on noise transmission through double-leaf aeroelastic plates. AIAA J 47(8):1939–1951
24. Graham WR (1996) Boundary layer induced noise in aircraft, Part I: The flat plate model. J Sound Vib 192(1):101–120
25. Graham WR (1996) Boundary layer induced noise in aircraft, Part II: The trimmed plat plate model. J Sound Vib 192(1):121–138

26. Ribner HS (1957) Reflection, transmission, and amplification of sound by a moving medium. J Acoust Soc Am 29(4):435–441
27. Franken PA, Ingard U (1956) Sound propagation into a moving medium. J Acoust Soc Am 28(1):126–127
28. Graham WR (1968) The effect of mean flow on the radiation efficiency of rectangular plates. Proc R Soc Lond Series A Math Phys Eng Sci 1998(454):111–137
29. Maury C, Elliott SJ (2005) Analytic solutions of the radiation modes problem and the active control of sound power. Proc R Soc Lond Series A Math Phys Eng Sci 2053(461):55–78
30. Dowell EH (1975) Aeroelasticity of plates and shells. Norrdhoff International Publishing, Leyden
31. Ingard U (1959) Influence of fluid motion past a plane boundary on sound reflection, absorption, and transmission. J Acoust Soc Am 31(7):1035–1036
32. Yeh C (1967) Reflection and transmission of sound waves by a moving fluid laver. J Acoust Soc Am 41(4A):817–821
33. Schmidt PL, Frampton KD (2009) Acoustic radiation from a simple structure in supersonic flow. J Sound Vib 328(3):243–258
34. Panneton R, Atalla N (1996) Numerical prediction of sound transmission through finite multilayer systems with poroelastic materials. J Acoust Soc Am 100(1):346–354
35. Chazot JD, Guyader JL (2007) Prediction of transmission loss of double panels with a patch-mobility method. J Acoust Soc Am 121(1):267–278
36. ASTM E 90-04 (2004) Standard test method for laboratory measurement of airborne sound transmission loss of building partitions and elements
37. Tam CKW, Auriault L (1998) Mean flow refraction effects on sound radiated from localized sources in a jet. J Fluid Mech 370:149–174
38. Lyle KH, Dowell EH (1994) Acoustic radiation damping of flat rectangular plates subjected to subsonic flows. I: Isotropic. J Fluids Struct 8(7):711–735
39. Heitman KE, Mixson JS (1986) Laboratory study of cabin acoustic treatments installed in an aircraft fuselage. J Aircr 23(1):32–38
40. Dowell EH (1969) Transmission of noise from a turbulent boundary layer through a flexible plate into a closed cavity. J Acoust Soc Am 46(1B):238–252
41. Gardonio P, Elliott SJ (1999) Active control of structure-borne and airborne sound transmission through double panel. J Aircr 36(6):1023–1032
42. Alujevic N, Frampton K, Gardonio P (2008) Stability and performance of a smart double panel with decentralized active dampers. AIAA J 46(7):1747–1756
43. Carneal JP, Fuller CR (1995) Active structural acoustic control of noise transmission through double panel systems. AIAA J 33(4):618–623
44. Unruh JF, Dobosz SA (1988) Fuselage structural-acoustic modeling for structure-borne interior noise transmission. J Vib Acoust Stress Reliab Des-Trans ASME 110(2):226–233
45. Grosveld F (1992) Plate acceleration and sound transmission due to random acoustic and boundary-layer excitation. AIAA J 30(3):601–607
46. Mixson JS, Roussos L (1983) Laboratory study of add-on treatments for interior noise control in light aircraft. J Aircr 20(6):516–522
47. Xin FX, Lu TJ (2011) Effects of core topology on sound insulation performance of lightweight all-metallic sandwich panels. Mater Manuf Process 26(9):1213–1221
48. Alujevic N, Gardonio P, Frampton KD (2008) Smart double panel with decentralized active dampers for sound transmission control. AIAA J 46(6):1463–1475
49. Carneal JP, Fuller CR (2004) An analytical and experimental investigation of active structural acoustic control of noise transmission through double panel systems. J Sound Vib 272(3–5):749–771
50. Mixson JS, Powell CA (1985) Review of recent research on interior noise of propeller aircraft. J Aircr 22(11):931–949
51. Wilby JF, Gloyna FL (1972) Vibration measurements of an airplane fuselage structure I. Turbulent boundary layer excitation. J Sound Vib 23(4):443–466

52. Wilby JF, Gloyna FL (1972) Vibration measurements of an airplane fuselage structure II. Jet noise excitation. J Sound Vib 23(4):467–486
53. Davies HG (1971) Sound from turbulent-boundary-layer-excited panels. J Acoust Soc Am 49(3B):878–889
54. Berry A, Guyader J-L, Nicolas J (1990) A general formulation for the sound radiation from rectangular, baffled plates with arbitrary boundary conditions. J Acoust Soc Am 88(6):2792–2802
55. Berry A (1994) A new formulation for the vibrations and sound radiation of fluid-loaded plates with elastic boundary conditions. J Acoust Soc Am 96(2):889–901
56. Maestrello L (1995) Responses of finite baffled plate to turbulent flow excitations. AIAA J 33(1):13–19
57. Beranek LL, Work GA (1949) Sound transmission through multiple structures containing flexible blankets. J Acoust Soc Am 21(4):419–428
58. London A (1950) Transmission of reverberant sound through double walls. J Acoust Soc Am 22(2):270–279
59. Wang J, Lu TJ, Woodhouse J et al (2005) Sound transmission through lightweight double-leaf partitions: theoretical modelling. J Sound Vib 286(4–5):817–847
60. Craik RJM, Smith RS (2000) Sound transmission through double leaf lightweight partitions part I: Airborne sound. Appl Acoust 61(2):223–245
61. Leppington FG, Broadbent EG, Butler GF (2006) Transmission of sound through a pair of rectangular elastic plates. Ima J Appl Math 71(6):940–955
62. Villot M, Guigou C, Gagliardini L (2001) Predicting the acoustical radiation of finite size multi-layered structures by applying spatial windowing on infinite structures. J Sound Vib 245(3):433–455
63. Kropp W, Rebillard E (1999) On the air-borne sound insulation of double wall constructions. Acta Acust Unit Acust 85:707–720
64. Antonio JMP, Tadeu A, Godinho L (2003) Analytical evaluation of the acoustic insulation provided by double infinite walls. J Sound Vib 263(1):113–129
65. Xin FX, Lu TJ, Chen CQ (2009) Sound transmission across lightweight all-metallic sandwich panels with corrugated cores. Chinese J Acoust 28(3):231–243
66. Price AJ, Crocker MJ (1970) Sound transmission through double panels using statistical energy analysis. J Acoust Soc Am 47(3A):683–693
67. Langley RS, Smith JRD, Fahy FJ (1997) Statistical energy analysis of periodically stiffened damped plate structures. J Sound Vib 208(3):407–426
68. Sgard FC, Atalla N, Nicolas J (2000) A numerical model for the low frequency diffuse field sound transmission loss of double-wall sound barriers with elastic porous linings. J Acoust Soc Am 108(6):2865–2872
69. Crighton DG (1989) The 1988 Rayleigh medal lecture: fluid loading–the interaction between sound and vibration. J Sound Vib 133(1):1–27
70. Yeh C (1968) A further note on the reflection and transmission of sound waves by a moving fluid layer. J Acoust Soc Am 43(6):1454–1455
71. Leissa AW (1993) Vibration of shells. Acoustical Society of America, Woodbury
72. Liu BL, Feng LP, Nilsson A (2007) Sound transmission through curved aircraft panels with stringer and ring frame attachments. J Sound Vib 300(3–5):949–973
73. Biot MA (1956) Theory of propagation of elastic waves in a fluid-saturated porous solid. I. Low-frequency range. J Acoust Soc Am 28(2):168–178
74. Biot MA (1956) Theory of propagation of elastic waves in a fluid-saturated porous solid. II. Higher frequency range. J Acoust Soc Am 28(2):179–191
75. Lighthill J (1979) Waves in fluids. Cambridge University Press, London
76. Hovem JM, Ingram GD (1979) Viscous attenuation of sound in saturated sand. J Acoust Soc Am 66(6):1807–1812
77. Stinson MR, Champoux Y (1992) Propagation of sound and the assignment of shape factors in model porous materials having simple pore geometries. J Acoust Soc Am 91(2):685–695

78. Selamet A, Easwaran V, Novak JM et al (1998) Wave attenuation in catalytic converters: reactive versus dissipative effects. J Acoust Soc Am 103(2):935–943
79. Delany ME, Bazley EN (1970) Acoustical properties of fibrous absorbent materials. Appl Acoust 3(2):105–116
80. Allard J-F, Champoux Y (1992) New empirical equations for sound propagation in rigid frame fibrous materials. J Acoust Soc Am 91(6):3346–3353
81. Muehleisen RT, Beamer CW, Tinianov BD (2005) Measurements and empirical model of the acoustic properties of reticulated vitreous carbon. J Acoust Soc Am 117(2):536–544
82. Bies DA, Hansen CN (1980) Flow resistance information for acoustical design. Appl Acoust 13(5):357–391
83. Lauriks W, Mees P, Allard JF (1992) The acoustic transmission through layered systems. J Sound Vib 155(1):125–132
84. Brouard B, Lafarge D, Allard JF (1995) A general method of modelling sound propagation in layered media. J Sound Vib 183(1):129–142
85. Chonan S, Kugo Y (1991) Acoustic design of a three? Layered plate with high sound interception. J Sound Vib 89(2):792–798
86. Kang HJ, Ih JG, Kim JS et al (2000) Prediction of sound transmission loss through multilayered panels by using Gaussian distribution of directional incident energy. J Acoust Soc Am 107(3):1413–1420
87. Bolton JS, Shiau NM, Kang YJ (1996) Sound transmission through multi-panel structures lined with elastic porous materials. J Sound Vib 191(3):317–347
88. Xin FX, Lu TJ (2010) Analytical modeling of sound transmission across finite aeroelastic panels in convected fluids. J Acoust Soc Am 128(3):1097–1107
89. Rebillard P, Allard JF, Depollier C et al (1992) The effect of a porous facing on the impedance and absorption coefficient of a layer of porous materials. J Sound Vib 156(3):541–555
90. Xin FX, Lu TJ (2011) Transmission loss of orthogonally rib-stiffened double-panel structures with cavity absorption. J Acoust Soc Am 129(4):1919–1934
91. Xin FX, Lu TJ (2010) Sound radiation of orthogonally rib-stiffened sandwich structures with cavity absorption. Compos Sci Technol 70(15):2198–2206
92. Morse PM, Ingard KU (1968) Theoretical acoustics. MacGraw-Hill, New York
93. Pellicier A, Trompette N (2007) A review of analytical methods, based on the wave approach, to compute partitions transmission loss. Appl Acoust 68(10):1192–1212
94. Xin FX, Lu TJ (2010) Analytical modeling of fluid loaded orthogonally rib-stiffened sandwich structures: sound transmission. J Mech Phys Solids 58(9):1374–1396
95. Yairi M, Sakagami K, Morimoto M et al (2003) Effect of acoustical damping with a porous absorptive layer in the cavity to reduce the structure-borne sound radiation from a double-leaf structure. Appl Acoust 64(4):365–384

Chapter 3
Vibroacoustics of Stiffened Structures in Mean Flow

Abstract This chapter is organized as two parts: in the first part, a theoretical modeling approach is proposed for noise radiated from aeroelastic skin plates of aircraft fuselage stiffened by orthogonally distributed rib-stiffeners and subjected to external jet noise in the presence of convected mean flow. The focus is placed upon quantifying the effects of external mean flow on the aeroelastic-acoustic characteristics of the rib-stiffened plate. The Euler-Bernoulli beam equation and the torsional wave equation governing separately the flexural and torsional motions of the rib-stiffeners are employed to accurately describe the force-moment coupling between the stiffeners and the plate. The external mean flow fluid is modeled using the convected wave equation. Given the periodicity of the considered structure, the resulting governing equations of the system are solved by applying the Poisson summation formula and the Fourier transformation technique. The radiated sound pressure is closely related to the plate displacement by means of the Helmholtz equation and the fluid-structure boundary conditions. To highlight the radiation characteristics of the periodically stiffened structure as well as the mean flow effects, the final radiated sound pressure is presented in the form of decibels with reference to that of a bare plate immersed in mean flow. Systematic parametric studies are conducted to evaluate the effects of external mean flow speed, noise incident angle, and periodic spacings on the aeroelastic-acoustic performance of the rib-stiffened plate.

In the second part, this chapter investigates the sound transmission loss of aeroelastic plates reinforced by two sets of orthogonal rib-stiffeners in the presence of external mean flow. Built upon the periodicity of the structure, a comprehensive theoretical model is developed by considering the convection effect of mean flow. The rib-stiffeners are modeled by employing the Euler-Bernoulli beam theory and the torsional wave equation. While the solution for the transmission loss of the structure based on plate displacement and acoustic pressures is given in the form of space-harmonic series, the corresponding coefficients are obtained from the solution of a system of linear equations derived from the plate-beam coupling vibration governing equation and Helmholtz equation. The model predictions are validated by comparing with existing theoretical and experimental results in the absence of mean flow. A parametric study is subsequently performed to quantify the effects of mean flow as well as structure geometrical parameters upon the transmission loss. It is demonstrated that the transmission loss of periodically rib-stiffened structure is increased significantly with increasing Mach number of mean flow over a wide frequency range. The STL value for the case of sound wave incident downstream is pronouncedly larger than that associated with sound wave incident upstream.

3.1 Noise Radiation from Orthogonally Rib-Stiffened Plates

3.1.1 Introduction

Recent research and development in aircraft design have reconcentrated on the long-lasting concerns about external flow interaction with structure responses and noise radiation into the aircraft interior, which is of paramount importance for designing supersonic (or high subsonic) civil and military aircrafts with lower interior noise level [1–19]. It has been widely regarded that the interior cabin noise is usually attributed to the direct incidence of engine exhaust noise and the high-speed turbulent boundary layer (TBL) flow over the exterior fuselage [11–17, 20–25], which generate high-level noise, thereby affecting the comforts of passengers. In particular, the ultrahigh-bypass turbofans have remarkably increased tip Mach numbers, resulting in enhanced low-frequency noise impinging on the exterior of aircraft fuselages [3, 26]. To reduce the cabin interior noise level, considerable efforts have been dedicated to address the increasingly pressing issue of external fluid flow coupling with structure dynamic response.

While early research on acoustic problems involving fluid flow concentrated on sound reflection and transmission at the idealized interface between a steady fluid medium and a moving fluid medium [27–29], numerous researches in the past decades focused on the aeroelastic-acoustic interaction problem of an aeroelastic plate coupled with fluid flow. Concerning sound transmission through aircraft fuselage in the presence of external mean flow, Koval [2] derived a theoretical model for the field-incidence transmission loss of a single-walled plate and calculated the effects of airflow, panel curvature, and internal fuselage pressurization. As an extension of Koval's model, Xin et al. [30] theoretically investigated the external mean flow effects on noise transmission through double-leaf plate structures. In this research, four different types of acoustic phenomenon (namely, mass-air-mass resonance, standing-wave resonance, standing-wave attenuation, and coincidence resonance) for a planar double-leaf plate as well as the ring frequency resonance for a curved double-leaf plate were identified, with closed-form formulas for the natural frequencies of these phenomena derived based upon physical principles. To evaluate the influence of mean flow on boundary layer-generated interior noise, Howe and Shah [20] presented an analytic model to solve the acoustic radiation in terms of prescribed turbulent boundary layer pressure fluctuation, which may

3.1 Noise Radiation from Orthogonally Rib-Stiffened Plates

be used to validate more general numerical schemes for fluid-structure interaction. With mean flow effects on forced vibroacoustic response of a baffled plate accounted for, Sgard et al. [31] proposed a coupled finite element method-boundary element method (FEM-BEM) approach to investigate the mean flow effects as well as the acoustic radiation pattern for a baffled plate with different kinds of boundary conditions; the mean flow effects were explicitly shown in terms of added mass, stiffness, and radiation damping. Considering the nonlinearities induced by in-plane forces and shearing forces due to the stretching of plate-bending motion, Wu and Maestrello [32] developed theoretical formulations to estimate the dynamic and acoustic responses of a finite baffled plate subject to turbulent boundary layer excitations. It was found that, in the presence of mean flow, the temporal instability can be induced by the added stiffness due to acoustic radiation and the effect of added stiffness increased quadratically with mean flow speed. More recently, the effects of mean flow on sound transmission across a simply supported rectangular aeroelastic panel were analytically solved [33]. Focusing upon aircraft sidewall structures, Legault and Atalla [34–36] investigated theoretically the sound transmission problems of such periodically stiffened structures, and their theoretical predictions agreed well with experimental results.

In addition to the aforementioned investigations, Frampton and Clark [4, 15, 16, 37–41] carried out comprehensive studies on aeroelastic plates interacting with aerodynamic loading, including theoretical modeling for sound radiation and transmission as well as acoustic control scheme. Besides, to effectively reduce the interior noise level of aircraft cabin fuselages over a wide frequency range, various active strategies [11–13, 21, 26, 42] were proposed to suppress the vibration and noise radiation of skin plates, providing alternative noise reduction solutions.

Although numerous experimental and theoretical studies concerning aeroelastic-acoustic problems of fuselage-like structures exist, at present there is a lack of a thorough and fundamental understanding of the physical mechanism associated with the interaction between fluid flow and a vibrating structure. Specifically, although the factual construction of aircraft fuselage is commonly made of thin-walled structural elements with periodic rib-stiffeners, the issue of aeroelastic-acoustic interaction for such structures has not been well addressed by existing studies. The focus of the present work is therefore placed upon the aeroelastic-acoustic problem of orthogonally rib-stiffened skin plates in the presence of external mean flow. A theoretical model is developed by combining the Kirchhoff thin plate theory with the convected wave equation, with the fluid momentum equation applied to satisfy the fluid-structure boundary condition. The Euler-Bernoulli beam equation and the torsional wave equation are employed to describe the force-moment coupling between the rib-stiffeners and the faceplate. The system of governing equations is solved by applying the Poisson summation formula and Fourier transformation technique. Based on the theoretical formulations, systematic parametric studies are carried out to quantify how the external mean flow affects sound radiation from periodically stiffened aeroelastic plates and to explore the physical mechanisms underlying aeroelastic-acoustic interaction.

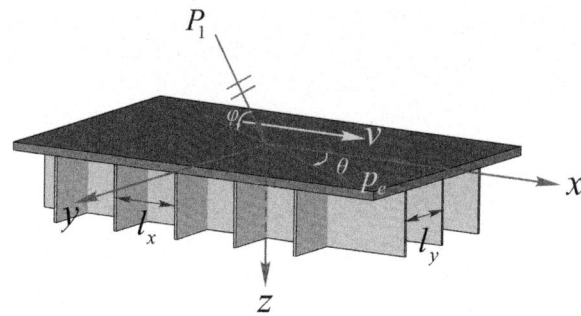

Fig. 3.1 Schematic of orthogonally rib-stiffened plate excited by convected harmonic pressure

3.1.2 Theoretical Formulation

3.1.2.1 Dynamic Responses to Convected Harmonic Pressure

Typical aircraft fuselages are made of periodically rib-stiffened plates and are often excited by external airflow and engine exhaust noise in cruise condition. To explore the dynamic response and sound radiation behavior of such rib-stiffened plates immersed in external airflow, a uniform plane sound wave varying harmonically in time is assumed to impinge on the plate from the external mean flow side, and the resulting sound radiation level within the interior static fluid side is examined. As shown in Fig. 3.1, the aeroelastic structure considered consists of a flat facesheet and two sets of orthogonally distributed rib-stiffeners. Both the facesheet and the stiffeners are made of homogenous and isotropic materials, and the whole structure is immersed in inviscid, irrotational fluid media. The upper and bottom fluid media separated by the faceplate occupy the spaces of $z < 0$ and $z > h$ and, in terms of mass density and sound speed, are characterized by (ρ_1, c_1) and $(\rho_2, \text{and } c_2)$, respectively. Let h denote the faceplate thickness, H denote the height of the stiffeners, l_x and l_y denote the periodic spacings of the stiffeners in the x- and y-directions, and t_x and t_y denote the thicknesses of the stiffeners in the x- and y-directions, respectively. The mean fluid flow with uniform speed v is assumed to move along the x-direction. The incident sound wave transmitting from the external mean flow side is characterized by elevation angle φ_1 and azimuth angle θ with respect to the coordinate system defined in Fig. 3.1. The impinged noise excitation is partially reflected and partially transmitted through the structure via the faceplate into the interior stationary fluid medium, which is strongly affected by the external mean flow and the rib-stiffeners.

To analyze theoretically the system of Fig. 3.1, a number of simplifying assumptions are made: (1) the faceplate is sufficiently thin so that it can be modeled using the classical Kirchhoff thin plate theory; (2) the fluid media are taken as inviscid, irrotational, and incompressible [4–6, 15, 16, 31, 38–40, 43]; and (3) the surface of the faceplate adjacent to the mean flow is sufficiently smooth so that it is appropriate to consider the fluid-plate interface as one of the streamlines of the fluid flow, i.e., the mean flow is tangential to the acoustically deformed boundary [2, 44].

3.1 Noise Radiation from Orthogonally Rib-Stiffened Plates

With the forces and moments exerted by the rib-stiffeners on the faceplate accounted for, the normal displacement w of the plate is governed by

$$D\nabla^4 w + m_p \frac{\partial^2 w}{\partial t^2} = -\sum_m q_s(x,y)\delta(x - ml_x) - \frac{\partial}{\partial x}\left[\sum_m \kappa_s(x,y)\delta(x - ml_x)\right]$$

$$-\sum_n q_t(x,y)\delta(y - nl_y) - \frac{\partial}{\partial y}\left[\sum_n \kappa_t(x,y)\delta(y - nl_y)\right]$$

$$+ P_i(x,y,0) + P_r(x,y,0) - P_t(x,y,h) \qquad (3.1)$$

where m_p is the surface mass; D is the flexural rigidity of the plate; q_s and κ_s are the forces and moments of y-wise stiffeners; q_t and κ_t are the forces and moments of x-wise stiffeners; P_i, P_r, and P_t are the incident, reflected, and radiated sound pressure; and m and n are the number of the x- and y-wise stiffeners, respectively.

The Euler-Bernoulli beam equation and torsional wave equation governing the flexural and torsional motions of the rib-stiffeners are shown below (note that the bending moments have been implicitly included in the Euler-Bernoulli equation, because the first-order partial derivative of the bending moments about the coordinate is equivalent to transverse force [45]):

1. x-wise stiffeners:

$$E_x I_x \frac{\partial^4 w}{\partial x^4} + m_x \frac{\partial^2 w}{\partial t^2} = q_t(x,y), \quad G_x J_x \frac{\partial^2 \theta_x}{\partial x^2} - \rho_x I_{px} \frac{\partial^2 \theta_x}{\partial t^2} = \kappa_t(x,y) \qquad (3.2)$$

2. y-wise stiffeners:

$$E_y I_y \frac{\partial^4 w}{\partial y^4} + m_y \frac{\partial^2 w}{\partial t^2} = q_s(x,y), \quad G_y J_y \frac{\partial^2 \theta_y}{\partial y^2} - \rho_y I_{py} \frac{\partial^2 \theta_y}{\partial t^2} = \kappa_s(x,y) \qquad (3.3)$$

where $E_x I_x$, $G_x J_x$, I_{px}, ρ_x, and m_x are the flexural stiffness, torsional stiffness, polar moment of inertia, density, and surface mass for the x-wise stiffeners, respectively; $E_y I_y$, $G_y J_y$, I_{py}, ρ_y, and m_y are the flexural stiffness, torsional stiffness, polar moment of inertia, density, and surface mass for the y-wise stiffeners, respectively; θ_x $(= \partial w/\partial y)$ and θ_y $(= \partial w/\partial x)$ are the torsion angles of the x- and y-wise stiffeners, respectively.

The incident sound pressure is taken as a traveling pressure wave excitation:

$$P_i(x,y,z) = p_e e^{i(\omega t - k_x x - k_y y - k_z z)} \qquad (3.4)$$

where the sound wavenumber components depend upon the incident sound elevation angle φ and azimuth angle θ as

$$k_x = k\cos\varphi\cos\theta, \quad k_y = k\cos\varphi\sin\theta, \quad k_z = k\sin\varphi \qquad (3.5)$$

With k_x and k_y replaced by α_0 and β_0 as the initial wavenumber components, the incident sound pressure at the interface between air and plate can be written as

$$P_i(x,y,0) = p_e e^{i(\omega t - \alpha_0 x - \beta_0 y)} \qquad (3.6)$$

The incident sound pressure in an inviscid and irrotational fluid moving in a plane parallel to the plate surface satisfies the convected wave equation [2, 46] as

$$\frac{D^2 P_i}{Dt^2} = \left(\frac{\partial}{\partial t} + \mathbf{V}\cdot\nabla\right)^2 P_i = c_0^2 \nabla^2 P_i \qquad (3.7)$$

In the case when the uniform flow of velocity V moves along the x-direction (Fig. 3.1), the wave equation in mean flow is simplified as

$$\left(\frac{\partial}{\partial t} + v\cdot\frac{\partial}{\partial x}\right)^2 P_i = c_0^2 \nabla^2 P_i \qquad (3.8)$$

Upon substituting Eq. (3.4) into Eq. (3.8), the wavenumber in mean flow is obtained as

$$k = \frac{\omega}{c_0(1 + M\cos\varphi\cos\theta)} \qquad (3.9)$$

where $M = v/c_0$ is the Mach number of the mean flow and c_0 being the sound speed in fluid. When the fluid is stationary (i.e., $M = 0$), $k = \omega/c_0$.

Taking advantage of the periodicity of the rib-stiffened plates examined in the present study and employing the Poisson summation formula, we can express the wave components in the form of space-harmonic series [47–52] as

$$\sum_m \delta(x - ml_x) = \frac{1}{l_x}\sum_m e^{-i(2m\pi/l_x)x}, \quad \sum_n \delta(y - nl_y) = \frac{1}{l_y}\sum_n e^{-i(2n\pi/l_y)y} \qquad (3.10)$$

The Fourier transform pairs of a function with respect to (x,y) and (α,β) are here defined as

$$w(x,y) = \int_{-\infty}^{+\infty}\int_{-\infty}^{+\infty} \tilde{w}(\alpha,\beta) e^{i(\alpha x + \beta y)} d\alpha d\beta \qquad (3.11)$$

$$\tilde{w}(\alpha,\beta) = \left(\frac{1}{2\pi}\right)^2 \int_{-\infty}^{+\infty}\int_{-\infty}^{+\infty} w(x,y) e^{-i(\alpha x + \beta y)} dx dy \qquad (3.12)$$

3.1 Noise Radiation from Orthogonally Rib-Stiffened Plates

Applying the Poisson summation formula and taking the Fourier transform of Eq. (3.1) lead to

$$\left[D\left(\alpha^2 + \beta^2\right)^2 - m\omega^2\right]\tilde{w}(\alpha, \beta) = -\frac{1}{l_x}\sum_m \left[\tilde{q}_s(\alpha_m, \beta) + i\alpha\tilde{\kappa}_s(\alpha_m, \beta)\right]$$

$$-\frac{1}{l_y}\sum_n \left[\tilde{q}_t(\alpha, \beta_n) + i\beta\tilde{\kappa}_t(\alpha, \beta_n)\right] + p_e\delta(\alpha + \alpha_0)\delta(\beta + \beta_0)$$

$$+ \tilde{P}_r(\alpha, \beta, 0) - \tilde{P}_t(\alpha, \beta, h) \quad (3.13)$$

where $\alpha_m = \alpha + 2m\pi/l_x$, $\beta_n = \beta + 2m\pi/l_y$, $\tilde{q}_s(\alpha_m, \beta)$, $\tilde{\kappa}_s(\alpha_m, \beta)$, $\tilde{q}_t(\alpha, \beta_n)$, and $\tilde{\kappa}_t(\alpha, \beta_n)$ can be obtained by taking the Fourier transform of (3.2) and (3.3) as

$$\tilde{q}_s(\alpha_m, \beta) = \left[E_y I_y \beta^4 - m_y \omega^2\right]\tilde{w}(\alpha_m, \beta) \quad (3.14)$$

$$\tilde{\kappa}_s(\alpha_m, \beta) = -\left[G_y J_y \beta^2 - \rho_y I_{py}\omega^2\right](i\alpha_m)\tilde{w}(\alpha_m, \beta) \quad (3.15)$$

$$\tilde{q}_t(\alpha, \beta_n) = \left[E_x I_x \alpha^4 - m_x \omega^2\right]\tilde{w}(\alpha, \beta_n) \quad (3.16)$$

$$\tilde{\kappa}_t(\alpha, \beta_n) = -\left[G_x J_x \alpha^2 - \rho_x I_{px}\omega^2\right](i\beta_n)\tilde{w}(\alpha, \beta_n) \quad (3.17)$$

The sound pressure in the incident side $P_i + P_r$ (including both the incident and reflected sound pressure) and the radiated sound pressure P_t satisfy the convected wave equation and the Helmholtz equation, respectively:

$$\left(\frac{\partial^2}{\partial x^2} + \frac{\partial^2}{\partial y^2} + \frac{\partial^2}{\partial z^2}\right)[P_i(x,y,z) + P_r(x,y,z)]$$
$$-\frac{1}{c_0^2}\left(\frac{\partial}{\partial t} + v \cdot \frac{\partial}{\partial x}\right)^2[P_i(x,y,z) + P_r(x,y,z)] = 0 \quad (3.18)$$

$$\left(\frac{\partial^2}{\partial x^2} + \frac{\partial^2}{\partial y^2} + \frac{\partial^2}{\partial z^2}\right)P_t(x,y,z) + \left(\frac{\omega}{c_0}\right)^2 P_t(x,y,z) = 0 \quad (3.19)$$

which together with the boundary condition

$$\left.\frac{\partial(P_i + P_r)}{\partial z}\right|_{z=0} = \omega^2 \rho_0 w, \quad \left.\frac{\partial P_t}{\partial z}\right|_{z=h} = \omega^2 \rho_0 w \quad (3.20)$$

ensures the equality of fluid velocity at the fluid-solid interface and plate velocity, ρ_0 being the fluid density. Transforming Eqs. (3.18), (3.19), and (3.20) yields

$$\tilde{P}_r(\alpha, \beta, z) = p_e\delta(\alpha + \alpha_0)\delta(\beta + \beta_0)e^{\gamma_1 z} + \omega^2\rho_0\tilde{w}(\alpha, \beta)e^{\gamma_1 z}/\gamma_1(\alpha, \beta) \quad (3.21)$$

$$\tilde{P}_t(\alpha, \beta, z) = -\frac{\omega^2 \rho_0 \tilde{w}(\alpha, \beta) e^{-\gamma_2 z + \gamma_2 h}}{\gamma_2(\alpha, \beta)} \qquad (3.22)$$

where

$$\gamma_1^2 = \alpha^2 + \beta^2 - (\omega - v\alpha)^2/c_0^2 \qquad (3.23)$$

$$\gamma_2^2 = \alpha^2 + \beta^2 - \omega^2/c_0^2 \qquad (3.24)$$

Incorporating Eqs. (3.14), (3.15), (3.16), (3.17) and Eqs. (3.21), (3.22) into Eq. (3.13) results in

$$\left[D(\alpha^2 + \beta^2)^2 - m\omega^2 - \omega^2 \rho_0/\gamma_1(\alpha, \beta) - \omega^2 \rho_0/\gamma_2(\alpha, \beta) \right] \tilde{w}(\alpha, \beta)$$

$$+ \frac{1}{l_x} \sum_m \left[E_y I_y \beta^4 - m_y \omega^2 + \alpha \alpha_m \left(G_y J_y \beta^2 - \rho_y I_{py} \omega^2 \right) \right] \cdot \tilde{w}(\alpha_m, \beta)$$

$$+ \frac{1}{l_y} \sum_n \left[E_x I_x \alpha^4 - m_x \omega^2 + \beta \beta_n \left(G_x J_x \alpha^2 - \rho_x I_{px} \omega^2 \right) \right] \cdot \tilde{w}(\alpha, \beta_n)$$

$$= 2 p_e \delta(\alpha + \alpha_0) \delta(\beta + \beta_0) \qquad (3.25)$$

To solve Eq. (3.25), (α, β) are replaced by (α'_m, β'_n), leading to a set of simultaneous algebraic equations as

$$\left[D(\alpha'^2_m + \beta'^2_n)^2 - m\omega^2 - \omega^2 \rho_0/\gamma_1(\alpha'_m, \beta'_n) - \omega^2 \rho_0/\gamma_2(\alpha'_m, \beta'_n) \right] \tilde{w}(\alpha'_m, \beta'_n)$$

$$+ \frac{1}{l_x} \sum_m \left[E_y I_y \beta'^4_n - m_y \omega^2 + \alpha'_m \alpha_m \left(G_y J_y \beta'^2_n - \rho_y I_{py} \omega^2 \right) \right] \cdot \tilde{w}(\alpha_m, \beta'_n)$$

$$+ \frac{1}{l_y} \sum_n \left[E_x I_x \alpha'^4_m - m_x \omega^2 + \beta'_n \beta_n \left(G_x J_x \alpha'^2_m - \rho_x I_{px} \omega^2 \right) \right] \cdot \tilde{w}(\alpha'_m, \beta_n)$$

$$= 2 p_e \delta(\alpha'_m + \alpha_0) \delta(\beta'_n + \beta_0) \qquad (3.26)$$

To facilitate subsequent numerical calculations, this equation is rewritten as

$$\left[D(\alpha'^2_m + \beta'^2_n)^2 - m\omega^2 - \omega^2 \rho_0/\gamma_1(\alpha'_m, \beta'_n) - \omega^2 \rho_0/\gamma_2(\alpha'_m, \beta'_n) \right] \tilde{w}(\alpha'_m, \beta'_n)$$

$$+ \frac{1}{l_x} \left(E_y I_y \beta'^4_n - m_y \omega^2 \right) \sum_m \tilde{w}(\alpha_m, \beta'_n) + \frac{1}{l_x} \alpha'_m \left(G_y J_y \beta'^2_n - \rho_y I_{py} \omega^2 \right) \sum_m \alpha_m \tilde{w}(\alpha_m, \beta'_n)$$

$$+ \frac{1}{l_y} \left(E_x I_x \alpha'^4_m - m_x \omega^2 \right) \sum_n \tilde{w}(\alpha'_m, \beta_n) + \frac{1}{l_y} \beta'_n \left(G_x J_x \alpha'^2_m - \rho_x I_{px} \omega^2 \right) \sum_n \beta_n \tilde{w}(\alpha'_m, \beta_n)$$

$$= 2 p_e \delta(\alpha'_m + \alpha_0) \delta(\beta'_n + \beta_0) \qquad (3.27)$$

3.1 Noise Radiation from Orthogonally Rib-Stiffened Plates

The amplitudes of the plate in wavenumber space should satisfy an infinite set of simultaneous equations, which contain a doubly infinite number of unknowns $\tilde{w}(\alpha'_m, \beta'_n)$ for $m' = -\infty, \infty$ and $n' = -\infty, \infty$. To perform numerical calculations, the equations can be truncated to retain a finite number of unknowns with $m' = -\hat{m}, \hat{m}$ and $n' = -\hat{n}, \hat{n}$ (insofar as the solutions converge). To be concise, the resulting simultaneous equations containing a finite number (i.e., $\overline{M}\,\overline{N}$, where $\overline{M} = 2\hat{m} + 1$, $\overline{N} = 2\hat{n} + 1$) of unknowns can be expressed in the matrix form as

$$T_{\overline{M}\,\overline{N} \times \overline{M}\,\overline{N}} U_{\overline{M}\,\overline{N} \times 1} = Q_{\overline{M}\,\overline{N} \times 1} \tag{3.28}$$

where $T_{\overline{M}\,\overline{N} \times \overline{M}\,\overline{N}}$ denotes the generalized stiffness matrix, $U_{\overline{M}\,\overline{N} \times 1}$ represents the displacement matrix, and $Q_{\overline{M}\,\overline{N} \times 1}$ signifies the generalized force matrix. Detailed expressions for these matrices can be found in Appendix A. The resulting set of simultaneous equations for a total of $\overline{M}\,\overline{N}$ unknowns is then numerically solved to obtain the solution for $\tilde{w}(\alpha, \beta)$.

3.1.2.2 The Radiated Sound Pressure

As aforementioned, the displacement of the orthogonally rib-stiffened plate can be obtained by solving the governing equations as

$$\tilde{w}(\alpha, \beta) = \sum_{m'n'} W_{m'n'} \cdot 2p_e \delta(\alpha'_m + \alpha_0) \delta(\beta'_n + \beta_0) \tag{3.29}$$

where $W_{m'n'}$ is associated with the inverse form of the generalized stiffness matrix $T_{\overline{M}\,\overline{N} \times \overline{M}\,\overline{N}}$. In fact, this expression gives the series form solution of Eq. (3.28).

Once the plate displacements are determined, the radiated sound pressure induced by plate vibration can be obtained by employing Eq. (3.22) as

$$\tilde{P}_{ts}(\alpha, \beta, h) = -\frac{\omega^2 \rho_0 \tilde{w}(\alpha, \beta)}{\gamma_2(\alpha, \beta)} \tag{3.30}$$

The radiated sound pressure in real physical space is calculated by applying the Fourier transform as

$$\begin{aligned}
P_{ts}(x, y) &= \int_{-\infty}^{+\infty}\int_{-\infty}^{+\infty} \tilde{P}_{ts}(\alpha, \beta) \cdot e^{i(\alpha x + \beta y)} d\alpha d\beta \\
&= \int_{-\infty}^{+\infty}\int_{-\infty}^{+\infty} \sum_{m'n'} W_{m'n'}(-\omega^2 \rho_0) \cdot 2p_e \delta(\alpha'_m + \alpha_0) \delta(\beta'_n + \beta_0) / \gamma_2(\alpha, \beta) \\
&\quad \cdot e^{i(\alpha x + \beta y)} d\alpha d\beta \\
&= \sum_{m'n'} W_{m'n'}(-\omega^2 \rho_0) \cdot 2p_e / \gamma_2(-\alpha_0 - 2m'\pi/l_x, -\beta_0 - 2n'\pi/l_y) \\
&\quad \cdot e^{-i[(\alpha_0 + 2m'\pi/l_x)x + (\beta_0 + 2n'\pi/l_y)y]}
\end{aligned} \tag{3.31}$$

The truncation manipulation of Eq. (3.27) into a finite range actually implies that the infinite extent structure is replaced by a finite extent structure with geometrical dimensions of $\overline{M}l_x \times \overline{N}l_y$. It has been established that sufficiently large values chosen for $\overline{M}$ and $\overline{N}$ can ensure the solution convergence in subsequent numerical calculations. Correspondingly, the total radiated sound power can be evaluated by [10, 40, 53, 54]

$$\begin{aligned}\Pi_s &= \frac{1}{2}\mathrm{Re}\left\{\int_{-\hat{n}l_y}^{\hat{n}l_y}\int_{-\hat{m}l_x}^{\hat{m}l_x} P_{ts}(x,y)\cdot v_{ts}^*(x,y)\,\mathrm{d}x\mathrm{d}y\right\}\\ &= \frac{1}{2\rho_0 c_0}\int_{-\hat{n}l_y}^{\hat{n}l_y}\int_{-\hat{m}l_x}^{\hat{m}l_x}\left|\sum_{m'n'}W_{m'n'}(-\omega^2\rho_0)\cdot 2p_e/\gamma_2(\alpha_0+2m'\pi/l_x,\beta_0+2n'\pi/l_y)\right.\\ &\quad \left.\cdot e^{-i[(\alpha_0+2m'\pi/l_x)x+(\beta_0+2n'\pi/l_y)y]}\right|^2\mathrm{d}x\mathrm{d}y\\ &= \frac{1}{2\rho_0 c_0}\int_{-\hat{n}l_y}^{\hat{n}l_y}\int_{-\hat{m}l_x}^{\hat{m}l_x}\left|4\omega^4\rho_0^2 p_e^2\sum_{mn,kl}\frac{W_{mn}W_{kl}}{\gamma_2(\alpha_m,\beta_n)\gamma_2(\alpha_k,\beta_l)}\right.\\ &\quad \left.\cdot e^{-i(\alpha_m x+\beta_n y)}e^{-i(\alpha_k x+\beta_l y)}\right|\mathrm{d}x\mathrm{d}y\\ &= \frac{2}{\rho_0 c_0}\omega^4\rho_0^2 p_e^2\left|\sum_{mn,kl}\frac{W_{mn}W_{kl}}{\gamma_2(\alpha_m,\beta_n)\gamma_2(\alpha_k,\beta_l)}\cdot\frac{4}{(\alpha_m+\alpha_k)(\beta_n+\beta_l)}\right.\\ &\quad \left.\cdot \sin[(\alpha_m+\alpha_k)\hat{m}l_x]\sin[(\beta_n+\beta_l)\hat{n}l_y]\right| \end{aligned} \quad (3.32)$$

where the symbol * denotes complex conjugate and $v_{ts}(x,y) = P_{ts}(x,y)/(\rho_0 c_0)$ is the local acoustic velocity on the condition of plane waves assumption.

To highlight the radiation characteristics of periodically rib-stiffened structures as well as the mean flow effects, the solution for a bare faceplate without any rib-stiffeners in steady fluid is given below, which is also used as reference. The displacement of a bare plate can be easily obtained from Eq. (3.25) by disregarding the terms related to the rib-stiffeners as

$$\tilde{w}(\alpha,\beta) = \frac{2p_e\delta(\alpha+\alpha_0)\delta(\beta+\beta_0)}{D(\alpha^2+\beta^2)^2 - m\omega^2 - 2\omega^2\rho_0/\gamma_2(\alpha,\beta)} \quad (3.33)$$

The radiated sound pressure is

$$\tilde{P}_{tu}(\alpha,\beta,h) = -\omega^2\rho_0\tilde{w}(\alpha,\beta)/\gamma_2(\alpha,\beta) \quad (3.34)$$

3.1 Noise Radiation from Orthogonally Rib-Stiffened Plates

Incorporating Eq. (3.33) and taking the Fourier transform of Eq. (3.34) yield

$$P_{tu}(x, y) = \int_{-\infty}^{+\infty}\int_{-\infty}^{+\infty} \tilde{P}_{tu}(\alpha, \beta) \cdot e^{i(\alpha x + \beta y)} d\alpha d\beta$$

$$= \int_{-\infty}^{+\infty}\int_{-\infty}^{+\infty} \Delta(\alpha, \beta)\left(-\omega^2 \rho_0\right) \cdot 2p_e \delta(\alpha + \alpha_0)\delta(\beta + \beta_0) e^{i(\alpha x + \beta y)} d\alpha d\beta$$

$$= \Delta(-\alpha_0, -\beta_0)\left(-\omega^2 \rho_0\right) \cdot 2p_e e^{-i(\alpha_0 x + \beta_0 y)} \tag{3.35}$$

where

$$\Delta(\alpha, \beta) = \left\{\left[D(\alpha^2 + \beta^2)^2 - m\omega^2\right]\gamma_2(\alpha, \beta) - 2\omega^2\rho_0\right\}^{-1},$$

$$\gamma_2(\alpha, \beta) = \sqrt{\alpha^2 + \beta^2 - \omega^2/c_0^2} \tag{3.36}$$

The total radiated sound power by a bare plate having the same geometrical dimensions as the rib-stiffened plate is [10, 40, 53, 54]

$$\Pi_u = \frac{1}{2}\mathrm{Re}\left\{\int_{-\hat{n}l_y}^{\hat{n}l_y}\int_{-\hat{m}l_x}^{\hat{m}l_x} P_{tu}(x, y) \cdot v_{tu}^*(x, y) dxdy\right\}$$

$$= \frac{1}{2\rho_0 c_0}\int_{-\hat{n}l_y}^{\hat{n}l_y}\int_{-\hat{m}l_x}^{\hat{m}l_x}\left|\Delta(\alpha_0, \beta_0)\left(-\omega^2\rho_0\right) \cdot 2p_e e^{-i(\alpha_0 x + \beta_0 y)}\right|^2 dxdy$$

$$= \frac{2}{\rho_0 c_0}\omega^4 \rho_0^2 p_e^2 \int_{-\hat{n}l_y}^{\hat{n}l_y}\int_{-\hat{m}l_x}^{\hat{m}l_x}\left|\Delta^2(\alpha_0, \beta_0) e^{-2i(\alpha_0 x + \beta_0 y)}\right| dxdy$$

$$= \frac{2}{\rho_0 c_0}\omega^4 \rho_0^2 p_e^2 \left|\frac{\sin(2\alpha_0 \hat{m} l_x)\sin(2\beta_0 \hat{n} l_y)}{\alpha_0 \beta_0 \left\{\left[D(\alpha_0^2 + \beta_0^2)^2 - m\omega^2\right]\gamma_2(\alpha_0, \beta_0) - 2\omega^2\rho_0\right\}^2}\right| \tag{3.37}$$

where the symbol * denotes complex conjugate and $v_{tu}(x, y) = P_{tu}(x, y)/(\rho_0 c_0)$ is the local acoustic velocity on the condition of plane waves assumption.

Finally, with the bare plate taken as reference, the radiated sound power level (PWL) L_W of the orthogonally rib-stiffened plate is expressed in decibel scale as

$$L_W = 10\log_{10}\left(\frac{\Pi_s}{\Pi_u}\right) \tag{3.38}$$

The sound power level L_W thus defined is applied below to quantify how the external mean flow affects the process of noise transmission and reveal the underlying physical mechanisms.

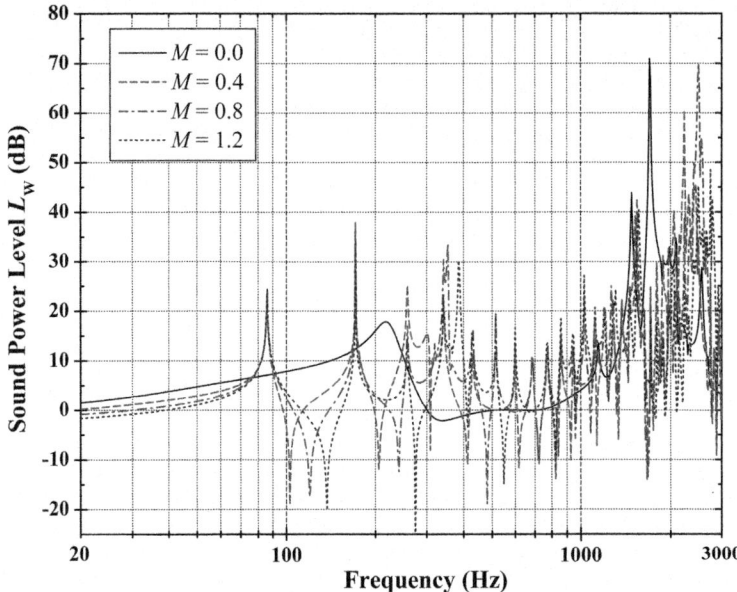

Fig. 3.2 Radiated sound power of the rib-stiffened plate (*re* that of the bare plate) plotted as a function of frequency for selected Mach numbers of mean flow ($\varphi = 60°$, $\theta = 0°$, $l_x = l_y = 0.2$ m)

3.1.3 Effect of Mach Number

Given that the primary objective of this investigation is to examine how external mean flow affects noise transmission trough a rib-stiffened aeroelastic plate into aircraft interior, the Mach number as a key parameter for external mean flow falls into the focal point category. To highlight the periodicity of the stiffened plate as well as the external mean flow effects, the radiated sound power of the structure immersed in mean flow is normalized by that of the bare faceplate in stationary fluid, i.e., Eq. (3.38). Figure 3.2 plots the predicted structure-radiated sound pressure as a function of frequency for selected Mach numbers (i.e., $M = 0$, 0.4, 0.8, and 1.2), with $\varphi = 60°$, $\theta = 0°$, and $l_x = l_y = 0.2$ m.

It is seen from Fig. 3.2 that the presence of external mean flow affects significantly sound radiated from the rib-stiffened plate, as evidenced by the dramatic difference between the overall tendency of sound power level (PWL) L_W versus frequency curve for the case $M = 0$ and that for other cases $M = 0.4$, 0.8, and 1.2. The curves associated with mean flow exhibit similar tendency, although the specific PWL values and peak (or dip) locations differ for different Mach numbers. The results of Fig. 3.2 show that the presence of mean flow reduces visually the PWL at low frequencies (<80 Hz) and enhances the modal behavior of the periodically rib-stiffened plate. Further, as the Mach number is increased, whereas the PWL decreases at low frequencies (<80 Hz), the location of the corresponding resonance

3.1 Noise Radiation from Orthogonally Rib-Stiffened Plates

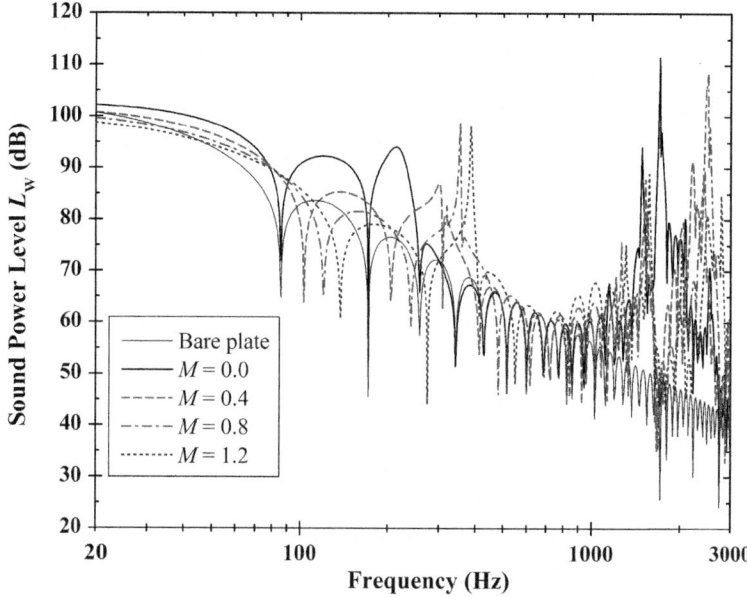

Fig. 3.3 Radiated sound power of the bare plate and the rib-stiffened plate ($re\ 10^{-12}$ W) plotted as a function of frequency for selected Mach numbers of mean flow ($\varphi = 60°$, $\theta = 0°$, $l_x = l_y = 0.2$ m)

peak remains approximately unchanged, while the antiresonance dips move to higher frequencies. These results imply the substantial influence of structural periodicity and external mean flow on sound radiation from rib-stiffened plates immersed in mean flow, because the choice of dimensionless PWL (with reference to that of a bare plate) eliminates other system factors.

To explore more details associated with the sound radiation properties of rib-stiffened plates in the presence of external mean flow, Fig. 3.3 presents the sound power level (PWL) in dB re 10^{-12} W of the rib-stiffened plate and the reference sound power level in dB re 10^{-12} W radiated from a bare plate under harmonic plane sound wave excitation of 1 Pa pressure. It is seen from Fig. 3.3 that the presence of mean flow leads to a modest decrease in the radiated sound power of the structure, increasing thereby the sound transmission loss. This feature is consistent with the existing results of Koval [2]. Moreover, as the Mach number is increased, the overall radiated sound power of the structure decreases, especially in the low-frequency range. Also, with the increase of the Mach number, the modal dips appearing in the PWL curve shift to higher frequencies, which implies that the presence of mean flow increases the modal frequency of the structure. If one notes that the location of the first dip in the PWL versus frequency curve of the stiffened plate at $M = 0$ is coincident with that of the first dip in the PWL curve of the bare plate at $M = 0$, it is understandable that no peak will appear at this location in the PWL curve of the stiffened plate at $M = 0$ due to the counteraction between the two. In contrast,

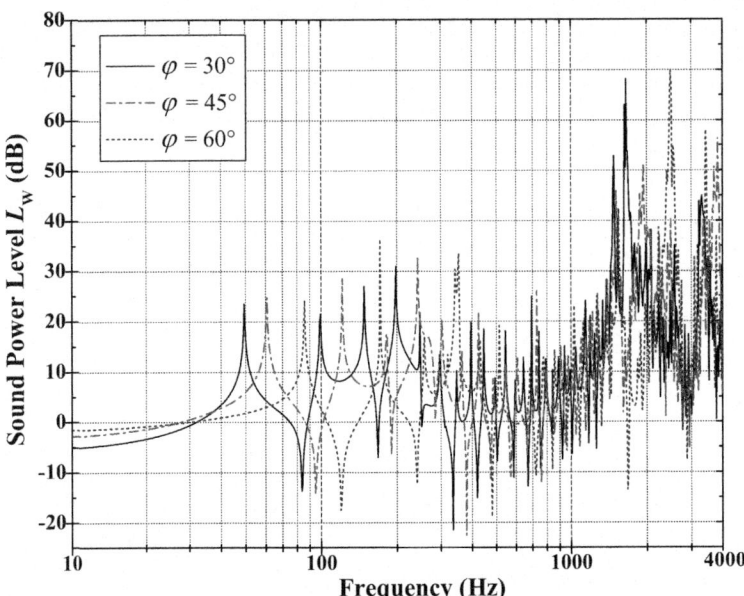

Fig. 3.4 Radiated sound power of the rib-stiffened plate (*re* that of the bare plate) plotted as a function of frequency for various incident angles ($\theta = 0°$, $M = 0.8$, $l_x = l_y = 0.2$ m)

while the first dips in the PWL curve of the stiffened plate at $M \neq 0$ move to higher frequencies, a peak will appear at this location in the PWL curve of the stiffened plate at $M \neq 0$.

3.1.4 Effect of Incidence Angle

For an aircraft in cruise condition, the noise induced by the jet engine or screw propeller may impact the fuselage skin structure at different incidence angles, depending upon the cruise speed and the skin structure location with respect to the source of noise. The effect of the noise incidence angle on sound radiation has, therefore, actual significance in the evaluation of aircraft interior noise at different cruise speeds as well as the design of specific skin structures. Figure 3.4 plots the radiated sound power level as a function of frequency for varying incidence angles, with $\theta = 0°$, $M = 0.8$ (selected for typical aircraft cruise speed; same below), and $l_x = l_y = 0.2$ m. A notable feature of Fig. 3.4 is that the peaks and dips in the PWL versus frequency curves shift to higher frequencies as the incidence angle is increased. Correspondingly, this induces changes in sound power level (PWL) at specific frequencies and the resultant alteration of the whole curve tendency. The sound wave with a more oblique sound incidence angle (i.e., a smaller incidence

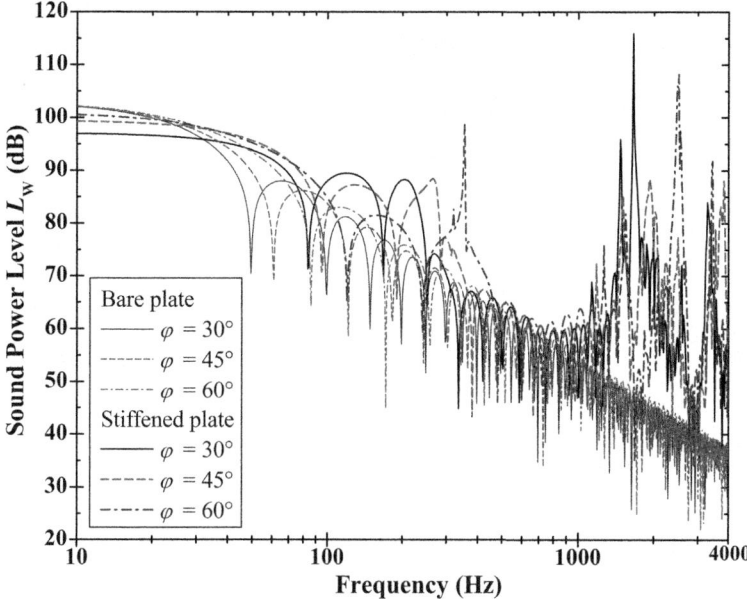

Fig. 3.5 Radiated sound power of the bare plate and the rib-stiffened plate (re 10^{-12} W) plotted as a function of frequency for various incident angles ($\theta = 0°$, $M = 0.8$, $l_x = l_y = 0.2$ m)

angle φ in the present coordinate of Fig. 3.1) is capable of exciting flexural bending waves with more frequency components in the plate, which is thus more likely to induce modal resonance and antiresonance over a wider frequency range. As a result, the peaks and dips move to lower frequencies as the incidence angle is decreased, as shown in Fig. 3.4.

In the presence of mean flow with Mach number $M = 0.8$, Fig. 3.5 presents both the PWL of the rib-stiffened plate and the reference PWL of the bare plate, in dB re 10^{-12} W. As the sound incidence angle is increased, the dips in the PWL curves of both the bare plate and the rib-stiffened plate are seen to shift to higher frequencies. Moreover, the PWL values decrease with increasing sound incidence angle in a wide frequency range, which is particularly significant in the mid-frequency range. This happens because the sound impedance of the structure is dependent of the sound incidence angle: larger sound impedance associated with a higher incidence angle will reduce noticeably the sound radiation of the structure.

3.1.5 Effect of Periodic Spacings

As a key parameter of the rib-stiffened structures considered here, the spacing between two adjacent stiffeners in the x- or y-direction characterizes the periodicity

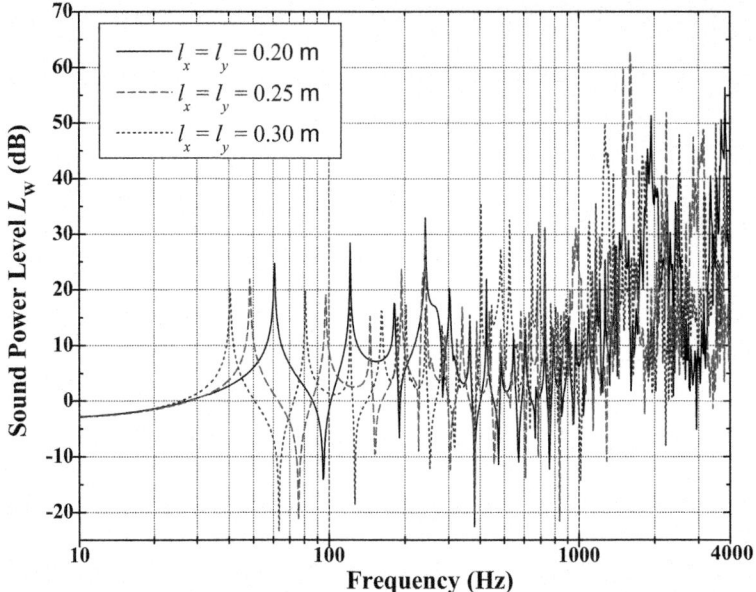

Fig. 3.6 Radiated sound power of the rib-stiffened plate (*re* that of the bare plate) plotted as a function of frequency for various periodic spacings ($\varphi = 45°$, $\theta = 0°$, $M = 0.8$)

of the structure. To explore the effect of structure periodicity on sound radiation, the radiated sound power is plotted in Fig. 3.6 as a function of frequency for several choices of periodic spacings, with $\varphi = 45°$, $\theta = 0°$, and $M = 0.8$. For simplicity, the stiffeners are taken as equally spaced along x- and y-directions, $l_x = l_y$. As the periodic spacing is increased while the whole tendency of the PWL versus frequency remains unchanged, the peaks and dips are noticeably shifted to lower frequencies (Fig. 3.6). This means that for periodically rib-stiffened plates immersed in external mean flow, the natural frequencies of the plate decrease with increasing periodic spacings, and relatively small alterations of the periodic spacings will not change broadly the periodicity nature of the structure. As the factual aircraft structures are often not perfectly periodic, the sound radiation behavior of locally nonperiodic structures is an interesting issue for aeroelastic-acoustic design of aircraft fuselages. This issue will be addressed in a separate study.

In the presence of mean flow, to gain more insights into the effect of periodic spacings on sound radiation, Fig. 3.7 plots the PWL of the rib-stiffened plate and the reference PWL of the bare plate for selected periodic spacings, with $M = 0.8$. As reference, the sound power of the bare plate is calculated by a truncation manipulation [Eq. (3.37)] so that it has the same dimensions ($\overline{M}l_x \times \overline{N}l_y$) as the rib-stiffened plate, which is therefore also related with periodic spacings, as shown in Fig. 3.7. It is seen that the peaks and dips in the PWL versus frequency curves all move to lower frequencies as the periodic spacing is increased. Moreover, with increasing periodic spacing, the radiation sound power decreases

3.1 Noise Radiation from Orthogonally Rib-Stiffened Plates

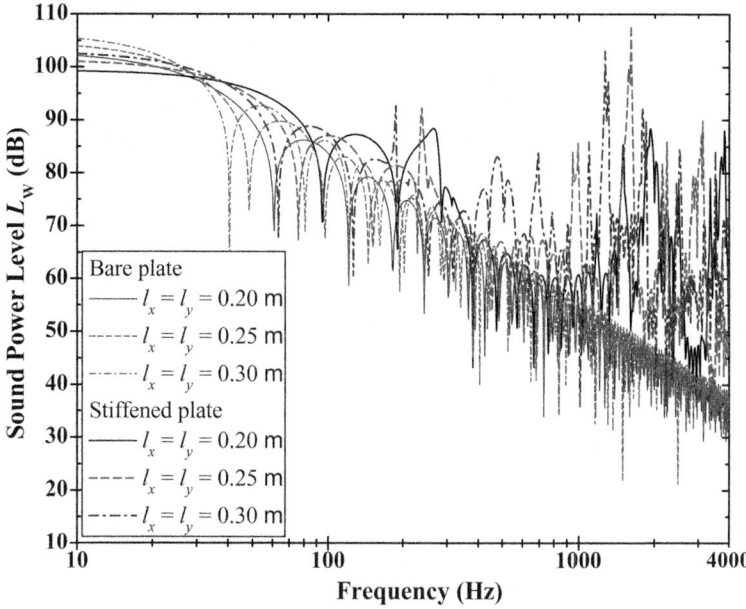

Fig. 3.7 Radiated sound power of the bare plate and the rib-stiffened plate ($re\ 10^{-12}$ W) plotted as a function of frequency for various periodic spacings ($\varphi = 45°$, $\theta = 0°$, $M = 0.8$)

in the low-frequency range (<100 Hz) and increases in the high-frequency range (>100 Hz) as far as the overall tendency is of concern. As a matter of fact, increasing the periodic spacing reduces the stiffness of the structure, which in turn causes the radiated sound power to decrease below 100 Hz and increase above 100 Hz.

3.1.6 Concluding Remarks

A theoretical model has been developed for sound radiation from aeroelastic plates stiffened by two sets of orthogonally distributed rib-stiffeners and subjected to external jet noise, with particular focus placed upon the influence of the presence of convected mean flow. The model is built upon the Kirchhoff thin plate theory and the convected wave equation, with the Euler-Bernoulli bean equation and the torsional wave equation applied to describe the flexural and torsional motions of the rib-stiffeners, respectively. In view of the periodic nature of the structure, the Poisson summation formula and the Fourier transformation technique are adopted to solve the aeroelastic-acoustic governing equations of the system. To highlight the effects of external mean flow and structural periodicity, the sound pressure radiated by the structure is given in the form of decibel scale with respect to that radiated by a bare plate immersed in stationary fluid.

To gain fundamental insights into the aeroelastic-acoustic behavior of rib-stiffened plates immersed in external mean flow, systematic numerical studies are carried out with the developed model to quantify the effects of mean flow speed, jet-noise incident angle, and stiffener spacings. It is established that the presence of mean flow affects significantly the sound radiation performance of the structure, reducing dramatically its PWL level at relatively low frequencies. As the mean flow Mach number is increased while the location of the resonance peak on the PWL versus frequency curve remains nearly unchanged, the antiresonance dips move to higher frequencies. As the sound incidence angle is increased, the peaks and dips on the PWL curves are remarkably shifted to higher frequencies, leading to changes in PWL value at specific frequencies and the resultant alteration of the whole curve tendency. As the periodic spacings are increased, the PWL peaks and dips all move to lower frequencies, while the whole tendency of the curve remains unchanged.

The theoretical model presented in this study is capable of giving reasonable predictions for sound radiation of rib-stiffened aircraft fuselage structures, which is helpful for the evaluation of aircraft interior noise level at different cruise speeds and the design of aircraft skin structures at different locations with respect to the jet engine.

3.2 Transmission Loss of Orthogonally Rib-Stiffened Plates

3.2.1 Introduction

As periodically rib-stiffened plates have been widely used in engineering structures such as aircraft fuselages and ship/submarine hulls [47–51, 55, 56], the sound transmission performance of such structures have attracted increasing attention. Particularly for high-speed transportation vehicles, the interior cabin noise mainly stems from the external turbulent boundary layer (TBL) and engine exhaust noise [16, 22, 30, 33, 37, 57–60]. The development of theoretical models that can provide fundamental insight and applicative guidance of noise reduction at the design stage of rib-stiffened fuselage structures is therefore of considerable practical significance. Regarding the dynamic and acoustic response of fluid-loaded periodically rib-stiffened structures, however, no such theory exists in the open literature that can be readily applied to predict the effects of external mean flow and structure geometrical parameters. This deficiency is addressed in this chapter.

The dynamic and acoustic response of fluid-loaded plates has been extensively investigated. For typical instance, concerning sound transmission through aircraft fuselage plates, Koval [2] formulated theoretical expressions for the field-incidence transmission loss of a single-walled plate by considering external airflow, panel curvature, and internal fuselage pressurization. Built upon Koval's work and with engine exhaust noise penetration through double-walled fuselage structures in mind, Xin and Lu [30] theoretically investigated the sound transmission characteristics of

3.2 Transmission Loss of Orthogonally Rib-Stiffened Plates

double-leaf plates in the presence of external mean flow. To simulate noise reduction of jet nozzle structures in conjunction with active cooling by convective fluid flow, they subsequently extended the work to study the acoustic behavior of finite simply supported aeroelastic plates immersed in convected fluids [33].

With emphasis placed upon boundary layer-induced aircraft noise, Graham [57] proposed a theoretical model to investigate the radiation of sound from a single, flat, elastic plate under turbulent boundary layer excitation, with insightful conclusions for this kind of problem obtained. Graham [22] extended this model to trimmed aircraft plates by studying a flat plate with its internal surface covered by two dissipative layers. It was found that the dissipative layers modify significantly the behavior of the system via two related effects: insulation damping and attenuation. With mean flow effects accounted for, Sgard et al. [31] developed a formulation using the coupled FEM-BEM approach for forced vibroacoustic response of baffled plates. The formulation showed explicitly the effects of mean flow in terms of added mass, stiffness, and radiation damping. Taking into account the structural nonlinearities induced by in-plane forces as well as shearing forces due to the stretching of plate-bending motion, Wu and Maestrello [32] derived theoretical formulations for the acoustic response of finite baffled plates subject to turbulent flow excitation. For arbitrary boundary conditions, Atalla and Nicolas [6] proposed a theoretical formulation for mean flow effects on sound radiation from rectangular baffled plates in inviscid, uniform subsonic flow. The effects of the mean flow in terms of added mass and radiation resistance were shown explicitly. Using the reverse flow reciprocal theorem to determine a Green's function, Howe and Shah [20] theoretically investigated the influence of mean flow on boundary layer-generated interior noise, with both simply supported edges and clamped edges considered. Particularly, it should be mentioned that the research group of Frampton and Clark [4, 15, 16, 37–41, 61] carried out comprehensive studies on sound radiation/transmission of aeroelastic plates excited by convected fluid flow. By adopting singular value decomposition, a method for the state-space modeling of aeroelastic plates subject to linearized potential flow aerodynamic loading was developed [4]. Subsequently, detailed analyses were performed to account for the TBL-induced noise coupling with aeroelastic plates [15, 38–41].

There also exist a number of active control studies associated with fluid-induced noise transmission through plates. For instance, Maury et al. [21] theoretically investigated the active control of sound transmitted through a TBL-excited elastic panel. Further, they presented a theoretical study concerned with the active control of airflow noise transmission through finite double-leaf fuselage structures [12].

Apparently, the vibroacoustic responses of fluid-loaded structures have been extensively studied, and a wide range of efficient theoretical techniques are available for dealing with aircraft fuselage interior noise problem. However, concerning external fluid loading on orthogonally rib-stiffened plates, effective and general guidelines helpful for decision-making at the early design stage of cabin structures appear to be lacking. This chapter attempts to address this deficiency. A theoretical model capable of providing fundamental insight into this issue is developed, with

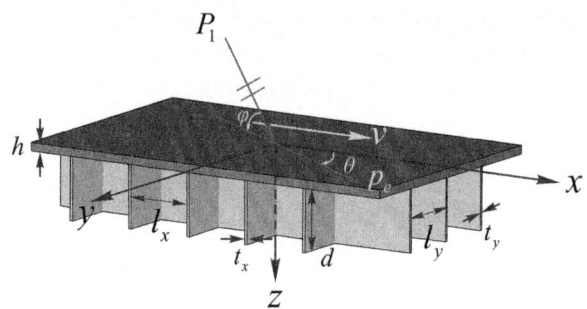

Fig. 3.8 Schematic of an orthogonally rib-stiffened plate excited by convected harmonic pressure (With permission from Acoustical Society of America)

particular focus placed upon the effects of mean flow and structure geometrical parameters on sound transmission loss across orthogonally rib-stiffened plates.

The description of the problem and the derivation of the theoretical model for sound transmission through orthogonally rib-stiffened plate are presented in Sect. 3.2.2. Numerical results and relevant discussions are described in Sects. 3.2.3–3.2.7, where the theoretical model validation is performed by comparing with available theoretical and experimental results in Sect. 3.2.3, together with systematic parameter investigations including effects of the Mach number, rib-stiffener spacings, thickness and height, the elevation, and azimuth angles of incident sound, respectively, in Sects. 3.2.4–3.2.7.

3.2.2 Theoretical Formulation

3.2.2.1 Structure Geometry and Model Definition

With reference to Fig. 3.8, consider an orthogonally rib-stiffened structure made of an infinite Kirchhoff thin plate lying in the plane of $z = 0$ and reinforced by periodically distributed rib-stiffeners along two orthogonal directions $x = ml_x$ and $y = nl_y$, with m and n being both positive or negative integers, and (l_x and l_y) representing rib-stiffener spacing in the x- and y-directions, respectively. Idealized line stiffeners are assumed, which are modeled using the Euler-Bernoulli beam theory and torsional wave equation. Let d represent the height of the rib-stiffeners in the two orthogonal directions; (t_x and t_y) signify the thickness of the x- and y-wise rib-stiffeners, respectively; and h denote the thickness of the thin surface plate. For simplicity, sound radiation from the rib-stiffeners will not be taken into account.

The surface plate divides the spatial region into two regimes: the incident field ($z < 0$) and the transmitted field ($z > h$). For convenience, the parameters associated with the two regimes are numbered by subscripts 1 and 2, respectively. A uniform mean flow is assumed to move parallel to the xy plane, along the x-axis in the incident field. An oblique plane acoustic wave varying harmonically in time is incident on the surface plate in the moving fluid, with elevation angle φ (angle

3.2 Transmission Loss of Orthogonally Rib-Stiffened Plates

about the xy plane) and azimuth angle θ (angle made by the projection of the wave vector in the xy plane about the x-axis) (see Fig. 3.8). The incident pressure wave induces the vibration of the structure, which creates a pressure disturbance in the surrounding fluid, resulting in a reflected sound pressure in the incident field and a transmitted sound pressure in the transmitted field. The pressure alteration caused by this disturbance in turn affects the structure vibration, generating the so-called aeroelastic coupling effect [4, 16, 33, 38].

3.2.2.2 Derivation of Aeroelastic Coupled Model

The oblique plane acoustic wave incident on the surface plate as shown in Fig. 3.8 may be expressed as

$$p(x, y, z, t) = I e^{-i(k_{1x}x + k_{1y}y + k_{1z}z - \omega t)} \tag{3.39}$$

In terms of aeroelastic coupling effect, the vibration of the plate induced by the incident sound pressure creates a pressure disturbance in the surrounding fluid, i.e., the incident field and the transmitted field. Therefore, a negative-going sound wave exists in the incident field, and a positive-going sound wave is present in the transmitted field. Correspondingly, with the structure-coupling effect between the rib-stiffeners and the surface plate [18, 30, 47, 49, 51] duly accounted for, the sound pressure may be written as

$$P_1(x, y, z, t) = I e^{-i(k_{1x}x + k_{1y}y + k_{1z}z - \omega t)}$$
$$+ \sum_{mn} \beta_{mn} e^{-i\left[(k_{1x} + 2m\pi/l_x)x + (k_{1y} + 2n\pi/l_y)y - k_{1z,mn}z - \omega t\right]} \tag{3.40}$$

$$P_2(x, y, z, t) = \sum_{mn} \varepsilon_{mn} e^{-i\left[(k_{2x} + 2m\pi/l_x)x + (k_{2xy} + 2n\pi/l_y)y + k_{2z,mn}z - \omega t\right]} \tag{3.41}$$

The summation indices m and n represent modal decomposition, as a result of the periodic rib-stiffener enclosed cavities, which justifies (through the Helmholtz equation given below) the present modal expressions of the reflected and transmitted sound waves. Note that while the time-dependence term $\exp(i\omega t)$ appearing in Eqs. (3.40) and (3.41) is omitted in subsequent formulations, it is considered implicitly. The wavenumber components in the two fields are associated with wavenumber k_1 (or k_2) and incident angles (φ_1, β) [or transmitted angles (φ_2, β)] as

$$k_{1x} = k_1 \cos\varphi_1 \cos\beta, \quad k_{1y} = k_1 \cos\varphi_1 \sin\beta, \quad k_{1z} = k_1 \sin\varphi_1 \tag{3.42}$$

$$k_{2x} = k_2 \cos\varphi_2 \cos\beta, \quad k_{2y} = k_2 \cos\varphi_2 \sin\beta, \quad k_{2z} = k_2 \sin\varphi_2 \tag{3.43}$$

In the incident field, there exists a uniform mean flow of velocity $\mathbf{V}$ moving parallel to the acoustically deformed boundary, i.e., the fluid-plate interface. The sound pressure in the moving flow should satisfy the convected equation [2, 27, 33, 41] given by

$$\frac{D^2 P_1}{Dt^2} = \left(\frac{\partial}{\partial t} + \mathbf{V} \cdot \nabla\right)^2 P_1 = c_1^2 \nabla^2 P_1 \tag{3.44}$$

As aforementioned, for convenience, the mean flow is assumed to move along the x-direction. As a consequence, Eq. (3.44) can be simplified as

$$\left(\frac{\partial}{\partial t} + v \cdot \frac{\partial}{\partial x}\right)^2 P_1 = c_1^2 \nabla^2 P_1 \tag{3.45}$$

Substitution of P_1 from Eq. (3.40) into Eq. (3.45) leads to

$$k_1 = \frac{\omega}{c_1 (1 + M \cos \varphi_1 \cos \beta)} \tag{3.46}$$

$$k_{1z,mn} = \sqrt{\left(\frac{\omega}{c_1} - \frac{v}{c_1}\frac{2m\pi}{l_x}\right)^2 - \left(k_{1x} + \frac{2m\pi}{l_x}\right)^2 - \left(k_{1y} + \frac{2n\pi}{l_y}\right)^2} \tag{3.47}$$

where $M = v/c_1$ denotes the Mach number of the mean flow. The above expression of wavenumber components in the z-direction manifests the cavity modal characteristics, resulting from the cavities enclosed by periodically distributed rib-stiffeners.

There is no flow in the transmitted field. In such a case the pressure P_2 obeys the classic wave equation, and the wavenumber in this field is given by

$$k_2 = \frac{\omega}{c_2} \tag{3.48}$$

Similar to the expression of Eq. (3.47), in view of the present structure periodicity and cavity modal characteristics, the wavenumber components in the z-direction for the sound wave in a static fluid can be simplified as

$$k_{2z,mn} = \sqrt{\left(\frac{\omega}{c_2}\right)^2 - \left(k_{2x} + \frac{2m\pi}{l_x}\right)^2 - \left(k_{2y} + \frac{2n\pi}{l_y}\right)^2} \tag{3.49}$$

Let c_t denote the trace wave speed in the surface plate. The trace wavenumber is then expressed as $k_t = \omega/c_t = 2\pi/\lambda_t$, λ_t being the wavelength of the plate trace wave. The fluid-plate interface condition dictates that the sound wave needs to fit with the plate, namely, the wavelength between the two must match with each other as

$$k_{1x} = k_t \cos \beta = k_{2x}, \quad k_{1y} = k_t \sin \beta = k_{2y} \tag{3.50}$$

3.2 Transmission Loss of Orthogonally Rib-Stiffened Plates

Meanwhile, given the periodicity of the structure, the displacement of the surface plate can be written in the following form [49]:

$$w(x,y) = \sum_{mn} a_{mn} e^{-i\left[(k_x+2m\pi/l_x)x+(k_y+2n\pi/l_y)y\right]} \tag{3.51}$$

The mnth modal amplitude for the plate displacement is related to the displacement as

$$a_{mn} = \frac{1}{l_x l_y} \int_0^{l_x} \int_0^{l_y} w(x,y) e^{i\left[(k_x+2m\pi/l_x)x+(k_y+2n\pi/l_y)y\right]} dx dy \tag{3.52}$$

where, according to the coherence condition between the bending wavelength in plate and acoustic wavelength in fluid, $k_x = k_{1x} = k_{2x}$ and $k_y = k_{1y} = k_{2y}$.

On the basis of Eq. (3.50), solving for φ_2, one gets

$$\varphi_2 = \arccos\left(\frac{c_2 \cos \varphi_1}{c_1 (1 + M \cos \varphi_1 \cos \beta)}\right) \tag{3.53}$$

Equation (3.53) gives the propagating angle of the transmitted wave, which actually represents the refraction law for wave transmission from one medium to another because one effect of the mean flow is to refract the wave at the surface plate. Note that the corresponding transmitted waves become evanescent wave, if

$$\left|\frac{c_2 \cos \varphi_1}{c_1 (1 + M \cos \varphi_1 \cos \beta)}\right| > 1 \tag{3.54}$$

In such a case, the total reflection of the incident sound wave occurs. Therefore, the vibration of the structure does not contribute to sound radiation in the transmitted field. In other words, the evanescent wave in the transmitted field will soon vanish in the decline exponential form with the distance in the z-direction.

Regarding aeroelastic coupling, the displacement continuity condition should be taken into account at the fluid-plate interfaces. To this end, let λ_1 and λ_2 be the displacements of fluid particles in the incident field and the transmitted field, respectively, both adjacent to the surface plate. These fluid particle displacements should be in coherence with the Navier-Stokes equation for inviscid and irrotational fluid, i.e.,

$$\frac{D^2 \lambda_1}{Dt^2} = \left(\frac{\partial}{\partial t} + \mathbf{V} \cdot \nabla\right)^2 \lambda_1 = -\frac{1}{\rho_1} \frac{\partial P_1}{\partial z}\bigg|_{z=0} \tag{3.55}$$

$$\frac{\partial^2 \lambda_2}{\partial t^2} = -\frac{1}{\rho_2} \frac{\partial P_2}{\partial z}\bigg|_{z=h_r} \tag{3.56}$$

For harmonic sound wave excitation, the fluid particle displacements can be written as

$$\bar{\lambda}_j = \sum_{mn} \bar{\lambda}_{j,mn} e^{-i\left[(k_{jx}+2m\pi/l_x)x+(k_{jy}+2n\pi/l_y)y\right]}, \quad (j = 1, 2) \tag{3.57}$$

Substitution of Eqs. (3.40), (3.41), and (3.57) into Eqs. (3.55) and (3.56) yields

$$-ik_{1z}Ie^{-i(k_{1x}x+k_{1y}y)} + \sum_{mn}\left[ik_{1z,mn}\beta_{mn} - \rho_1(\omega - vk_{1x,m})^2 \bar{\lambda}_{1,mn}\right]$$
$$e^{-i[k_{1x,m}x+k_{1y,n}y]} = 0 \tag{3.58}$$

$$\sum_{mn}\left[-ik_{2z,mn}\varepsilon_{mn}e^{-ik_{2z,mn}h} - \rho_2(\omega - vk_{2x,m})^2 \bar{\lambda}_{2,mn}\right]e^{-i[k_{2x,m}x+k_{2y,n}y]} = 0 \tag{3.59}$$

where

$$k_{jx,m} = k_{jx} + \frac{2m\pi}{l_x}, \quad k_{jy,n} = k_{jy} + \frac{2n\pi}{l_y}, \quad (j = 1, 2) \tag{3.60}$$

Because Eqs. (3.58) and (3.59) are valid for all values of x and y, one can derive the following relationships between the pressure modal amplitudes and fluid particle displacements:

$$\beta_{00} = I + \frac{\rho_1(\omega - vk_{1x})^2 \bar{\lambda}_{1,00}}{ik_{1z}} \tag{3.61}$$

$$\beta_{mn} = \frac{\rho_1(\omega - vk_{1x,m})^2 \bar{\lambda}_{1,mn}}{ik_{1z,mn}} \quad \text{at} \quad m \neq 0 \,\|\, n \neq 0 \tag{3.62}$$

$$\varepsilon_{mn} = -\frac{\rho_2(\omega - vk_{2x,m})^2 \bar{\lambda}_{2,mn}}{ik_{2z,mn}} e^{ik_{2z,mn}h} \tag{3.63}$$

Fully considering the plate displacement expression of Eq. (3.51) and the fluid particle displacement expression of Eq. (3.57), one can write the displacement continuity condition at fluid-plate interfaces in the modal amplitude form as

$$\bar{\lambda}_{1,mn} = a_{mn} = \bar{\lambda}_{2,mn} \tag{3.64}$$

Given the abovementioned assumptions for the coupled aeroelastic-acoustic problem, the surface plate vibration is modeled by applying the Kirchhoff thin plate theory. Consequently, by taking into account the equivalent forces and moments from the periodically distributed rib-stiffeners exerting on the plate, the governing equation for surface plate vibration may be obtained as

3.2 Transmission Loss of Orthogonally Rib-Stiffened Plates

$$D\nabla^4 w + m\frac{\partial^2 w}{\partial t^2} = -\sum_m q_s(x,y)\delta(x-ml_x) - \frac{\partial}{\partial x}\left[\sum_m \kappa_s(x,y)\delta(x-ml_x)\right]$$
$$-\sum_n q_t(x,y)\delta(y-nl_y) - \frac{\partial}{\partial y}\left[\sum_n \kappa_t(x,y)\delta(y-nl_y)\right]$$
$$+ P_1(x,y,0) - P_2(x,y,h) \qquad (3.65)$$

where $\nabla^4 = (\partial^2/\partial^2 x + \partial^2/\partial^2 y)^2$, w, D, and m are the displacement, bending stiffness, and area density of the surface plate, respectively, and $\delta(\cdot)$ signifies the Dirac delta function. Note in particular that the material loss η is introduced via the complex Young's modulus given by

$$D = \frac{Eh^3(1+i\eta)}{12(1-\nu^2)} \qquad (3.66)$$

where E is the Young's modulus and ν is the Poisson ratio of the plate material.

The periodically distributed rib-stiffeners exert equivalent forces and moments on the connected surface plate in terms of their tensional motion and rotational motion and have the same displacements as the attached plate. The displacement of individual rib-stiffener aligned with either x- or y-direction is modeled using the Euler-Bernoulli beam theory as

$$E_x I_x \frac{\partial^4 w}{\partial x^4} + m_x \frac{\partial^2 w}{\partial t^2} = q_t(x,y), \quad E_y I_y \frac{\partial^4 w}{\partial y^4} + m_y \frac{\partial^2 w}{\partial t^2} = q_s(x,y) \qquad (3.67)$$

where $(E_x I_x, E_y I_y)$, (m_x, m_y) and (q_t, q_s) are the flexural stiffness, area density, and equivalent line forces for x-wise and y-wise rib-stiffeners, respectively.

The rotation of individual rib-stiffeners is modeled with the torsional wave equation as

$$G_x J_x \frac{\partial^2 \theta_x}{\partial x^2} - \rho_x I_{px} \frac{\partial^2 \theta_x}{\partial t^2} = \kappa_t(x,y), \quad G_y J_y \frac{\partial^2 \theta_y}{\partial y^2} - \rho_y I_{py} \frac{\partial^2 \theta_y}{\partial t^2} = \kappa_s(x,y)$$
$$(3.68)$$

where $(G_x J_x, G_y J_y)$, $(\rho_x I_{px}, \rho_y I_{py})$ and (κ_t, κ_s) are the torsional stiffness, torsional inertial, and equivalent line moments for x-wise and y-wise rib-stiffeners, respectively, and (θ_x, θ_y) stands for the clockwise angle of rotation for x-wise and y-wise rib-stiffeners about their centroid, calculated by

$$\theta_x(x,y) = \frac{\partial w(x,y)}{\partial y}, \quad \theta_y(x,y) = \frac{\partial w(x,y)}{\partial x} \qquad (3.69)$$

Due to the periodic nature of the considered structure, the displacement of the surface plate includes a series of trace waves having different wave vectors:

$\mathbf{k}_{mn} = (k_x + 2m\pi/l_x)\hat{e}_x + (k_y + 2n\pi/l_y)\hat{e}_y$, $\hat{e}_x$ and $\hat{e}_y$ being the unit vector in the x- and y-directions, respectively. The displacement of the surface plate and the corresponding mnth modal amplitude have been given in Eqs. (3.51) and (3.52).

For simplicity, the new variations $k_{x,m}$ and $k_{y,n}$ are introduced as

$$k_{x,m} = k_x + \frac{2m\pi}{l_x}, \quad k_{y,n} = k_y + \frac{2n\pi}{l_y} \tag{3.70}$$

which represents wavenumber components in the x- and y-directions associated with the mnth trace wave in the plate.

Substitution of (3.51) into (3.67) and (3.68) yields

$$q_s(x, y) = \sum_{mn} \left[E_y I_y k_{y,n}^4 - m_y \omega^2 \right] \cdot a_{mn} e^{-i(k_{x,m}x + k_{y,n}y)} \tag{3.71}$$

$$\frac{\partial}{\partial x} \kappa_s(x, y) = \sum_{mn} \left[G_y J_y k_{x,m}^2 k_{y,n}^2 - \rho_y I_{py} k_{x,m}^2 \omega^2 \right] \cdot a_{mn} e^{-i(k_{x,m}x + k_{y,n}y)} \tag{3.72}$$

$$q_t(x, y) = \sum_{mn} \left[E_x I_x k_{x,m}^4 - m_x \omega^2 \right] \cdot a_{mn} e^{-i(k_{x,m}x + k_{y,n}y)} \tag{3.73}$$

$$\frac{\partial}{\partial y} \kappa_t(x, y) = \sum_{mn} \left[G_x J_x k_{x,m}^2 k_{y,n}^2 - \rho_x I_{px} k_{y,n}^2 \omega^2 \right] \cdot a_{mn} e^{-i(k_{x,m}x + k_{y,n}y)} \tag{3.74}$$

Since the considered structure is spatially periodic in both x- and y-directions, the modal amplitudes a_{mn} can be solved by utilizing the principle of virtual work for each element of the periodic structure. As close relationships between plate modal amplitudes a_{mn} and sound pressure amplitudes β_{mn} and ε_{mn} have been given in Eqs. (3.61), (3.62), (3.63), and (3.64), the sound pressures can be straightforwardly obtained once the former is determined. The principle of virtual work states that the virtual work of the whole system stemming from the virtual displacements must equal to zero, while the imposed virtual displacement may be written as

$$\delta w^* = \delta a_{mn} e^{-i[(k_x + 2m\pi/l_x)x + (k_y + 2n\pi/l_y)y]} \tag{3.75}$$

In view of the periodic nature of the structure, only one periodic element needs to be considered. The equation governing the surface plate vibration is expressed as

$$D\nabla^4 w + m \frac{\partial^2 w}{\partial t^2} - P_1(x, y, 0) + P_2(x, y, h) = 0 \tag{3.76}$$

Therefore, for a given virtual displacement, the virtual work contributed by the surface plate element is given by

3.2 Transmission Loss of Orthogonally Rib-Stiffened Plates

$$\delta\prod_p = \int_0^{l_x}\int_0^{l_y}\left[D\nabla^4 w + m\frac{\partial^2 w}{\partial t^2} - P_1(x,y,0) + P_2(x,y,h)\right]\cdot \delta w^* dxdy \tag{3.77}$$

The virtual works done by the equivalent forces and moments of x-wise and y-wise rib-stiffeners are

$$\delta\prod_x = \int_0^{l_x}\left[q_t(x,0) + \frac{\partial}{\partial y}\kappa_t(x,0)\right]\cdot \delta a_{kl} e^{ik_{x,k}x} dx \tag{3.78}$$

$$\delta\prod_y = \int_0^{l_y}\left[q_s(0,y) + \frac{\partial}{\partial x}\kappa_s(0,y)\right]\cdot \delta a_{kl} e^{ik_{y,l}y} dy \tag{3.79}$$

Accordingly, the principle of virtual work dictates that

$$\delta\prod_p + \delta\prod_x + \delta\prod_y = 0 \tag{3.80}$$

To arrive at a simple combination of governing equation, one needs to firstly derive the final expressions of Eqs. (3.77), (3.78), and (3.79), which are established as detailed below:

$$\delta\prod_p = \int_0^{l_x}\int_0^{l_y}\left[D\nabla^4 w + m\frac{\partial^2 w}{\partial t^2} - P_1(x,y,0) + P_2(x,y,h)\right]\cdot \delta w^* dxdy$$

$$= \int_0^{l_x}\int_0^{l_y}\left\{\sum_{mn}\left[D\left(k_{x,m}^2 + k_{y,n}^2\right)^2 - m\omega^2\right]\alpha_{mn}e^{-i(k_{x,m}x+k_{y,n}y)} - 2Ie^{-i(k_x x + k_y y)} \right.$$
$$\left. - \sum_{mn}\left[\frac{\rho_1(\omega-vk_{x,m})^2\bar{\lambda}_{1,mn}}{ik_{1z,mn}} + \frac{\rho_2(\omega-vk_{x,m})^2\bar{\lambda}_{2,mn}}{ik_{2z,mn}}\right]e^{-i(k_{x,m}x+k_{y,n}y)}\right\}$$

$$\times \delta a_{kl} e^{i(k_{x,k}x+k_{y,l}y)} dxdy$$

$$= \left\{\left[D\left(k_{x,k}^2 + k_{y,l}^2\right)^2 - m\omega^2\right]a_{kl} - \left[\frac{\rho_1(\omega-vk_{x,k})^2\bar{\lambda}_{1,kl}}{ik_{1z,kl}} + \frac{\rho_2(\omega-vk_{x,k})^2\bar{\lambda}_{2,kl}}{ik_{2z,kl}}\right]\right\}$$

$$\cdot l_x l_y \cdot \delta a_{kl} - \int_0^{l_x}\int_0^{l_y} 2Ie^{-i(k_x x + k_y y)} \cdot e^{i(k_{x,k}x+k_{y,l}y)} dxdy \cdot \delta a_{kl} \tag{3.81}$$

$$\delta\prod_x = \int_0^{l_x}\left[q_t(x,0) + \frac{\partial}{\partial y}\kappa_t(x,0)\right]\cdot \delta a_{kl} e^{ik_{x,k}x} dx$$

$$= \int_0^{l_x}\sum_{mn}\left[E_x I_x k_{x,m}^4 - m_x\omega^2 + G_x J_x k_{x,m}^2 k_{y,n}^2 - \rho_x I_{px} k_{y,n}^2 \omega^2\right]\alpha_{mn}$$

$$\times e^{-ik_{x,m}x} \cdot \delta a_{kl} e^{ik_{x,k}x} dx$$

$$= \sum_n \left[E_x I_x k_{x,k}^4 - m_x\omega^2 + G_x J_x k_{x,k}^2 k_{y,n}^2 - \rho_x I_{px} k_{y,n}^2 \omega^2\right]\cdot a_{kn} l_x \cdot \delta a_{kl} \tag{3.82}$$

$$\delta \Pi_y = \int_0^{l_y} \left[q_s(0,y) + \frac{\partial}{\partial x} \kappa_s(0,y) \right] \cdot \delta a_{kl} e^{i k_{y,l} y} dy$$

$$= \int_0^{l_y} \sum_{mn} \left[E_y I_y k_{y,n}^4 - m_y \omega^2 + G_y J_y k_{x,m}^2 k_{y,n}^2 - \rho_y I_{py} k_{x,m}^2 \omega^2 \right] \alpha_{mn}$$

$$\times e^{-i k_{y,n} y} \cdot \delta a_{kl} e^{i k_{y,l} y} dy$$

$$= \sum_m \left[E_y I_y k_{y,l}^4 - m_y \omega^2 + G_y J_y k_{x,m}^2 k_{y,l}^2 - \rho_y I_{py} k_{x,m}^2 \omega^2 \right] \cdot a_{ml} l_y \cdot \delta a_{kl}$$

(3.83)

Substituting Eqs. (3.81), (3.82), and (3.83) into Eq. (3.80) and noting that the virtual displacement is arbitrary, one gets

$$\left\{ \left[D \left(k_{x,k}^2 + k_{y,l}^2 \right)^2 - m \omega^2 \right] a_{kl} - \left[\frac{\rho_1 (\omega - v k_{x,k})^2 \overline{\lambda}_{1,kl}}{i k_{1z,kl}} + \frac{\rho_2 (\omega - v k_{x,k})^2 \overline{\lambda}_{2,kl}}{i k_{2z,kl}} \right] \right\} \cdot l_x l_y$$

$$+ \sum_n \left[E_x I_x k_{x,k}^4 - m_x \omega^2 + G_x J_x k_{x,k}^2 k_{y,n}^2 - \rho_x I_{px} k_{y,n}^2 \omega^2 \right] \cdot a_{kn} l_x$$

$$+ \sum_m \left[E_y I_y k_{y,l}^4 - m_y \omega^2 + G_y J_y k_{x,m}^2 k_{y,l}^2 - \rho_y I_{py} k_{x,m}^2 \omega^2 \right] \cdot a_{ml} l_y$$

$$= \begin{cases} 2 l_x l_y & \text{when} \quad k = 0 \ \& \quad l = 0 \\ 0 & \text{when} \quad k \neq 0 \ \| \quad l \neq 0 \end{cases}$$

(3.84)

To separate the different sum indices, upon taking necessary algebraic manipulations, the resultant governing equation for the system can be rewritten as

$$\left\{ \left[D \left(k_{x,k}^2 + k_{y,l}^2 \right)^2 - m \omega^2 \right] a_{kl} - \left[\frac{\rho_1 (\omega - v k_{x,k})^2 \overline{\lambda}_{1,kl}}{i k_{1z,kl}} + \frac{\rho_2 (\omega - v k_{x,k})^2 \overline{\lambda}_{2,kl}}{i k_{2z,kl}} \right] \right\} \cdot l_x l_y$$

$$+ \left[E_x I_x k_{x,k}^4 - m_x \omega^2 \right] \cdot l_x \sum_n a_{kn} + \left[G_x J_x k_{x,k}^2 - \rho_x I_{px} \omega^2 \right] \cdot l_x \sum_n k_{y,n}^2 a_{kn}$$

$$+ \left[E_y I_y k_{y,l}^4 - m_y \omega^2 \right] \cdot l_y \sum_m a_{ml} + \left[G_y J_y k_{y,l}^2 - \rho_y I_{py} \omega^2 \right] \cdot l_y \sum_m k_{x,m}^2 a_{ml}$$

$$= \begin{cases} 2 l_x l_y & \text{when} \quad k = 0 \ \& \quad l = 0 \\ 0 & \text{when} \quad k \neq 0 \ \| \quad l \neq 0 \end{cases}$$

(3.85)

To solve the above infinite set of coupled algebraic simultaneous equations, one needs to truncate the equation into a finite system consisting of series of assumed modes, insofar as the solution converges. That is, the sum indices (m, n) take values in a finite range, i.e., $m = -\hat{m}$ to $\hat{m}$ and $n = -\hat{n}$ to $\hat{n}$. After going through straightforward algebraic manipulations (Appendix B), the finite governing equation with dimensions $\overline{M} \, \overline{N}$ ($\overline{M} = 2\hat{m}+1$ and $\overline{N} = 2\hat{n}+1$) can be expressed in a matrix form as

$$\mathbf{T}_{\overline{M}\,\overline{N} \times \overline{M}\,\overline{N}} \mathbf{U}_{\overline{M}\,\overline{N} \times 1} = \mathbf{Q}_{\overline{M}\,\overline{N} \times 1} \quad (3.86)$$

3.2 Transmission Loss of Orthogonally Rib-Stiffened Plates

where $T_{\overline{MN} \times \overline{MN}}$ represents the generalized stiffness matrix, $U_{\overline{MN} \times 1}$ is the plate displacement matrix, and $Q_{\overline{MN} \times 1}$ is the generalized force matrix. The resulting set of simultaneous equations for a total of $\overline{M}\,\overline{N}$ unknowns can be numerically solved to obtain the solution for plate modal amplitude a_{mn}, which is then used to calculate the sound pressure amplitudes β_{mn} and ε_{mn}.

To evaluate the sound energy penetrating through the periodic structure in the presence of mean flow, the transmission coefficient is defined here as the ratio of the transmitted sound power to the incident sound power [49, 51, 62] as

$$\tau(\varphi_1, \beta) = \frac{\sum_{m=1}^{+\infty}\sum_{n=1}^{+\infty} |\varepsilon_{mn}|^2 \mathrm{Re}\,(k_{2z,mn})}{|I|^2 k_{1z}} \tag{3.87}$$

The diffuse sound transmission coefficient is then calculated in an averaged form over all possible incident angles [49, 51] as

$$\tau_{\mathrm{diff}} = \frac{\int_0^\pi \int_0^{\varphi_{\mathrm{lim}}} \tau(\varphi, \theta) \sin\varphi \cos\varphi \, d\varphi \, d\theta}{\int_0^\pi \int_0^{\varphi_{\mathrm{lim}}} \sin\varphi \cos\varphi \, d\varphi \, d\theta} \tag{3.88}$$

Finally, the sound transmission loss (STL) expressed in decibel scale [9, 10] is obtained as

$$\mathrm{STL} = 10 \log_{10}\left(\frac{1}{\tau}\right) \tag{3.89}$$

3.2.3 Model Validation

Since the resultant solution is expressed in the form of space-harmonic series, a sufficient number of terms should be adopted to ensure a satisfactory level of convergence and accuracy of the solution. A convergence check study is firstly carried out following the convergence check scheme proposed in our earlier works [18, 47, 49, 51]. It is established that the space-harmonic series solution requires at least 441 terms (i.e., the indices m and n both range from -10 to 10) to ensure convergence at 10 kHz. The same number (441 terms) is also applied to predict STL values below 10 kHz, and it is found that this can ensure accuracy within the error bound of 0.1 dB.

No experimental measurements or theoretical predictions exist for the effects of external mean flow on sound transmission loss of orthogonally rib-stiffened plates. Since the present theoretical model can be favorably degraded to the case of no mean flow by setting the Mach number $M = 0$, for validation, the model predictions

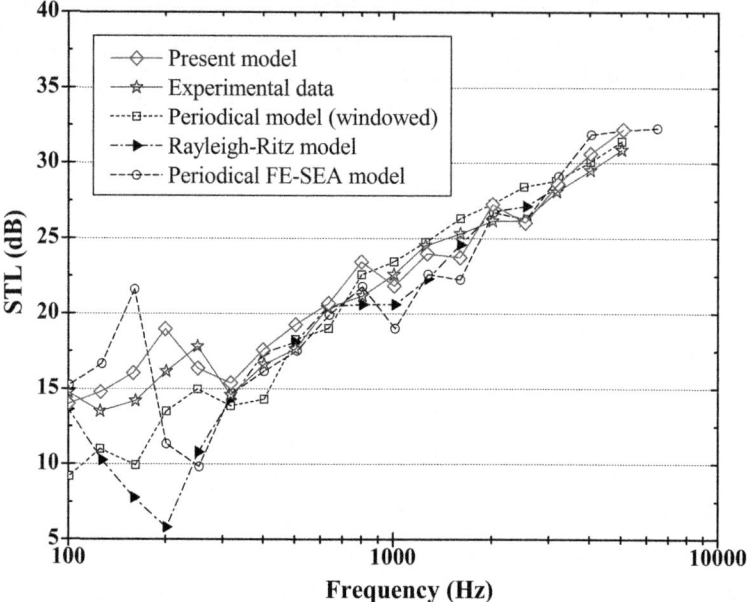

Fig. 3.9 Diffuse STL plotted as a function of incident frequency: comparison between present model predictions with experimental and theoretical results of Mejdi and Atalla [36, 63] (With permission from Acoustical Society of America)

are compared with experimental measurements and theoretical results of Mejdi and Atalla [36, 63] for orthogonally rib-stiffened plates in the absence of mean flow, as shown in Fig. 3.2. In Mejdi and Atalla's paper, the spatially windowed periodic model only accounts for finiteness on sound radiation, while the Rayleigh-Ritz model accounts for the reflected wave field generated at the boundaries. The structural dimensions and material properties applied are identical to those taken by Mejdi and Atalla. The diffuse field excitation with field incidence ($\varphi_{\lim} = 78°$) is employed in the calculation. Further, the transmissibility is computed in 1/24 octave bands and is averaged in 1/3 octave bands over the frequency range of 100–5,000 Hz.

As can be observed in Fig. 3.9, an overall agreement has been achieved between the present model predictions and the theoretical/experimental results of Mejdi and Atalla [36, 63], especially at mid and high frequencies. Particularly, the excellent agreement achieved for frequencies exceeding 300 Hz is attributed to the fact that the wavelength becomes comparable to the stiffener spacing, implying that the theoretical model is capable of accurately capturing the features arising from structure periodicity. However, noticeable discrepancy does exist at low frequencies, which may be due to two possible reasons: (a) boundary conditions in experimental measurements while none considered in the present model and (b) the size of reverberation room limits the low-frequency range below 200 Hz and thus the corresponding experimental results may be questionable [36].

3.2 Transmission Loss of Orthogonally Rib-Stiffened Plates

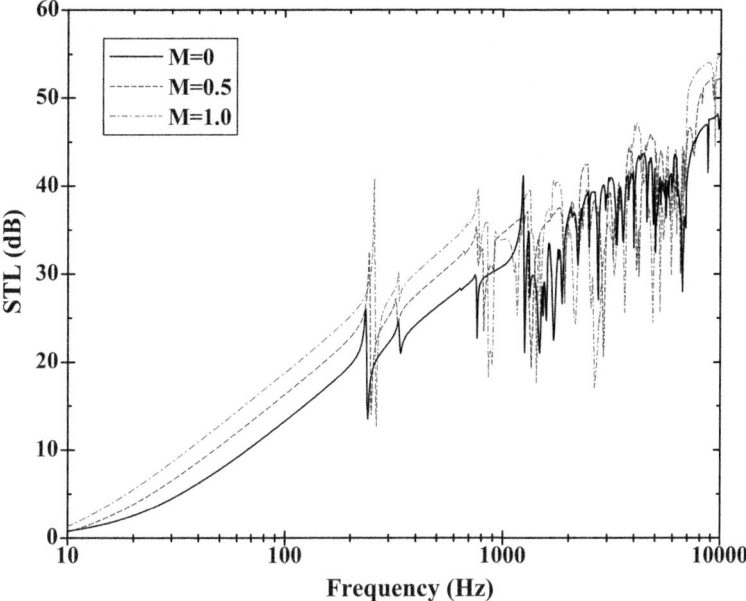

Fig. 3.10 STL variation with incident frequency for orthogonally rib-stiffened aeroelastic plate in the case of ($\varphi_1 = 60°$, $\theta = 0°$): effects of Mach number (With permission from Acoustical Society of America)

3.2.4 Effects of Mach Number of Mean Flow

Since the focus of this research is placed upon external mean flow effects on sound transmission loss of orthogonally rib-stiffened plates, it is of great interest to numerically explore the STL curve tendency at different Mach numbers of mean flow. Figure 3.10 plots the STL variation of the structure with incident frequency for selected Mach numbers (i.e., $M = 0, 0.5, 1$) for $\varphi_1 = 60°$ and $\theta = 0°$. The STL value is seen to increase with increasing Mach number over a wide frequency range, particularly at relatively low frequencies. The existing results for flat plates with no stiffeners have demonstrated the similar tendency, namely, the STL value for flat plates is also increased with increasing Mach number in a wide frequency range [15, 16, 30, 33]. In comparison, the peaks and dips in the STL curves arising from the constraint of the attached rib-stiffeners (liking simple boundary conditions) only slightly change their frequency locations. This actually implies that external mean flow has a weak influence on the modal frequency of the constrained surface plate, although its vibration amplitudes are noticeably changed by mean flow. In other words, it may be deduced from the present results that while the rib-stiffeners play a significant role in the STL curve tendency in terms of the modal behavior of the sub-plates, the added-mass effect of light fluid loading (i.e., airflow considered here) can be ignored, as it has been overwhelmed by the influence of periodic rib-stiffeners.

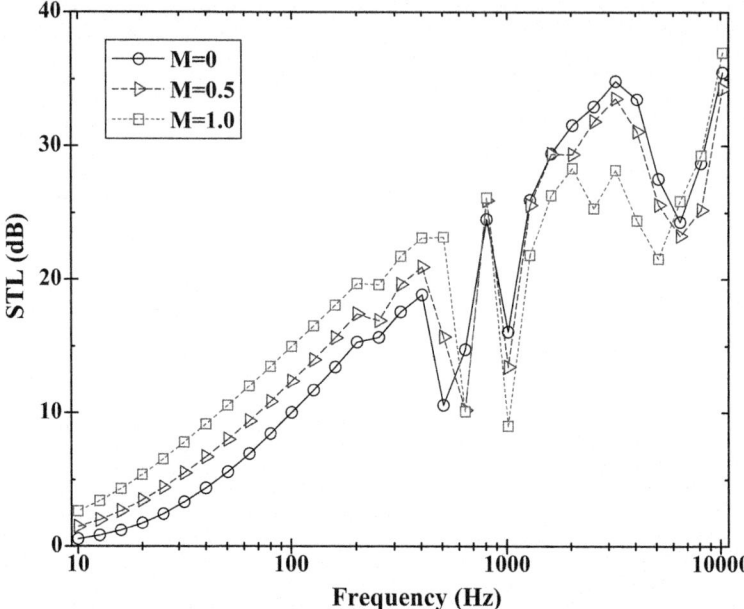

Fig. 3.11 Diffuse STL variation with incident frequency for orthogonally rib-stiffened aeroelastic plate for $\varphi_{\lim} = 78°$: effects of Mach number (With permission from Acoustical Society of America)

To clarify the dense peaks and dips of the plate modal behavior, Fig. 3.11 presents the predicted STL variation with incident frequency under a 3D (three-dimensional) diffuse field incidence for $\varphi_{\lim} = 78°$ and different Mach numbers. As the Mach number is increased while the overall tendency of the STL versus frequency curves remains approximately unaffected, the magnitude of STL increases considerably in the frequency range below 500 Hz. In sharp contrast, within the high-frequency range (approximately 1,000–7,000 Hz), the STL value decreases with increasing Mach number. The theoretical predictions capture the main feature of the STL versus frequency curve in terms of two transition dips [36] at mid frequencies and one coincidence dip at high frequency. The appearance of these dips significantly alter the tendency of the STL curve and thus should be the cause of increased STL below 500 Hz and decreased STL in the frequency range of 1,000–7,000 Hz as the Mach number is increased.

3.2.5 Effects of Rib-Stiffener Spacings

The periodically rib-stiffeners introduce extra reinforced stiffness of the surface plate, inducing significant changes of its vibroacoustic performance especially in the presence of external mean flow. The spacings of the rib-stiffeners are chosen here a

3.2 Transmission Loss of Orthogonally Rib-Stiffened Plates

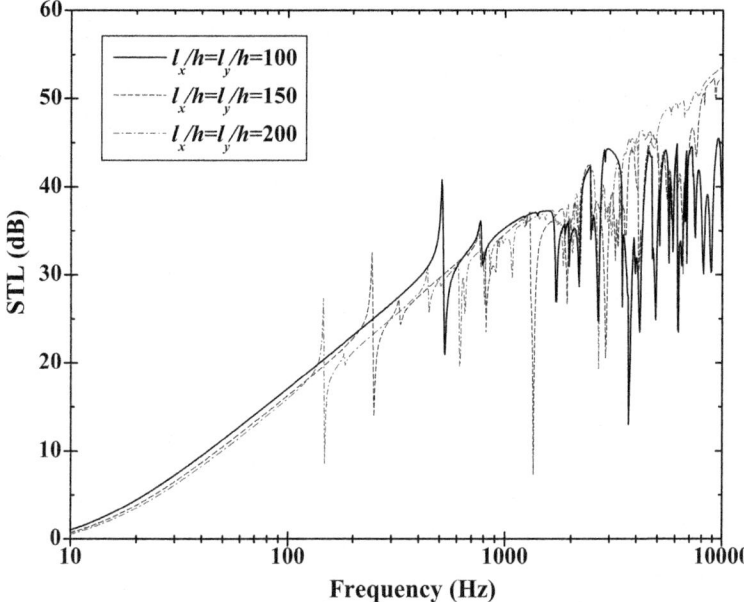

Fig. 3.12 STL variation with incident frequency for orthogonally rib-stiffened aeroelastic plate: effects of rib-stiffener spacings, with $M = 0.5$, $\varphi_1 = 60°$, and $\theta = 0°$ (With permission from Acoustical Society of America)

key parameter to evaluate the influence of the rib-stiffeners on sound transmission loss in the presence of external mean flow. Figure 3.12 plots the STL variation against incident frequency for selected rib-stiffener spacings (i.e., $l_x/h = l_y/h = $ 100, 150, and 200), with $M = 0.5$, $\varphi_1 = 60°$, and $\theta = 0°$.

The dense dips appearing in the STL curves of Fig. 3.12 correspond to the frequencies at which the incident wave undergoes a kind of resonance with the bending wave propagating in the plate. This effect is analogous to the familiar "coincidence resonance," but the spatial harmonics created by wave reflection at the rib-stiffeners introduce multiple possibilities for wavelength matching and coincidence [62]. While the appearance of STL peaks should be arising from the case that the excitation at a structural resonance frequency is without significant wavenumber coincidence between the excitation and the structural wave, and vice versa. In another viewpoint, the constraints exerted on the surface plate by the rib-stiffeners may be regarded as simple boundary condition to some extent. Thereby, the dense dips in the STL curve may be taken as the modal behavior of the subplates. As illustrated in Fig. 3.12, the modal resonance dips of the subplates shift to lower frequencies as the rib-stiffener spacings are increased (particularly so for the first modal resonance dip), implying that the natural frequencies of these subplates decrease when their sizes are increased.

To explore further the influence of rib-stiffener spacings on structure sound transmission loss, Fig. 3.13 plots the diffuse STL variation (averaged in 1/3 octave

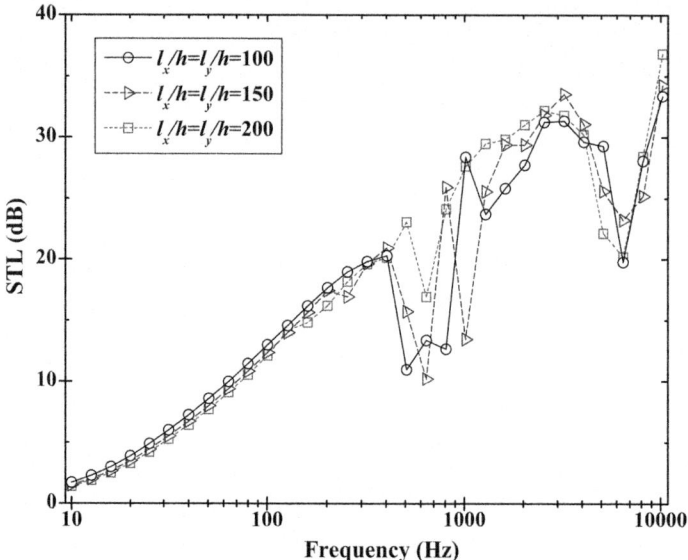

Fig. 3.13 Diffuse STL variation with incident frequency for orthogonally rib-stiffened aeroelastic plate: effects of rib-stiffener spacings, with $M=0.5$ and $\varphi_{\lim}=78°$ (With permission from Acoustical Society of America)

bands) as a function of incident frequency for $M=0.5$ and $\varphi_{\lim}=78°$. Consistent with Fig. 3.12, the modal resonance dips of the subplates shift downward as the rib-stiffener spacings are increased, although this is not clearly displayed in the 1/3 octave bands plot of Fig. 3.13. As a matter of fact, this trend can be well captured in a higher resolution plot, such as the 1/24 octave band. Since Fig. 3.12 has provided enough details, the 1/3 octave band plot of Fig. 3.13 is employed here to show the overall trend of the rib-stiffener effects. As can be seen from Fig. 3.13, changes in rib-stiffener spacings do not alter significantly the tendency of the STL curves, indicated by the consistence of the corresponding transition dips and coincidence dips. It should be pointed out that discrepancies among shifts of modal resonance dips do exist as shown in Fig. 3.12, but this is not captured by the relatively low-resolution plot (i.e., 1/3 octave band) of Fig. 3.13.

Figures 3.12 and 3.13 present the alteration of STL curve with changes in bidirectional spacings (in both x- and y-directions) of the rib-stiffeners. When only one directional spacing (e.g., x-direction) is varied, the corresponding STL variation is presented in Fig. 3.14 for plane sound wave incident case with ($M=0.5$, $\varphi_1=60°$, $\theta=0°$) and in Fig. 3.15 for diffuse field incident case with ($M=0.5$, $\varphi_{\lim}=78°$).

Upon comparing Fig. 3.14 with Fig. 3.12, it is found that the overall trend of the STL versus frequency curves is similar. However, one noticeable difference is that the shifts of modal resonance dips appearing in Fig. 3.14 are not as significant as those in Fig. 3.12. This is readily understandable, as the sizes of the subplates are varied in both directions in Fig. 3.12, while these are altered in only one direction in Fig. 3.14.

3.2 Transmission Loss of Orthogonally Rib-Stiffened Plates

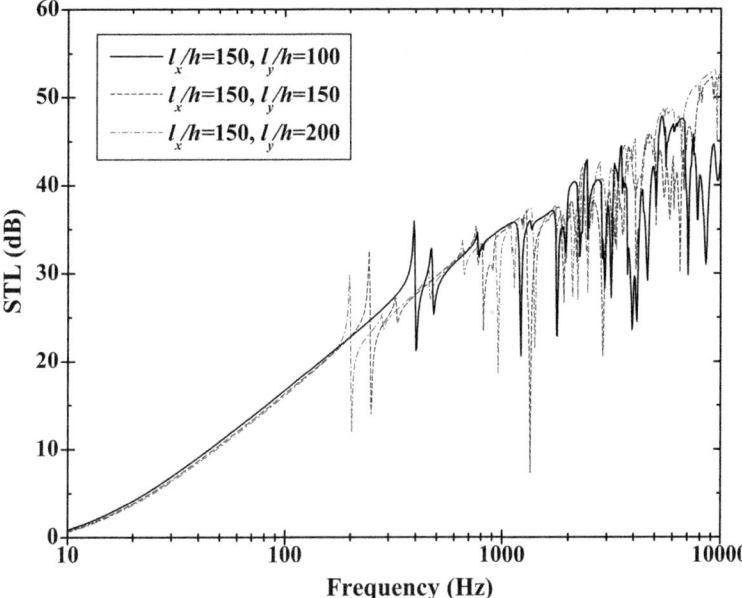

Fig. 3.14 STL variation with incident frequency for orthogonally rib-stiffened aeroelastic plate: effects of rib-stiffener spacings, with $M = 0.5$, $\varphi_1 = 60°$ and $\theta = 0°$ (With permission from Acoustical Society of America)

The STL variations against frequency for diffuse field incident case shown in Fig. 3.15 present an overall tendency of the STL curve in 1/3 octave band. Analogous to Fig. 3.13, the main feature of the STL curve has been well captured, including the coincidence dip and the transition dips appearing between global modes at relatively low frequencies and periodic modes in mid and high frequencies.

3.2.6 Effects of Rib-Stiffener Thickness and Height

The rib-stiffeners affect the vibroacoustic performance of the plate to which they are attached by introducing added mass and reinforced stiffness. In addition to the periodic spacings discussed in the previous section, the thickness and height of the rib-stiffeners are varied below to explore their influence on the STL characteristics of the structure.

Figures 3.16 and 3.17 plot the STL as a function of incident frequency for plane wave incident ($\varphi_1 = 60°$, $\theta = 0°$) and diffuse field incident ($\varphi_{\lim} = 78°$) at $M = 0.5$, respectively, for selected stiffener thicknesses ($t_x/h = t_y/h = 0.25, 0.50$, and 0.75). It is seen that the STL peaks and dips become more remarkable as the thickness is

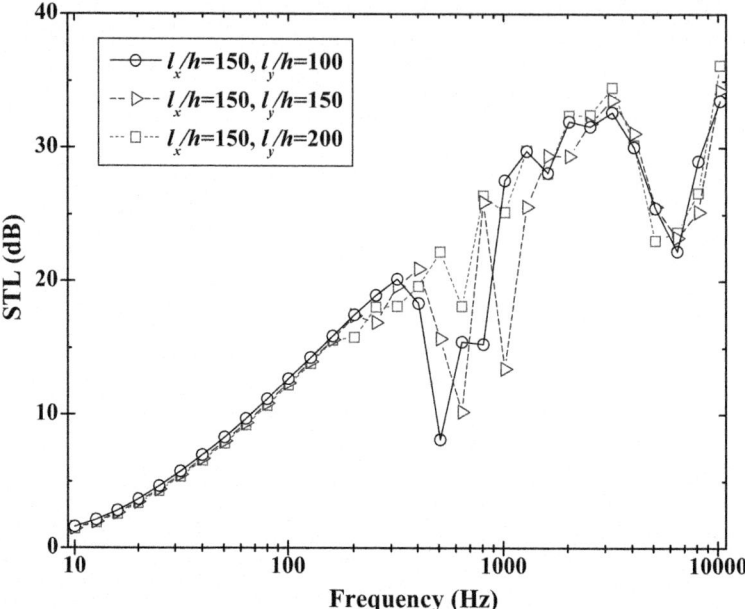

Fig. 3.15 Diffuse STL variation with incident frequency for orthogonally rib-stiffened aeroelastic plate: effects of rib-stiffeners spacings, with $M = 0.5$ and $\varphi_{\lim} = 78°$ (With permission from Acoustical Society of America)

increased, resulting from the added inertial effect of the rib-stiffeners. However, the overall trend of the STL curve does not change significantly.

As another key parameter of rib-stiffener, the effects of rib-stiffener height are examined in Figs. 3.18 and 3.19 for plane wave incident ($\varphi_1 = 60°$, $\theta = 0°$) and diffuse field incident ($\varphi_{\lim} = 78°$) at $M = 0.5$. Similar to the effects of thickness, increasing the rib-stiffener height also causes more remarkable STL peaks and dips. Moreover, increase in rib-stiffener height causes another phenomenon: the peaks and dips shift to higher frequencies, which is not observed in Fig. 3.18 for the case of increasing thickness. This is attributable to the significant change of flexural and torsional stiffness of the rib-stiffeners when their height is altered.

3.2.7 Effects of Elevation and Azimuth Angles of Incident Sound

Previous studies have shown that sound incident angle affects significantly the sound transmission performance of a structure for plane sound wave incident. This also holds in the present case of fluid-loaded orthogonal rib-stiffened plates, and there

3.2 Transmission Loss of Orthogonally Rib-Stiffened Plates

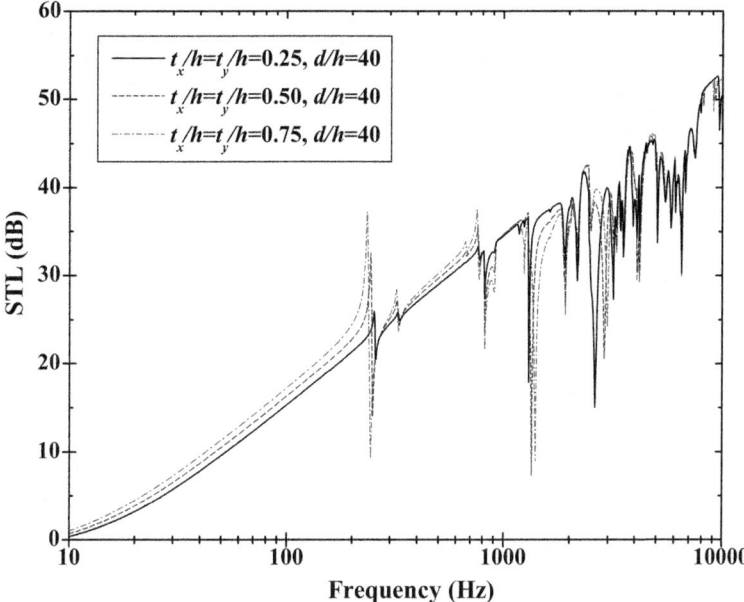

Fig. 3.16 STL variation with incident frequency for orthogonally rib-stiffened aeroelastic plate: effects of rib-stiffener thickness, with $M = 0.5$, $\varphi_1 = 60°$, and $\theta = 0°$ (With permission from Acoustical Society of America)

exist a number of differences with respect to the case of no mean flow. As can be observed in Fig. 3.20, the STL value slightly increases with increasing incident elevation angle (see Fig. 3.8 for definition), especially in the low-frequency range. The varying incident elevation angle also exerts a noticeable influence on the modal behavior of the subplates in the high-frequency range. While the overall variation tendency in the presence of external mean flow is the same as the case of no mean flow, the increase in STL value is not as significant as the latter. This is caused by the refraction effect of the mean flow, as the presence of which has considerably affected the plane wave incident.

Figure 3.21 plots the STL variation against incident frequency for selected incident azimuth angles ($\theta = 0°$, 45°, 135°, and 180°) in the presence of mean flow ($M = 0.5$). Given that the flow is directed along the x-direction, the cases of $\theta = 0°$ and 45° belong to the downstream category, while $\theta = 135°$ and 180° belong to the upstream category. This induces the trend exhibited in Fig. 3.21 that the two STL curves of $\theta = 0°$ and 45° are relatively far away from the two others of $\theta = 135°$ and 180°. In other words, the STL value for the case of sound wave incident downstream is pronouncedly larger than that associated with sound wave incident upstream.

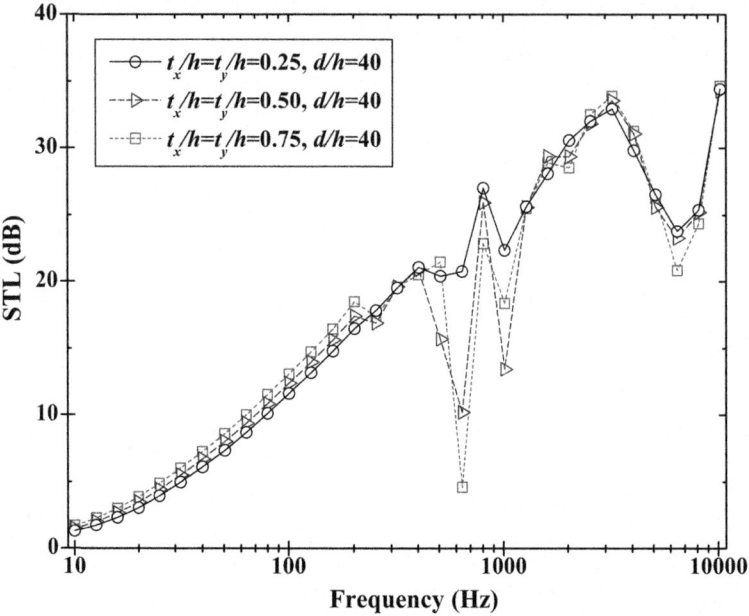

Fig. 3.17 Diffuse STL variation with incident frequency for orthogonally rib-stiffened aeroelastic plate: effects of rib-stiffener thickness, with $M = 0.5$ and $\varphi_{\lim} = 78\,^\circ$ (With permission from Acoustical Society of America)

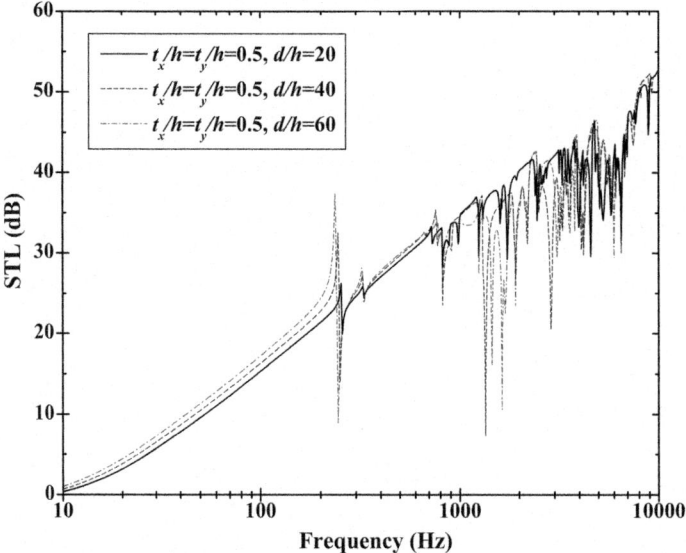

Fig. 3.18 STL variation with incident frequency for orthogonally rib-stiffened aeroelastic plate: effects of rib-stiffener height, with $M = 0.5$, $\varphi_1 = 60\,^\circ$, and $\theta = 0\,^\circ$ (With permission from Acoustical Society of America)

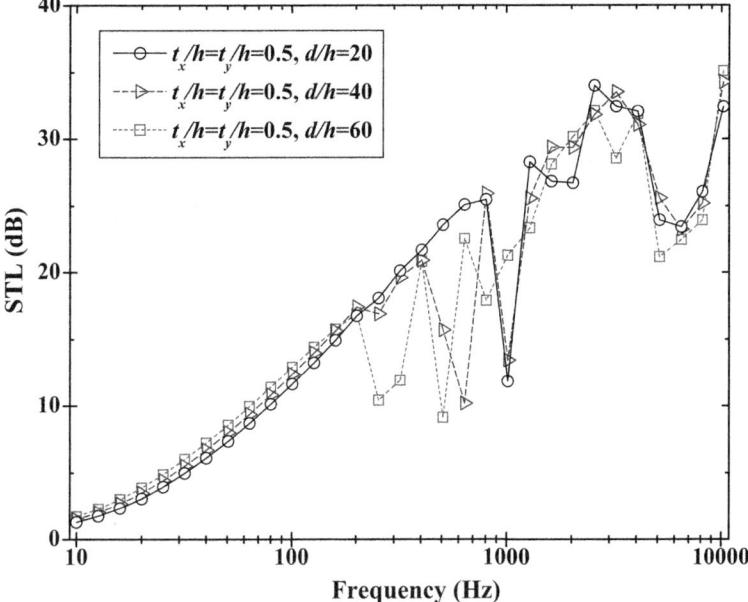

Fig. 3.19 Diffuse STL variation with incident frequency for orthogonally rib-stiffened aeroelastic plate: effects of rib-stiffener height, with $M = 0.5$ and $\varphi_{\lim} = 78°$ (With permission from Acoustical Society of America)

3.2.8 Conclusions

An analytic model has been developed for the effects of external mean flow and structure geometrical parameters on sound transmission across orthogonally rib-stiffened aeroelastic plates, from which the transmission loss of the periodic structure under both plane sound wave incident and diffuse field incident can be predicted. The proposed model is verified by comparing with existing theoretical and experimental results for the case of no mean flow. The model predictions agree well with experimental results, especially at high frequencies where the trace wavelength in the plate is smaller than the rib-stiffener spacings. In the presence of mean flow, comprehensive parametric studies subsequently carried out with the model demonstrate the remarkable influence of structure geometries on transmission loss. It is demonstrated that the transmission loss of periodically rib-stiffened structure is increased significantly with increasing Mach number of mean flow over a wide frequency range. The resonance dips shift to lower frequencies as the rib-stiffener spacings are increased. While, increasing the sir-stiffener thickness and height causes more remarkable peaks and dips. The STL value for the case of sound wave incident downstream is pronouncedly larger than that associated with sound wave incident upstream. These main results indicate that a carefully

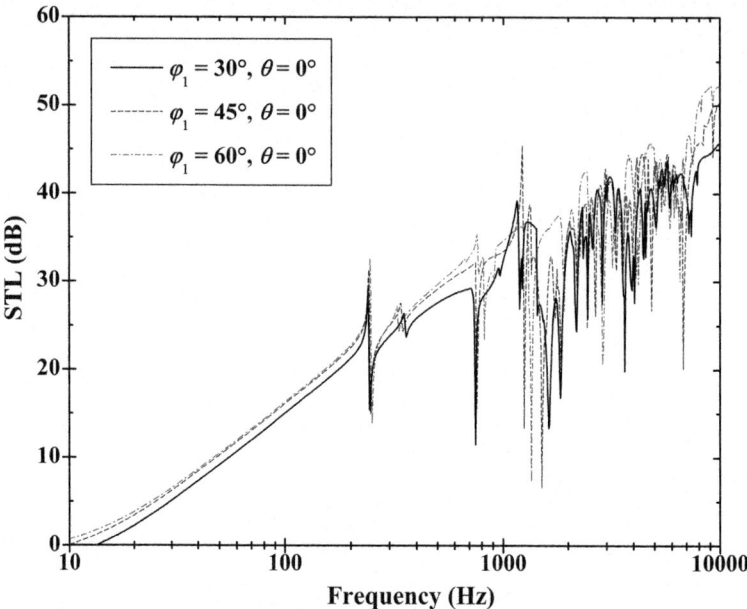

Fig. 3.20 STL variation with incident frequency for orthogonally rib-stiffened aeroelastic plate: effects of incident elevation angle, with $M = 0.5$ (With permission from Acoustical Society of America)

chosen combination of rib-stiffener geometrical parameters (e.g., periodic spacings, thickness, and height) is capable of improving the sound insulation behavior of such structures.

Appendices

Appendix A

The displacement components of the faceplate in wavenumber space are

$$\{U_{m'n'}\} = \begin{bmatrix} U_{11} & U_{21} & \cdots & U_{\overline{M}1} & U_{12} & U_{22} & \cdots & U_{\overline{M}2} & \cdots & U_{\overline{M}\,\overline{N}} \end{bmatrix}^{\mathrm{T}}_{\overline{M}\,\overline{N} \times 1} \quad (3.A.1)$$

where

$$U_{m'n'} = \tilde{w}_1\left(\alpha'_m, \beta'_n\right) \quad (3.A.2)$$

The right-hand side of Eq. (3.28) represents the generalized force

$$\{Q_{m'n'}\} = \begin{bmatrix} Q_{11} & Q_{21} & \cdots & Q_{\overline{M}1} & Q_{12} & Q_{22} & \cdots & Q_{\overline{M}2} & \cdots & Q_{\overline{M}\,\overline{N}} \end{bmatrix}^{\mathrm{T}}_{\overline{M}\,\overline{N} \times 1} \quad (3.A.3)$$

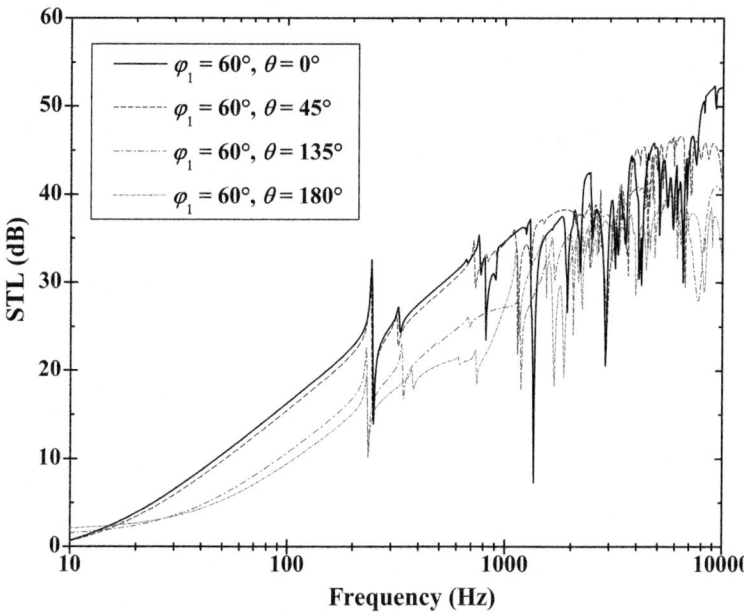

Fig. 3.21 STL variation with incident frequency for orthogonally rib-stiffened aeroelastic plate: effects of incident azimuth angle, with $M = 0.5$ (With permission from Acoustical Society of America)

where

$$Q_{m'n'} = 2p_e \delta \left(\alpha'_m + \alpha_0 \right) \delta \left(\beta'_n + \beta_0 \right) \quad (3.A.4)$$

$$\lambda_{1,m'n'} = D \left(\alpha'^2_m + \beta'^2_n \right)^2 - m\omega^2 - \omega^2 \rho_0/\gamma_1 \left(\alpha'_m, \beta'_n \right) - \omega^2 \rho_0/\gamma_2 \left(\alpha'_m, \beta'_n \right) \quad (3.A.5)$$

$$T_1 = \begin{bmatrix} \lambda_{1,11} & & & & & & & \\ & \lambda_{1,21} & & & & & & \\ & & \ddots & & & & & \\ & & & \lambda_{1,\overline{M}1} & & & & \\ & & & & \lambda_{1,12} & & & \\ & & & & & \lambda_{1,22} & & \\ & & & & & & \ddots & \\ & & & & & & & \lambda_{1,\overline{M}2} \\ & & & & & & & & \ddots \\ & & & & & & & & & \lambda_{1,\overline{M}\,\overline{N}} \end{bmatrix}_{\overline{M}\,\overline{N} \times \overline{M}\,\overline{N}} \quad (3.A.6)$$

$$\lambda_{2,\overline{M}n'}$$

$$= \frac{1}{l_x} \begin{bmatrix} \left(E_y I_y \beta'^4_n - m_y \omega^2\right) & \left(E_y I_y \beta'^4_n - m_y \omega^2\right) & \cdots & \left(E_y I_y \beta'^4_n - m_y \omega^2\right) \\ \left(E_y I_y \beta'^4_n - m_y \omega^2\right) & \left(E_y I_y \beta'^4_n - m_y \omega^2\right) & \cdots & \left(E_y I_y \beta'^4_n - m_y \omega^2\right) \\ \vdots & \vdots & \ddots & \vdots \\ \left(E_y I_y \beta'^4_n - m_y \omega^2\right) & \left(E_y I_y \beta'^4_n - m_y \omega^2\right) & \cdots & \left(E_y I_y \beta'^4_n - m_y \omega^2\right) \end{bmatrix}_{\overline{M}\times\overline{N}}$$
(3.A.7)

$$T_2 = \begin{bmatrix} \lambda_{2,\overline{M}1} & & & \\ & \lambda_{2,\overline{M}2} & & \\ & & \ddots & \\ & & & \lambda_{2,\overline{M}\,\overline{N}} \end{bmatrix}_{\overline{M}\,\overline{N}\times\overline{M}\,\overline{N}}$$
(3.A.8)

$$\lambda_{3,\overline{M}n'} = \frac{1}{l_x}\left(G_y J_y \beta'^2_n - \rho_y I_{py}\omega^2\right) \begin{bmatrix} \alpha_1 \alpha'_1 & \alpha_2 \alpha'_1 & \cdots & \alpha_{\overline{M}} \alpha'_1 \\ \alpha_1 \alpha'_2 & \alpha_2 \alpha'_2 & \cdots & \alpha_{\overline{M}} \alpha'_2 \\ \vdots & \vdots & \ddots & \vdots \\ \alpha_1 \alpha'_{\overline{M}} & \alpha_2 \alpha'_{\overline{M}} & \cdots & \alpha_{\overline{M}} \alpha'_{\overline{M}} \end{bmatrix}_{\overline{M}\times\overline{N}}$$
(3.A.9)

$$T_3 = \begin{bmatrix} \lambda_{3,\overline{M}1} & & & \\ & \lambda_{3,\overline{M}2} & & \\ & & \ddots & \\ & & & \lambda_{3,\overline{M}\,\overline{N}} \end{bmatrix}_{\overline{M}\,\overline{N}\times\overline{M}\,\overline{N}}$$
(3.A.10)

$$\lambda_{4,\overline{M}\,\overline{N}}$$

$$= \frac{1}{l_y} \begin{bmatrix} \left(E_x I_x \alpha'^4_1 - m_x \omega^2\right) & & & \\ & \left(E_x I_x \alpha'^4_2 - m_x \omega^2\right) & & \\ & & \ddots & \\ & & & \left(E_x I_x \alpha'^4_{\overline{M}} - m_x \omega^2\right) \end{bmatrix}_{\overline{M}\times\overline{N}}$$
(3.A.11)

$$T_4 = \begin{bmatrix} \lambda_{4,\overline{M}\,\overline{N}} & \lambda_{4,\overline{M}\,\overline{N}} & \lambda_{4,\overline{M}\,\overline{N}} & \lambda_{4,\overline{M}\,\overline{N}} \\ \lambda_{4,\overline{M}\,\overline{N}} & \lambda_{4,\overline{M}\,\overline{N}} & \lambda_{4,\overline{M}\,\overline{N}} & \lambda_{4,\overline{M}\,\overline{N}} \\ \lambda_{4,\overline{M}\,\overline{N}} & \lambda_{4,\overline{M}\,\overline{N}} & \lambda_{4,\overline{M}\,\overline{N}} & \lambda_{4,\overline{M}\,\overline{N}} \\ \lambda_{4,\overline{M}\,\overline{N}} & \lambda_{4,\overline{M}\,\overline{N}} & \lambda_{4,\overline{M}\,\overline{N}} & \lambda_{4,\overline{M}\,\overline{N}} \end{bmatrix}_{\overline{M}\,\overline{N}\times\overline{M}\,\overline{N}}$$
(3.A.12)

$\lambda_{5,\overline{M}n',n}$

$$= \frac{1}{l_y}\beta'_n\beta_n \begin{bmatrix} \left(G_xJ_x\alpha'^2_1-\rho_xI_{px}\omega^2\right) & & & \\ & \left(G_xJ_x\alpha'^2_2-\rho_xI_{px}\omega^2\right) & & \\ & & \ddots & \\ & & & \left(G_xJ_x\alpha'^2_{\overline{M}}-\rho_xI_{px}\omega^2\right) \end{bmatrix}_{\overline{M}\times\overline{N}}$$
(3.A.13)

$$T_5 = \begin{bmatrix} \lambda_{5,\overline{M}1,1} & \lambda_{5,\overline{M}1,2} & \cdots & \lambda_{5,\overline{M}1,\overline{N}} \\ \lambda_{5,\overline{M}2,1} & \lambda_{5,\overline{M}2,2} & \cdots & \lambda_{5,\overline{M}2,\overline{N}} \\ \vdots & \vdots & \ddots & \vdots \\ \lambda_{5,\overline{M}\,\overline{N},1} & \lambda_{5,\overline{M}\,\overline{N},2} & \cdots & \lambda_{5,\overline{M}\,\overline{N},\overline{N}} \end{bmatrix}_{\overline{M}\,\overline{N}\times\overline{M}\,\overline{N}}$$
(3.A.14)

Employing the definition of the sub-matrices presented above, one obtains

$$T = T_1 + T_2 + T_3 + T_4 + T_5 \quad (3.A.15)$$

Appendix B

The modal amplitudes of the surface plate displacement are

$$\{U_{mn}\} = \begin{bmatrix} a_{11} & a_{21} & \cdots & a_{\overline{M}1} & a_{12} & a_{22} & \cdots & a_{\overline{M}2} & \cdots & a_{\overline{M}\,\overline{N}} \end{bmatrix}^T_{\overline{M}\,\overline{N}\times 1} \quad (3.B.1)$$

The right-hand side of Eq. (3.86) represents the generalized force

$$\{Q_{mn}\} = \begin{bmatrix} Q_{11} & Q_{21} & \cdots & Q_{\overline{M}1} & Q_{12} & Q_{22} & \cdots & Q_{\overline{M}2} & \cdots & Q_{\overline{M}\,\overline{N}} \end{bmatrix}^T_{\overline{M}\,\overline{N}\times 1} \quad (3.B.2)$$

where

$$Q_{mn} = \begin{cases} 2Il_xl_y & \text{at } m = \frac{\overline{M}+1}{2} \,\&\, n = \frac{\overline{N}+1}{2} \\ 0 & \text{at } m \neq \frac{\overline{M}+1}{2} \,\|\, n \neq \frac{\overline{N}+1}{2} \end{cases} \quad (3.B.3)$$

$$\lambda_{1,mn} = \left[D\left(k^2_{x,m}+k^2_{y,n}\right)^2 - m\omega^2 - \frac{\rho_1(\omega-vk_{x,m})^2}{ik_{1z,mn}} - \frac{\rho_2(\omega-vk_{x,m})^2}{ik_{2z,mn}}\right] \cdot l_xl_y$$
(3.B.4)

$$T_1 = \text{diag}\begin{bmatrix} \lambda_{1,11} & \lambda_{1,21} & \cdots & \lambda_{1,\overline{M}1} & \lambda_{1,12} & \lambda_{1,22} & \cdots & \lambda_{1,\overline{M}2} & \cdots & \lambda_{1,\overline{M}\,\overline{N}} \end{bmatrix}_{\overline{M}\,\overline{N}\times\overline{M}\,\overline{N}} \quad (3.B.5)$$

$$\lambda_{2,\overline{MN}} = l_x \cdot \text{diag}\big[\big(E_xI_xk_{x,1}^4 - m_x\omega^2\big)\big(E_xI_xk_{x,2}^4 - m_x\omega^2\big)\cdots$$
$$\times \big(E_xI_xk_{x,\overline{M}}^4 - m_x\omega^2\big)\big]_{\overline{M}\times\overline{N}} \tag{3.B.6}$$

$$T_2 = \begin{bmatrix} \lambda_{2,\overline{MN}} & \lambda_{2,\overline{MN}} & \lambda_{2,\overline{MN}} & \lambda_{2,\overline{MN}} \\ \lambda_{2,\overline{MN}} & \lambda_{2,\overline{MN}} & \lambda_{2,\overline{MN}} & \lambda_{2,\overline{MN}} \\ \lambda_{2,\overline{MN}} & \lambda_{2,\overline{MN}} & \lambda_{2,\overline{MN}} & \lambda_{2,\overline{MN}} \\ \lambda_{2,\overline{MN}} & \lambda_{2,\overline{MN}} & \lambda_{2,\overline{MN}} & \lambda_{2,\overline{MN}} \end{bmatrix}_{\overline{MN}\times\overline{MN}} \tag{3.B.7}$$

$$\lambda_{3,\overline{M}n} = l_x k_{y,n}^2 \cdot \text{diag}\big[\big(G_xJ_xk_{x,1}^2 - \rho_xI_{px}\omega^2\big)\big(G_xJ_xk_{x,2}^2 - \rho_xI_{px}\omega^2\big)\cdots$$
$$\times \big(G_xJ_xk_{x,\overline{M}}^2 - \rho_xI_{px}\omega^2\big)\big]_{\overline{M}\times\overline{N}} \tag{3.B.8}$$

$$T_3 = \begin{bmatrix} \lambda_{3,\overline{M}1} & \lambda_{3,\overline{M}2} & \cdots & \lambda_{3,\overline{M}1,\overline{N}} \\ \lambda_{3,\overline{M}1} & \lambda_{3,\overline{M}2} & \cdots & \lambda_{3,\overline{MN}} \\ \vdots & \vdots & \ddots & \vdots \\ \lambda_{3,\overline{M}1} & \lambda_{3,\overline{M}2} & \cdots & \lambda_{3,\overline{MN}} \end{bmatrix}_{\overline{MN}\times\overline{MN}} \tag{3.B.9}$$

$$\lambda_{4,\overline{M}n}$$
$$= l_y \cdot \begin{bmatrix} \big(E_yI_yk_{y,n}^4-m_y\omega^2\big) & \big(E_yI_yk_{y,n}^4-m_y\omega^2\big) & \cdots & \big(E_yI_yk_{y,n}^4-m_y\omega^2\big) \\ \big(E_yI_yk_{y,n}^4-m_y\omega^2\big) & \big(E_yI_yk_{y,n}^4-m_y\omega^2\big) & \cdots & \big(E_yI_yk_{y,n}^4-m_y\omega^2\big) \\ \vdots & \vdots & \ddots & \vdots \\ \big(E_yI_yk_{y,n}^4-m_y\omega^2\big) & \big(E_yI_yk_{y,n}^4-m_y\omega^2\big) & \cdots & \big(E_yI_yk_{y,n}^4-m_y\omega^2\big) \end{bmatrix}_{\overline{M}\times\overline{N}} \tag{3.B.10}$$

$$T_4 = \text{diag}\big[\lambda_{4,\overline{M}1}\ \lambda_{4,\overline{M}2}\ \cdots\ \lambda_{4,\overline{MN}}\big]_{\overline{MN}\times\overline{MN}} \tag{3.B.11}$$

$$\lambda_{5,\overline{M}n} = l_y\big(G_yJ_yk_{y,n}^2 - \rho_yI_{py}\omega^2\big) \begin{bmatrix} k_{x,1}^2 & k_{x,2}^2 & \cdots & k_{x,\overline{M}}^2 \\ k_{x,1}^2 & k_{x,2}^2 & \cdots & k_{x,\overline{M}}^2 \\ \vdots & \vdots & \ddots & \vdots \\ k_{x,1}^2 & k_{x,2}^2 & \cdots & k_{x,\overline{M}}^2 \end{bmatrix}_{\overline{M}\times\overline{N}} \tag{3.B.12}$$

$$T_5 = \text{diag}\big[\lambda_{5,\overline{M}1}\ \lambda_{5,\overline{M}2}\ \cdots\ \lambda_{5,\overline{MN}}\big]_{\overline{MN}\times\overline{MN}} \tag{3.B.13}$$

Employing the definition of the sub-matrices presented above, one obtains

$$T = T_1 + T_2 + T_3 + T_4 + T_5 \tag{3.B.14}$$

References

1. Dowell EH (1969) Transmission of noise from a turbulent boundary layer through a flexible plate into a closed cavity. J Acoust Soc Am 46(1B):238–252
2. Koval LR (1976) Effect of air flow, panel curvature, and internal pressurization on field-incidence transmission loss. J Acoust Soc Am 59(6):1379–1385
3. Carneal JP, Fuller CR (1995) Active structural acoustic control of noise transmission through double panel systems. AIAA J 33(4):618–623
4. Frampton KD, Clark RL, Dowell EH (1996) State-space modeling for aeroelastic panels with linearized potential flow aerodynamic loading. J Aircr 33(4):816–822
5. Frampton KD, Clark RL (1996) State-space modeling of aerodynamic forces on plate using singular value decomposition. AIAA J 34(12):2627–2630
6. Atalla N, Nicolas J (1995) A formulation for mean flow effects on sound radiation from rectangular baffled plates with arbitrary boundary conditions. J Vib Acoust 117(1):22–29
7. Unruh JF, Dobosz SA (1988) Fuselage structural-acoustic modeling for structure-borne interior noise transmission. J Vib Acoust Stress Reliab Des-Trans ASME 110(2):226–233
8. Mixson JS, Roussos L (1983) Laboratory study of add-on treatments for interior noise control in light aircraft. J Aircr 20(6):516–522
9. Xin FX, Lu TJ (2009) Analytical and experimental investigation on transmission loss of clamped double panels: implication of boundary effects. J Acoust Soc Am 125(3):1506–1517
10. Xin FX, Lu TJ, Chen CQ (2008) Vibroacoustic behavior of clamp mounted double-panel partition with enclosure air cavity. J Acoust Soc Am 124(6):3604–3612
11. Gardonio P, Elliott SJ (1999) Active control of structure-borne and airborne sound transmission through double panel. J Aircr 36(6):1023–1032
12. Maury C, Gardonio P, Elliott SJ (2002) Model for active control of flow-induced noise transmitted through double partitions. AIAA J 40(6):1113–1121
13. Alujevic N, Frampton K, Gardonio P (2008) Stability and performance of a smart double panel with decentralized active dampers. AIAA J 46(7):1747–1756
14. Heitman KE, Mixson JS (1986) Laboratory study of cabin acoustic treatments installed in an aircraft fuselage. J Aircr 23(1):32–38
15. Frampton KD, Clark RL (1997) Power flow in an aeroelastic plate backed by a reverberant cavity. J Acoust Soc Am 102(3):1620–1627
16. Clark RL, Frampton KD (1997) Aeroelastic structural acoustic coupling: implications on the control of turbulent boundary-layer noise transmission. J Acoust Soc Am 102(3):1639–1647
17. Grosveld F (1992) Plate acceleration and sound transmission due to random acoustic and boundary-layer excitation. AIAA J 30(3):601–607
18. Xin FX, Lu TJ, Chen CQ (2010) Sound transmission through simply supported finite double-panel partitions with enclosed air cavity. J Vib Acoust 132(1):011008: 011001–011011
19. Xin FX, Lu TJ (2011) Analytical modeling of sound transmission through clamped triple-panel partition separated by enclosed air cavities. Eur J Mech A/Solids 30(6):770–782
20. Howe MS, Shah PL (1996) Influence of mean flow on boundary layer generated interior noise. J Acoust Soc Am 99(6):3401–3411

21. Maury C, Gardonio P, Elliott SJ (2001) Active control of the flow-induced noise transmitted through a panel. AIAA J 39(10):1860–1867
22. Graham WR (1996) Boundary layer induced noise in aircraft, Part II: The trimmed plat plate model. J Sound Vib 192(1):121–138
23. Mixson JS, Powell CA (1985) Review of recent research on interior noise of propeller aircraft. J Aircr 22(11):931–949
24. Wilby JF, Gloyna FL (1972) Vibration measurements of an airplane fuselage structure I. Turbulent boundary layer excitation. J Sound Vib 23(4):443–466
25. Wilby JF, Gloyna FL (1972) Vibration measurements of an airplane fuselage structure II. Jet noise excitation. J Sound Vib 23(4):467–486
26. Gardonio P (2002) Review of active techniques for aerospace vibro-acoustic control. J Aircr 39(2):206–214
27. Yeh C (1967) Reflection and transmission of sound waves by a moving fluid laver. J Acoust Soc Am 41(4A):817–821
28. Ribner HS (1957) Reflection, transmission, and amplification of sound by a moving medium. J Acoust Soc Am 29(4):435–441
29. Yeh C (1968) A further note on the reflection and transmission of sound waves by a moving fluid layer. J Acoust Soc Am 43(6):1454–1455
30. Xin FX, Lu TJ, Chen CQ (2009) External mean flow influence on noise transmission through double-leaf aeroelastic plates. AIAA J 47(8):1939–1951
31. Sgard F, Atalla N, Nicolas J (1994) Coupled FEM-BEM approach for mean flow effects on vibro-acoustic behavior of planar structures. AIAA J 32(12):2351–2358
32. Wu SF, Maestrello L (1995) Responses of finite baffled plate to turbulent flow excitations. AIAA J 33(1):13–19
33. Xin FX, Lu TJ (2010) Analytical modeling of sound transmission across finite aeroelastic panels in convected fluids. J Acoust Soc Am 128(3):1097–1107
34. Legault J, Atalla N (2009) Numerical and experimental investigation of the effect of structural links on the sound transmission of a lightweight double panel structure. J Sound Vib 324(3–5):712–732
35. Legault J, Atalla N (2010) Sound transmission through a double panel structure periodically coupled with vibration insulators. J Sound Vib 329(15):3082–3100
36. Mejdi A, Atalla N (2010) Dynamic and acoustic response of bidirectionally stiffened plates with eccentric stiffeners subject to airborne and structure-borne excitations. J Sound Vib 329(21):4422–4439
37. Frampton KD, Clark RL (1997) Sound transmission through an aeroelastic plate into a cavity. AIAA J 35(7):1113–1118
38. Clark RL, Frampton KD (1999) Aeroelastic structural acoustic control. J Acoust Soc Am 105(2):743–754
39. Frampton KD (2003) Radiation efficiency of convected fluid-loaded plates. J Acoust Soc Am 113(5):2663–2673
40. Frampton KD (2005) The effect of flow-induced coupling on sound radiation from convected fluid loaded plates. J Acoust Soc Am 117(3):1129–1137
41. Schmidt PL, Frampton KD (2009) Acoustic radiation from a simple structure in supersonic flow. J Sound Vib 328(3):243–258
42. Alujevic N, Gardonio P, Frampton KD (2008) Smart double panel with decentralized active dampers for sound transmission control. AIAA J 46(6):1463–1475
43. Maestrello L (1995) Responses of finite baffled plate to turbulent flow excitations. AIAA J 33(1):13–19
44. Ingard U (1959) Influence of fluid motion past a plane boundary on sound reflection, absorption, and transmission. J Acoust Soc Am 31(7):1035–1036
45. Kessissoglou NJ, Pan J (1997) An analytical investigation of the active attenuation of the plate flexural wave transmission through a reinforcing beam. J Acoust Soc Am 102(6):3530–3541
46. Morse PM, Ingard KU (1968) Theoretical acoustics. McGraw-Hill, New York

References

47. Xin FX, Lu TJ (2011) Analytical modeling of wave propagation in orthogonally rib-stiffened sandwich structures: sound radiation. Comput Struct 89(5–6):507–516
48. Xin FX, Lu TJ (2010) Sound radiation of orthogonally rib-stiffened sandwich structures with cavity absorption. Compos Sci Technol 70(15):2198–2206
49. Xin FX, Lu TJ (2010) Analytical modeling of fluid loaded orthogonally rib-stiffened sandwich structures: sound transmission. J Mech Phys Solids 58(9):1374–1396
50. Mace BR (1980) Sound radiation from a plate reinforced by two sets of parallel stiffeners. J Sound Vib 71(3):435–441
51. Xin FX, Lu TJ (2011) Transmission loss of orthogonally rib-stiffened double-panel structures with cavity absorption. J Acoust Soc Am 129(4):1919–1934
52. Rumerman ML (1975) Vibration and wave propagation in ribbed plates. J Acoust Soc Am 57(2):370–373
53. Fahy F (1985) Sound and structural vibration: radiation, transmission and response. Academic, London
54. Cunefare KA (1992) Effect of modal interaction on sound radiation from vibrating structures. AIAA J 30(12):2819–2828
55. Mace BR (1980) Periodically stiffened fluid-loaded plates, I: Response to convected harmonic pressure and free wave propagation. J Sound Vib 73(4):473–486
56. Mace BR (1980) Periodically stiffened fluid-loaded plates, II: Response to line and point forces. J Sound Vib 73(4):487–504
57. Graham WR (1996) Boundary layer induced noise in aircraft, Part I: The flat plate model. J Sound Vib 192(1):101–120
58. Xin FX, Lu TJ (2013) External mean flow effects on noise radiation from orthogonally rib-stiffened aeroelastic plates. AIAA J 51(2):406–415
59. Xin FX, Lu TJ (2012) Sound radiation of parallelly stiffened plates under convected harmonic pressure excitation. Sci China-Technol Sci 55(2):496–500
60. Meng H, Xin FX, Lu TJ (2012) External mean flow effects on sound transmission through acoustic absorptive sandwich structure. AIAA J 50(10):2268–2276
61. Schmidt PL, Frampton KD (2009) Effect of in-plane forces on sound radiation from convected, fluid loaded plates. J Vib Acoust 131(2):021001–021008
62. Wang J, Lu TJ, Woodhouse J et al (2005) Sound transmission through lightweight double-leaf partitions: theoretical modelling. J Sound Vib 286(4–5):817–847
63. Legault J, Mejdi A, Atalla N (2011) Vibro-acoustic response of orthogonally stiffened panels: the effects of finite dimensions. J Sound Vib 330(24):5928–5948

Chapter 4
Sound Transmission Across Sandwich Structures with Corrugated Cores

Abstract An analytic study of sound transmission through all-metallic, two-dimensional, periodic sandwich structures having corrugated core is presented. The space-harmonic method is employed, and an equivalent structure containing one translational spring and one rotational spring per unit cell is proposed to simplify the analysis of the vibroacoustic problem. It is demonstrated that the core geometry exerts a significant effect on the sound insulation performance of the sandwich, so that one may tailor the core topology for specified acoustic applications. Subsequent analysis of the STL (sound transmission loss) and dispersion curves of the structure leads to fundamental insight into the physical mechanisms behind the appearance of various peaks and dips on the STL versus frequency curves. As the weight, stiffness, and acoustic property of the sandwich structures all change with the alteration of core configuration and geometry, it is further demonstrated that it is possible to explore the multifunctionality of the structure by optimally designing the core topology.

4.1 Introduction

Lightweight sandwich structures comprising two facesheets and an (fluid through) open core have attractive structural load-bearing and heat dissipation attributes [1, 2] and hence have found increasingly wide applications in high-speed transportation, aerospace and aeronautical aircrafts, ships, and other transportation vehicles where weight saving is of major concern [3–16]. Among these applications, an important issue is noise transmission from exterior of cabin into the interior, inasmuch as that has a particular significance for the safety and comfort of civil or military vehicles. However, to date, only a few investigations have focused on the fluid-structure coupling and dynamic responses of the sandwich structure as well as the physical mechanisms of sound transmission across the structure (see, e.g., Refs. [3, 4]).

Fig. 4.1 Illustration of a typical sandwich panel with corrugated core [13, 19] (With permission from Taylor & Francis Ltd.)

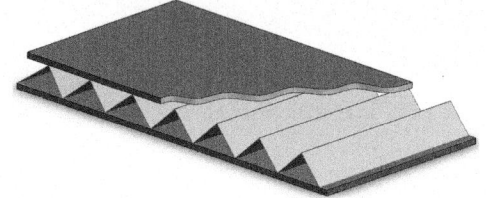

A typical lightweight all-metallic sandwich panel, which is shown in Fig. 4.1, has found successful applications in modern express locomotive and ship constructions. The main attractiveness of this type of sandwich is its simple two-dimensional (2D) corrugated core, which can be either connected to the facesheets via laser welding or formed together with the facesheets as one integral structure using the method of extrusion. In addition to carry structural loads, the openness of the corrugated core also allows the sandwich to be used as an effective heat dissipation medium, either separately or simultaneously [17, 18]. For practical applications such as express locomotives and ships, however, it is also necessary to investigate the sound insulation capability of the sandwich structure, especially the influence of core geometry on sound transmission across the structure. This task is performed in the current study. Together with load-bearing and/or heat dissipation properties, the knowledge thus acquired on sound insulation capability would be useful for the design and manufacture of sandwich structures having corrugated cores.

Extensive investigations over the past decades have been dedicated to studying the sound insulation performances of periodic beams [20, 21], single plates [22–24], and double panels with air cavity [25–34] or structural connections [27, 35–52]. For example, the sound transmission loss (*STL*) of a structure containing two impervious layers, an airspace, and two acoustic blankets was studied by Beranek et al. [27] for normal incident sound. London [28] studied the transmission of sound across a double wall consisting of two identical single walls coupled by an airspace by considering the impedance of a single wall and achieved good agreement between model predictions and experimental measurements. The problem of the transmission of a plane wave through two infinite parallel plates connected by periodically spaced frames was solved by Lin and Garrelick [36] while that for an infinite sandwich panel with a constrained viscoelastic damping layer was analyzed by Narayanan and Shanbhag [46]. Takahashi [37] studied the problem of sound radiation from periodically connected (e.g., point connected, point connected with rib-stiffening, and rib-connected) infinite double-plate structures excited by a harmonic point force. Using the multiple-reflection theory, Cummings et al. [25] investigated random incidence transmission loss of a thin double aluminum panel with glass fiber absorbent around the edge of the cavity. Mathur et al. [53] investigated the *STL* through periodically stiffened panels and double-leaf strictures and proposed a theoretical model using the space-harmonic method. Lee and Kim [35] solved the vibroacoustic problem of a single stiffened plate subjected to a plane sound wave using the space-harmonic approach proposed by Mead and Pujara [21].

Wang et al. [38] extended the space-harmonic approach to double-leaf partitions stiffened by periodic parallel studs and provided detailed physical interpretations for the mechanisms of sound transmission through the structure. However, few studies have addressed the issue of sound transmission across 2D sandwich structures with corrugated core.

The purpose of this chapter is to analytically study the sound insulation performance of all-metallic sandwich structure with corrugated core as shown in Fig. 4.1, with focus placed upon the influence of core geometry as well as the physical insight of the associated vibroacoustic phenomena. In Sect. 4.2, the theoretical development of the acoustic model is presented. Based on model predictions, the effects of core geometry on the vibroacoustic performance of the sandwich are quantified in Sect. 4.3. Physical interpretations of the appearances of various peaks and dips on *STL* versus frequency curves are presented in Sect. 4.4. Finally, this chapter concludes with a summary of current findings and a suggestion of future investigations required for multifunctional design of the sandwich.

4.2 Development of Theoretical Model

The transmission of sound through an infinite (in x- and z-directions) 2D sandwich panel with corrugated core is schematically illustrated in Fig. 4.2. A harmonic sound wave (with angular frequency ω) impinges on the left-side facesheet with incidence angle θ. While part of the sound is reflected back, the remaining portion transmits into the right side of the sandwich panel via two paths: corrugated core as structural route and air cavity as airborne route. The sandwich panel with corrugated core is modeled as two parallel plates (facesheets) structurally connected by uniformly distributed (equivalent) translational springs and rotational springs, and the mass of the core is considered as lumped mass attached to the two facesheets, as shown in Fig. 4.2. Due to periodicity, a unit cell of the sandwich structure is shown in Fig. 4.3a, whereas its equivalent structure is presented in Fig. 4.3b. For simplicity, the facesheets and the core are both made of aluminum and have geometrical and material properties as follows: core thickness t_0 and depth l, panel thickness h_1 and h_2 (subscripts "1" and "2" denote left- and right-side facesheets, respectively), unit cell length L, inclination angle between panel and core sheet $\pm \alpha$, Young's modulus E, and Poisson ratio ν.

With the assumption of small deflections, the stiffness of the translational spring K_F and the stiffness of the rotational spring K_M per unit length can be obtained analytically (details can be found in Xin et al. [13]). Let the surface density of the left- and right-side facesheets be denoted by m_1 and m_2, respectively. The mass of the core for one unit cell is $2M$, which is equivalent to two lumped mass M separately attached to the inner surfaces of the two facesheets (Fig. 4.3b). Under these conditions, the governing equations for the vibration of the sandwich structure may be written as [13, 19]

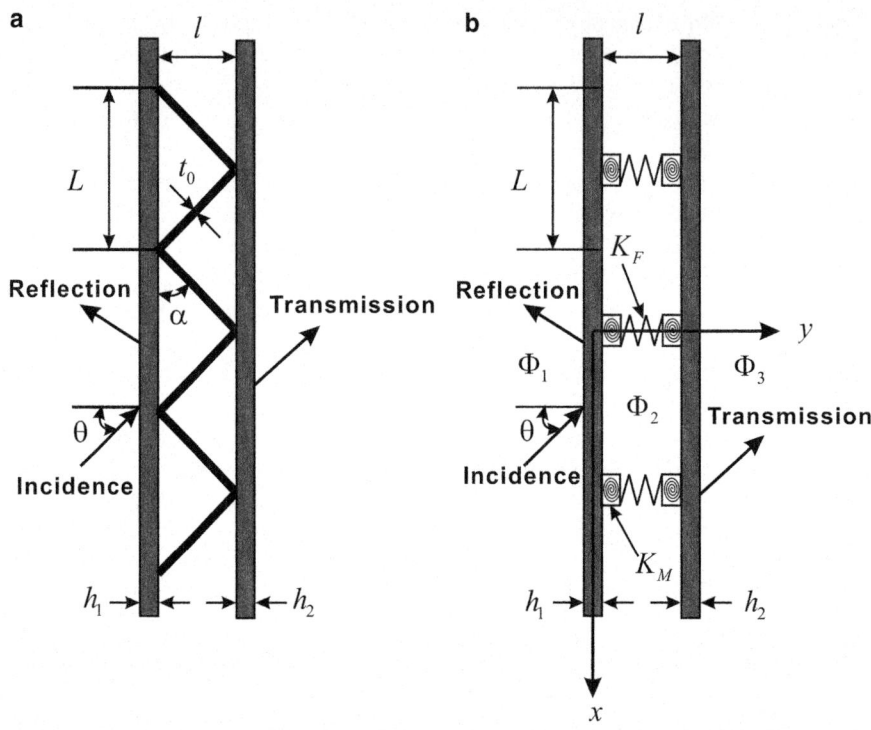

Fig. 4.2 (a) Side view of sandwich panel with corrugated core and (b) schematic of equivalent structure for space-harmonic modeling [13, 19] (With permission from Taylor & Francis Ltd.)

Fig. 4.3 (a) Unit cell of sandwich panel with corrugated core and (b) equivalent unit cell [13, 19] (With permission from Taylor & Francis Ltd.)

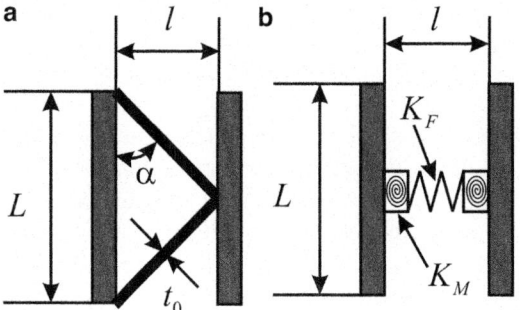

$$D_1 \frac{\partial^4 w_1(x,t)}{\partial x^4} + \left(m_1 + \frac{M}{L}\right) \frac{\partial^2 w_1(x,t)}{\partial t^2} + K_F \left(w_1(x,t) - w_2(x,t)\right) \\ - K_M \frac{\partial^2}{\partial x^2} \left(w_1(x,t) - w_2(x,t)\right) - j\omega\rho_0 \left(\Phi_1 - \Phi_2\right) = 0 \quad (4.1)$$

$$D_2 \frac{\partial^4 w_2(x,t)}{\partial x^4} + \left(m_2 + \frac{M}{L}\right) \frac{\partial^2 w_2(x,t)}{\partial t^2} + K_F \left(w_2(x,t) - w_1(x,t)\right) \\ - K_M \frac{\partial^2}{\partial x^2} \left(w_2(x,t) - w_1(x,t)\right) - j\omega\rho_0 \left(\Phi_2 - \Phi_3\right) = 0 \quad (4.2)$$

4.2 Development of Theoretical Model

where ρ_0 is air density, w_i ($i = 1, 2$) is the transverse deflection of the two facesheets, $j = \sqrt{-1}$, and Φ_i ($i = 1, 2, 3$) is the velocity potential of the acoustic field (see Fig. 4.2b) which is related to the corresponding velocity by $\hat{\mathbf{u}}_i = -\nabla \Phi_i$. With the loss factor of facesheet material denoted by η_i, the flexural rigidity of the facesheet D_i ($i = 1, 2$) can be written using complex Young's modulus $E_i(1 + j\eta_i)$ as

$$D_i = \frac{E_i h_i^3 (1 + j\eta_i)}{12 (1 - v_i^2)} \tag{4.3}$$

The velocity potentials of acoustic fields, i.e., the incident field, the air cavity field in between the two facesheets, and the transmitted field (Fig. 4.2b), are separately defined as follows [13, 19, 38]:

$$\Phi_1(x, y, t) = I e^{-j[k_x x + k_y y - \omega t]} + \sum_{-\infty}^{\infty} \beta_n e^{-j[(k_x + 2n\pi/L)x - k_{yn} y - \omega t]} \tag{4.4}$$

$$\Phi_2(x, y, t) = \sum_{-\infty}^{\infty} \varepsilon_n e^{-j[(k_x + 2n\pi/L)x + k_{yn} y - \omega t]} + \sum_{-\infty}^{\infty} \zeta_n e^{-j[(k_x + 2n\pi/L)x - k_{yn} y - \omega t]} \tag{4.5}$$

$$\Phi_3(x, y, t) = \sum_{-\infty}^{\infty} \xi_n e^{-j[(k_x + 2n\pi/L)x + k_{yn} y - \omega t]} \tag{4.6}$$

where I and β_n are the amplitudes of the incident (i.e., positive-going) sound wave and the reflected (i.e., negative-going) sound wave, respectively. Similarly, symbols ε_n and ζ_n represent separately the amplitude of the positive-going wave and the negative-going wave in the air cavity field. In the transmitted field, there is no negative-going waves; thus, the velocity potential is only for transmitted wave with amplitude ξ_n.

The wavenumbers k_x and k_y that appear in Eqs. (4.4), (4.5), and (4.6) are determined by sound incidence angle θ as

$$k_x = k \sin \theta, \quad k_y = k \cos \theta \tag{4.7}$$

where $k = \omega/c_0$, and c_0 is sound speed in air and k_{yn} is the wavenumber in the y-direction, defined as [13]

$$k_{yn} = \sqrt{\left(\frac{\omega}{c}\right)^2 - \left(k_x + \frac{2n\pi}{L}\right)^2} \tag{4.8}$$

When $\omega/c < |k_x + 2n\pi/L|$, the pressure waves become evanescent waves, and correspondingly, k_{yn} is given by

$$k_{yn} = j\sqrt{\left(k_x + \frac{2n\pi}{L}\right)^2 - \left(\frac{\omega}{c}\right)^2} \qquad (4.9)$$

For harmonic sound excitations of the infinitely large and periodic sandwich structure, the deflections of the two facesheets can be expressed as [13, 38]

$$w_1(x,t) = \sum_{n=-\infty}^{\infty} \alpha_{1,n} e^{-j[k_x + 2n\pi/L]x} e^{j\omega t} \qquad (4.10)$$

$$w_2(x,t) = \sum_{n=-\infty}^{\infty} \alpha_{2,n} e^{-j[k_x + 2n\pi/L]x} e^{j\omega t} \qquad (4.11)$$

At the interface of the air and the facesheet, the continuity of velocity should be satisfied, i.e., the normal velocity of the facesheet matches with that of its adjacent air particle:

$$-\frac{\partial \Phi_1}{\partial z} = j\omega w_1, \quad -\frac{\partial \Phi_2}{\partial z} = j\omega w_1; \quad \text{at} \quad y = 0 \qquad (4.12)$$

$$-\frac{\partial \Phi_2}{\partial z} = j\omega w_2, \quad -\frac{\partial \Phi_3}{\partial z} = j\omega w_2; \quad \text{at} \quad y = H \qquad (4.13)$$

Substitution of Eqs. (4.4), (4.5), and (4.6) and Eqs. (4.10) and (4.11) into Eqs. (4.12) and (4.13) leads to

$$\beta_0 = I - \frac{\omega \alpha_{1,0}}{k_y} \qquad (4.14)$$

$$\beta_n = -\frac{\omega \alpha_{1,n}}{k_{yn}}, \quad \text{when} \quad n \neq 0 \qquad (4.15)$$

$$\varepsilon_n = \frac{\omega \left(\alpha_{2,n} e^{jk_{yn}H} - \alpha_{1,n} e^{2jk_{yn}H}\right)}{k_{yn}\left(1 - e^{2jk_{yn}H}\right)} \qquad (4.16)$$

$$\zeta_n = \frac{\omega \left(\alpha_{2,n} e^{jk_{yn}H} - \alpha_{1,n}\right)}{k_{yn}\left(1 - e^{2jk_{yn}H}\right)} \qquad (4.17)$$

$$\xi_n = \frac{\omega \alpha_{2,n} e^{jk_{yn}H}}{k_{yn}} \qquad (4.18)$$

Substituting Eqs. (4.4), (4.5), and (4.6) and Eqs. (4.10) and (4.11) into the governing equations and incorporating Eqs. (4.14), (4.15), (4.16), (4.17), and (4.18), one obtains [13]

4.2 Development of Theoretical Model

$$\left[D_1 k_m^4 - m_1\omega^2 - \frac{2j\omega^2 \rho_0 e^{jk_{yn}l}}{k_{yn}\left(1-e^{2jk_{yn}l}\right)} \right]\alpha_{1,m} + \frac{K_F - \omega^2 M}{L}\left(\sum_{n=-\infty}^{\infty}\alpha_{1,n}\right)$$

$$+ \frac{K_M}{L}\left(\sum_{n=-\infty}^{\infty}\alpha_{1,n}k_n\right)k_m - \frac{K_F}{L}\left(\sum_{n=-\infty}^{\infty}\alpha_{2,n}\right) - \frac{K_M}{L}\left(\sum_{n=-\infty}^{\infty}\alpha_{2,n}k_n\right)k_m$$

$$+ \frac{2j\omega^2 \rho_0 e^{jk_{yn}l}}{k_{yn}\left(1-e^{2jk_{yn}l}\right)}\alpha_{2,m} = \begin{cases} 2j\omega^2\rho_0 I, & m=0 \\ 0, & m\neq 0 \end{cases} \qquad (4.19)$$

$$\left[D_2 k_m^4 - m_2\omega^2 - \frac{2j\omega^2 \rho_0 e^{jk_{yn}l}}{k_{yn}\left(1-e^{2jk_{yn}l}\right)} \right]\alpha_{2,m} + \frac{K_F - \omega^2 M}{L}\left(\sum_{n=-\infty}^{\infty}\alpha_{2,n}\right)$$

$$+ \frac{K_M}{L}\left(\sum_{n=-\infty}^{\infty}\alpha_{2,n}k_n\right)k_m - \frac{K_F}{L}\left(\sum_{n=-\infty}^{\infty}\alpha_{1,n}\right) - \frac{K_M}{L}\left(\sum_{n=-\infty}^{\infty}\alpha_{1,n}k_n\right)k_m$$

$$+ \frac{2j\omega^2 \rho_0 e^{jk_{yn}l}}{k_{yn}\left(1-e^{2jk_{yn}l}\right)}\alpha_{1,m} = 0 \qquad (4.20)$$

where the unknown coefficients $\alpha_{1,n}$ and $\alpha_{2,n}$ can be obtained by simultaneously solving the above algebraic equations. Once these coefficients are known, other parameters, β_n, ε_n, ζ_n, and ξ_n, can be obtained by applying Eqs. (4.14), (4.15), (4.16), (4.17), and (4.18).

Finally, the power transmission coefficient is defined as the ratio of incident sound power to transmitted sound power [13, 19] as

$$\tau(\theta) = \frac{\sum_{n=-\infty}^{\infty}|\xi_n|^2 \mathrm{Re}\left(k_{yn}\right)}{|I|^2 k_y} \qquad (4.21)$$

Correspondingly, the sound transmission loss (*STL*) is defined in decibel scale as

$$\mathrm{STL} = -10\log_{10}\tau \qquad (4.22)$$

Numerical simulations are performed in this section to investigate the effects of core topology (determined by inclination angle α between facesheet and core sheet) on the sound insulation performance of an all-metallic sandwich panel with corrugated core. In the subsequent analysis of the core topology effects on STL (i.e., Sect. 4.4) and optimal design (i.e., Sect. 4.5), the two sound transmission routes (i.e., airborne route and structure-borne route) are both considered, just as the theoretical model previously derived. However, to avoid the singularity problem in mathematics, the physical interpretations for the existence of peaks and dips in

STL curves in Sect. 4.4 neglect the airborne route of sound transmission and only consider the predominant structural route so that a clear understanding of the peaks and dips can be obtained.

4.3 Effects of Core Topology on Sound Transmission Across the Sandwich Structure

To highlight the effects of core topology on sound transmission across the sandwich structure, the inclination angle α is systematically varied, while all other parameters are held constant. The *STL* values numerically calculated with the space-harmonic method are plotted as functions of frequency in Fig. 4.4 for selected values of α: 20°, 30°, 45°, and 60°. Due to the similar qualitative aspect of the STL versus frequency curves for different incidence angles [35] and for the purpose of sufficiently exciting the structure vibration modes, the sound incident angle is fixed at $\theta = 45°$.

At frequencies corresponding separately to the "mass-air-mass" resonance, the standing-wave resonance, the "coincidence" resonance, and the structure natural resonance, the sandwich structure vibrates intensely. This significantly decreases the sound insulation capability of the structure, resulting in the existence of various dips in Fig. 4.4. The "mass-air-mass" resonance appears in the process of

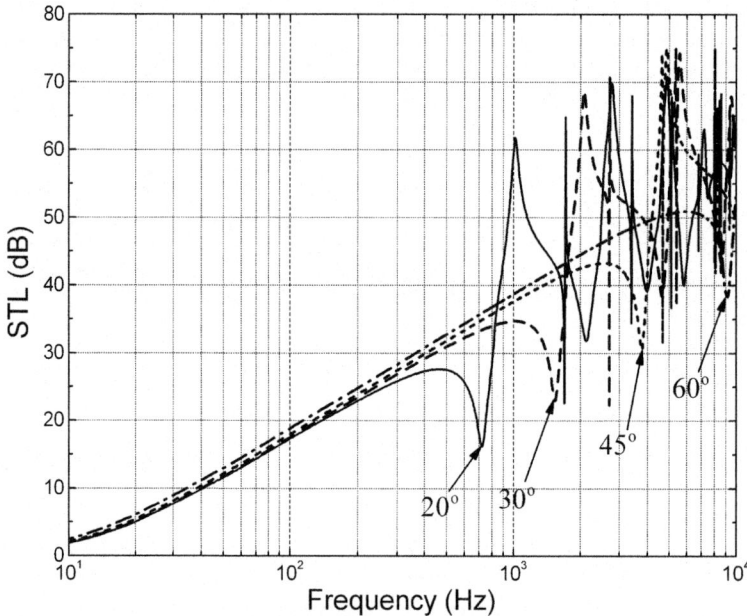

Fig. 4.4 *STL* of sandwich panels with different inclination angles (as shown in Fig. 4.2) in the case of incidence angle $\theta = 45°$ (With permission from Taylor & Francis Ltd.)

sound transmission through a "panel-air-panel"-type structure, at which the two panels move in opposite phase, with the air in between bouncing like springs. This resonance is active only when the frequency of the incident sound matches the natural frequency of the "mass-air-mass" resonance. Similarly, the acoustic standing-wave resonance is also a feature of the "panel-air-panel"-type structure, which occurs when the air cavity depth is integer numbers of the half wavelengths of incident sound. The "coincidence" frequency is relevant to the condition that the trace wavelength of the incidence sound matches the wavelength of bending wave in the facesheet. For the sandwich panel considered here, when sound is normally incident, the mass-air-mass resonance occurs at around 280 Hz, the first-order standing-wave resonance occurs at about 8,000 Hz, while the critical frequency beyond which the coincident resonance will occur is around 6,000 Hz (in comparison, the coincidence frequency is about 12,000 Hz for a typical sound incidence with angle $\theta = 45°$). In addition, due to the complexity and periodicity of the sandwich structure as well as fluid-structure coupling, the "mass-air-mass" resonance, the standing-wave resonance, the "coincidence" resonance, and the structure natural resonance are mutually coupled, creating a series of peaks and dips on the *STL* curves as shown in Fig. 4.4. A more detailed physical interpretation for the existence of those peaks and dips in structure-borne STL curves which are associated with the structure natural resonance only is presented in the next section.

It can be seen from Fig. 4.4 that the geometry of the corrugated core plays a significant role in the process of sound transmission through the sandwich. As the inclination angle is increased, not only the peaks and dips of the *STL* versus frequency curve are all shifted to higher frequencies, but also the smooth range of the curve at lower frequencies is stretched toward higher frequencies. The primary reason is that the change of inclination angle alters the equivalent stiffness of the translational and rotational springs as well as the unit cell length (see Xin et al. [13]), and hence the vibroacoustic behavior of the whole sandwich structure is changed. In other words, the sandwich structure with corrugated core can be regarded as a structure with inherent vibroacoustic properties that can be favorably altered by tuning its core geometry in terms of factual application requirements.

4.4 Physical Interpretation for the Existence of Peaks and Dips on STL Curves

The corrugated core in between the two facesheets stiffens the sandwich structure for enhanced load-bearing capability. However, its existence adds one structural route for sound propagation, resulting in the decrease of *STL* and the intense peaks and dips on *STL* versus frequency curves relative to double-leaf panel structures with air cavity. These peaks and dips caused by the core geometry as well as periodicity of the sandwich structure have a significant effect on its sound insulation performance, so it is necessary to explore their physical origins.

As mentioned above, the appearance of dips on *STL* curves is attributed to the effects of vibroacoustic resonances, the most likely being the familiar "coincidence" resonance discussed in the previous section. While the bending waves induced by the incident sound transmit in the facesheet, the reflected waves from the core sheets also act on the facesheet, which thence generate multiple possibilities for the wavenumbers of the two waves matching and coinciding with each other. To emphasize the main effects and investigate how the "fake coincidence resonances" affect *STL* curves, material damping is ignored here, and only the panel-core-panel route for wave transmission is considered. Furthermore, for simplicity, the coupling effects of air cavity between the two facesheets are ignored, and the mass and the equivalent rotational stiffness of the corrugated core are not taken into account. These assumptions are generally reasonable inasmuch as the structural coupling of the corrugated core is much stronger than the acoustic coupling of air cavity and the rotational stiffness of the core has a negligible effect with respect to its translational stiffness. The governing equations for facesheet vibration can therefore be simplified as

$$D_1 \frac{\partial^4 w_1(x,t)}{\partial x^4} + m_1 \frac{\partial^2 w_1(x,t)}{\partial t^2} + K_F(w_1(x,t) - w_2(x,t)) = 0 \quad (4.23)$$

$$D_2 \frac{\partial^4 w_2(x,t)}{\partial x^4} + m_2 \frac{\partial^2 w_2(x,t)}{\partial t^2} + K_F(w_2(x,t) - w_1(x,t)) = 0 \quad (4.24)$$

As shown in Fig. 4.2b, the equivalent of the sandwich structure with corrugated core is a symmetrical construction. The transmission of bending wave in the structure can be divided into two parts. One is the symmetrical vibration of the construction, in which the two facesheets move in opposite phase, both in or both out at the same time, analogous to the "breathing" motion of animals. Here, the corrugated core has a strong constraint influence on the two facesheets. The other is antisymmetric motion, in which the structure vibrates in step (the two facesheets move in the same phase) with the core carried by the facesheets. In this case, the core exerts no constraint on the movement of the facesheets. In other words, the sandwich structure vibrating in antisymmetric motion can be considered as a single panel in movement.

As for the antisymmetric motion, it can be observed that its dispersion curve (dash-dot line shown in Fig. 4.5a) is a continuous parabola, so sinusoidal vibration waves with any frequency can transmit along the length direction in the sandwich panel, while the periodic corrugated core in between the facesheets has a significant effect on the dispersion relation of the construction for symmetrical motion, whose dispersion curves (calculated with the transfer matrix method proposed by Mead [54]) consist of a series of periodic distributions of "passband" and "stop-band." At the frequency range corresponding to the "passband," for a given frequency, there exist infinite wavenumbers that are relevant to this frequency: the traveling wave identified by the correlative frequency and wavenumber can transmit along

4.4 Physical Interpretation for the Existence of Peaks and Dips on STL Curves 217

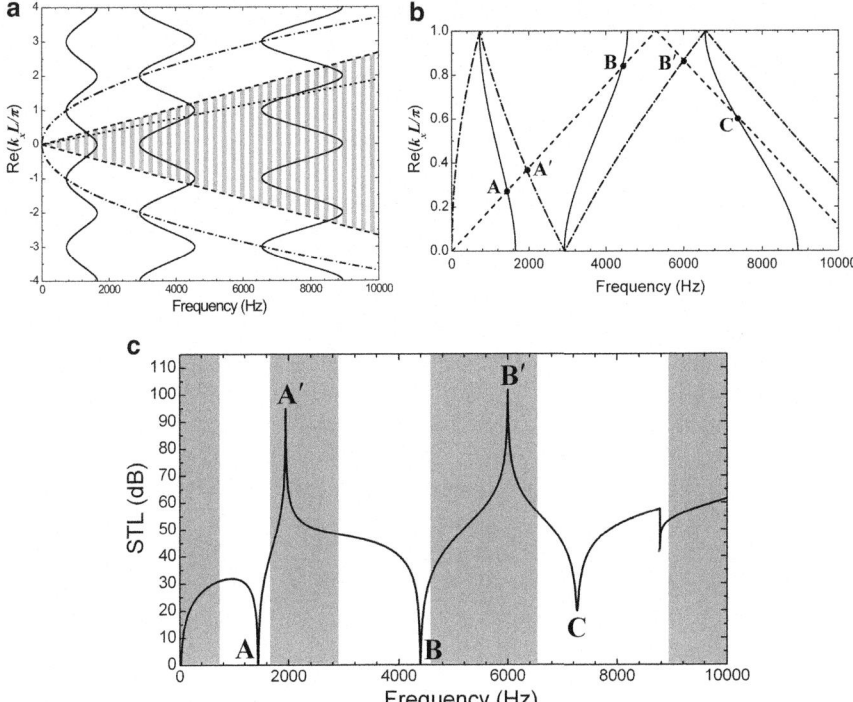

Fig. 4.5 Normalized wavenumber versus frequency. (**a**) *Solid line*, "passband" of symmetrical motion; *dash-dot line*, dispersion curve of antisymmetric motion; *dash line*, dispersion curve corresponding to incident sound at $\theta = \pm 90°$, with the wedge bounded in between the two lines called sound radiation area; *short dash line*, dispersion curve corresponding to incident sound at $\theta = 45°$. (**b**) Real part of normalized wavenumber versus frequency curve (i.e., "folding" version of Fig. 4.5a). *Solid line*, dispersion curve of symmetrical motion; *dash-dot line*, dispersion curve of antisymmetric motion; *dash line*, dispersion curve corresponding to incident sound at $\theta = 90°$; *short dash line*, dispersion curve corresponding to incident sound at $\theta = 45°$. (**c**) Relevant *STL* curve. *Shading area*, "stop-band"; *blank area*, "passband" (With permission from Taylor & Francis Ltd.)

the length direction in the structure without any attenuation. Whereas in the areas of the "stop-band," there is no wavenumber relevant to a given frequency: the bending waves transmitting in the structure with these frequencies are quickly evanescent in exponent form along the length direction in the structure. These "passbands" and "stop-bands" are primarily responsible for the appearances of peaks and dips on the *STL* versus frequency curves shown in Fig. 4.4.

Note that the short dash line in Fig. 4.5a represents the dispersion relation for incident sound with angle $\theta = 45°$ and two dash lines represent the dispersion curve for incidence sound with angle $\theta = \pm 90°$, respectively. Thus, the wedge bounded in between the two lines is the frequency/wavenumber area in which a wave in the sandwich structure can radiate sound.

It can be observed from Fig. 4.5a that there is no intersection between the dispersion curve of antisymmetric motion and the sound radiation area, implying that the antisymmetric motion caused by the incident sound cannot reach enough intensity to radiate sound waves except at the coincidence frequency. For the symmetrical motion of the sandwich, however, there are intersections of the dispersion curve for sound wave with a certain incidence angle, one per passband. At the frequencies corresponding to these intersections, the incident sound can excite structure vibration strong enough to radiate sound, causing the appearance of *STL* dips, as shown in Fig. 4.5b, c. Actually, Fig. 4.5b is the "folding" version of Fig. 4.5a in the range from 0 to 1. As a compact diagram, this fold manipulation is often employed in theoretical studies of sound propagation across periodic structures [19]. The *STL* versus frequency curve of the simplified problem (in which the mass and rotational stiffness of the core as well as air cavity coupling effects are all ignored) is presented in Fig. 4.5c using the same frequency scale as Fig. 4.5a, b. It can be readily seen that the intersections labeled by **A**, **B**, and **C** in Fig. 4.5b correspond with the *STL* dips labeled by the same symbols in Fig. 4.5c.

The intersections labeled by **A′** and **B′** in Fig. 4.5b are closely related to the two steep peaks in Fig. 4.5c. With reference to Fig. 4.5a, the intersections **A′** and **B′** do not really exist, whose appearances are owing to the "folding" manipulation from Fig. 4.5a to b. However, at the frequencies corresponding to these intersections, the transverse deflection of the vibratory facesheet only takes two terms in Eq. (4.10): one for the driven term $n = 0$ and the other in accord with the resonance condition (obtained from Eq. (4.23)) of the driven facesheet [38]:

$$D_1 \left(\frac{k_x + 2N\pi}{L} \right)^4 - m_1 \omega^2 = 0 \qquad (4.25)$$

The transverse deflection of the facesheet is a summation of these two terms, namely,

$$W_1 = I j \omega \rho_0 e^{-j(k_x x - \omega t)} \left[\frac{1 - e^{-j 2N\pi x/L}}{D_1 k_x^4 - m_1 \omega^2} \right] \qquad (4.26)$$

It can be obtained from Eq. (4.26) that the joints which connect the equivalent springs and the driven facesheet are just the nodal points of standing wave in the facesheet. Consequently, the joints have no displacement, and the springs do not transfer any force to the other facesheet. As a result, there is no sound radiation because the radiating facesheet is stationary without any movement. Naturally, at these frequencies relevant to the "fake" interactions in Fig. 4.5b, the sharp peaks on the *STL* curve will occur.

4.5 Optimal Design for Combined Sound Insulation and Structural Load Capacity

It has thus far been demonstrated that core topology plays a significant role in the sound insulation performance of the sandwich structure. Similarly, the structural load capacity of the sandwich is also strongly dependent upon the core geometry [1, 55]. It is therefore of interest to investigate the role of core geometry when the acoustic and structural properties of the sandwich structure are considered simultaneously. To this end, several dimensionless parameters are introduced.

The first dimensionless parameter introduced is the normalized mass of the sandwich (i.e., ratio of the mass of one unit cell to that of the solid filling the whole volume of the unit cell):

$$\overline{m} = \frac{(h_1 + h_2) L + 2lt_0/\sin\alpha}{L(l + h_1 + h_2)} \tag{4.27}$$

As the core for a load-bearing sandwich panel, the in-plane shear modulus is the most important, due to the fact that upon loading the panel deflects as a result of combined bending and shear deformation [1]. The in-plane shear modulus G of the corrugated core can be calculated following Eq. (35) in Ref. [55] and normalized as

$$\tilde{G} = \frac{G}{E} \tag{4.28}$$

Together with the above-defined dimensionless parameters and the sound insulation index *STL*, an integrated index for optimal design toward a lightweight sandwich with superior load-bearing and sound isolation properties may be defined as

$$\gamma_{SGM} = \frac{STL \times \tilde{G}}{\tilde{m}} \tag{4.29}$$

Note that *STL*, $\tilde{G}$, and $\tilde{m}$ are all functions of inclination angle α of the corrugated core.

Figure 4.6 plots γ_{SGM} as a function of frequency for selected core inclination angles (20°, 30°, 45°, 60°, and 70°), with the sound incidence angle fixed at $\theta = 45°$. It is seen from Fig. 4.6 that when the inclination angle α is smaller than 45°, the index γ_{SGM} increases with the increase of α, whereas in the range of $45° < \alpha < 60°$, the index γ_{SGM} maintains the same value over a wide frequency range, except for discrepancies at frequencies higher than 3×10^3 Hz. As the inclination angle is increased beyond 60°, the magnitude of γ_{SGM} decreases. Note that the higher the index γ_{SGM} is, the more superior of the combined acoustic and structural performance of the sandwich will be. Therefore, it can be concluded from Fig. 4.6 that a core inclination angle in the range of 45° to 60° is the preferred selection for the sandwich structure.

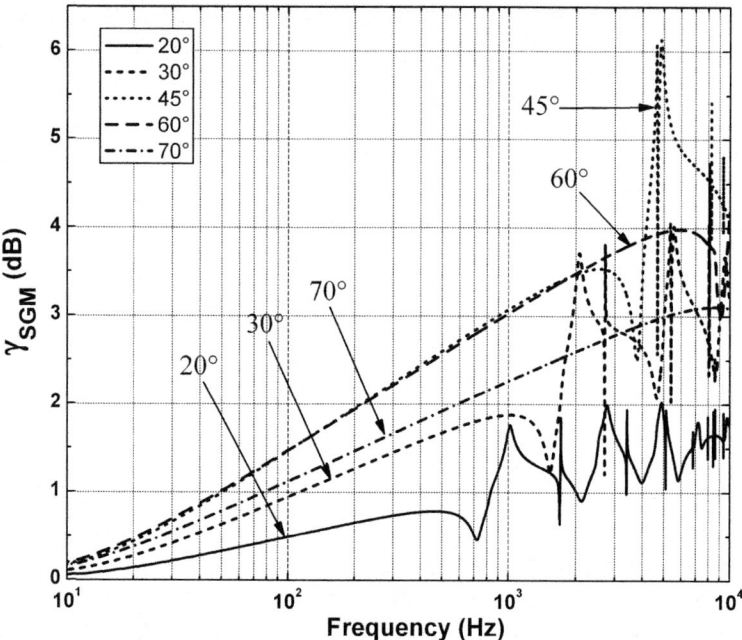

Fig. 4.6 Tendency plot of γ_{SGM} versus frequency for selected core inclination angles (as shown in Fig. 4.2) for fixed sound incidence angle of $\theta = 45°$ (With permission from Taylor & Francis Ltd.)

4.6 Conclusion

The influence of the core topology on the sound insulation capability of a two-dimensional periodic sandwich configuration with corrugated core is theoretically analyzed by the space-harmonic method. In the theoretical model, an equivalent structure containing one translational spring and one rotational spring per unit cell is proposed to simplify the analysis of the vibroacoustic problem. Obtained results demonstrate that the core topology significantly influences the sound insulation performance of the structure, so that one may tailor the core topology for specified acoustic applications. Subsequent analysis of the *STL* and dispersion curves of the sandwich structure leads to fundamental insight into the physical mechanisms behind the appearances of various peaks and dips on the *STL* curves.

As the inclination angle between the facesheet and the core sheet is increased, the peaks and dips on *STL* curves are shifted toward higher frequencies, and the *STL* values increase on the whole while the smooth portion of the *STL* curve in low frequencies is stretched longer.

For a given incident sound angle, the *STL* dips occur at frequencies corresponding to the interactions between the dispersion curve and the symmetrical motion of the structure. The peaks on *STL* occur because the junctions which connect the equivalent springs and the driven facesheet are just the nodal points of the standing or bending wave traveling in the incident facesheet.

As the weight, stiffness, and acoustic property of the sandwich structures all change with the alteration of core configuration and geometry, it is possible to explore the multifunctionality of the structure by optimally designing the core topology.

References

1. Gu S, Lu TJ, Evans AG (2001) On the design of two-dimensional cellular metals for combined heat dissipation and structural load capacity. Int J Heat Mass Transf 44(11):2163–2175
2. Lu TJ, Chen C, Zhu G (2001) Compressive behaviour of corrugated board panels. J Compos Mater 35(23):2098–2126
3. Franco F, Cunefare KA, Ruzzene M (2007) Structural-acoustic optimization of sandwich panels. J Vib Acoust 129(3):330–340
4. Spadoni A, Ruzzene M (2006) Structural and acoustic behavior of chiral truss-core beams. J Vib Acoust 128(5):616–626
5. Wu SF, Wu G, Puskarz MM et al (1997) Noise transmission through a vehicle side window due to turbulent boundary layer excitation. J Vib Acoust 119(4):557–562
6. Xin FX, Lu TJ, Chen CQ (2009) External mean flow influence on noise transmission through double-leaf aeroelastic plates. AIAA J 47(8):1939–1951
7. Carneal JP, Fuller CR (2004) An analytical and experimental investigation of active structural acoustic control of noise transmission through double panel systems. J Sound Vib 272(3–5):749–771
8. Grosveld F (1992) Plate acceleration and sound transmission due to random acoustic and boundary-layer excitation. AIAA J 30(3):601–607
9. Xin FX, Lu TJ, Chen CQ (2009) Dynamic response and acoustic radiation of double-leaf metallic panel partition under sound excitation. Comput Mater Sci 46(3):728–732
10. Maury C, Gardonio P, Elliott SJ (2002) Model for active control of flow-induced noise transmitted through double partitions. AIAA J 40(6):1113–1121
11. Thamburaj P, Sun JQ (2001) Effect of material and geometry on the sound and vibration transmission across a sandwich beam. J Vib Acoust 123(2):205–212
12. Lyle K, Mixson J (1987) Laboratory study of sidewall noise transmission and treatment for a light aircraft fuselage. J Aircr 24(9):660–665
13. Xin FX, Lu TJ, Chen CQ (2009) Sound transmission across lightweight all-metallic sandwich panels with corrugated cores. Chin J Acoust 28(3):231–243
14. Xin FX, Lu TJ, Chen CQ (2010) Sound transmission through simply supported finite double-panel partitions with enclosed air cavity. J Vib Acoust 132(1):011008: 011001–011011
15. Xin FX, Lu TJ (2010) Analytical modeling of fluid loaded orthogonally rib-stiffened sandwich structures: sound transmission. J Mech Phys Solids 58(9):1374–1396
16. Xin FX, Lu TJ (2010) Analytical modeling of sound transmission across finite aeroelastic panels in convected fluids. J Acoust Soc Am 128(3):1097–1107
17. Lu TJ, Valdevit L, Evans AG (2005) Active cooling by metallic sandwich structures with periodic cores. Prog Mater Sci 50(7):789–815
18. Liu T, Deng ZC, Lu TJ (2007) Bi-functional optimization of actively cooled, pressurized hollow sandwich cylinders with prismatic cores. J Mech Phys Solids 55(12):2565–2602
19. Xin FX, Lu TJ, Chen C (2008) Sound transmission through lightweight all-metallic sandwich panels with corrugated cores. Multi-Funct Mater Struct Parts 1 and 2 47–50:57–60
20. Mead DJ (1970) Free wave propagation in periodically supported, infinite beams. J Sound Vib 11(2):181–197

21. Mead DJ, Pujara KK (1971) Space-harmonic analysis of periodically supported beams: response to convected random loading. J Sound Vib 14(4):525–532
22. Lomas NS, Hayek SI (1977) Vibration and acoustic radiation of elastically supported rectangular plates. J Sound Vib 52(1):1–25
23. Renji K (2005) Sound transmission loss of unbounded panels in bending vibration considering transverse shear deformation. J Sound Vib 283(1–2):478–486
24. Sewell EC (1970) Transmission of reverberant sound through a single-leaf partition surrounded by an infinite rigid baffle. J Sound Vib 12(1):21–32
25. Cummings A, Mulholland KA (1968) The transmission loss of finite sized double panels in a random incidence sound field. J Sound Vib 8(1):126–133
26. Mulholland KA, Parbrook HD, Cummings A (1967) The transmission loss of double panels. J Sound Vib 6(3):324–334
27. Beranek LL, Work GA (1949) Sound transmission through multiple structures containing flexible blankets. J Acoust Soc Am 21(4):419–428
28. London A (1950) Transmission of reverberant sound through double walls. J Acoust Soc Am 22(2):270–279
29. Sewell EC (1970) Two-dimensional solution for transmission of reverberant sound through a double partition. J Sound Vib 12(1):33–57
30. White PH, Powell A (1966) Transmission of random sound and vibration through a rectangular double wall. J Acoust Soc Am 40(4):821–832
31. Hongisto V (2000) Sound insulation of doors-part 1: prediction models for structural and leak transmission. J Sound Vib 230(1):133–148
32. Chazot JD, Guyader JL (2007) Prediction of transmission loss of double panels with a patch-mobility method. J Acoust Soc Am 121(1):267–278
33. Xin FX, Lu TJ, Chen CQ (2008) Vibroacoustic behavior of clamp mounted double-panel partition with enclosure air cavity. J Acoust Soc Am 124(6):3604–3612
34. Xin FX, Lu TJ (2009) Analytical and experimental investigation on transmission loss of clamped double panels: Implication of boundary effects. J Acoust Soc Am 125(3):1506–1517
35. Lee JH, Kim J (2002) Analysis of sound transmission through periodically stiffened panels by space-harmonic expansion method. J Sound Vib 251(2):349–366
36. Lin G-F, Garrelick JM (1977) Sound transmission through periodically framed parallel plates. J Acoust Soc Am 61(4):1014–1018
37. Takahashi D (1983) Sound radiation from periodically connected double-plate structures. J Sound Vib 90(4):541–557
38. Wang J, Lu TJ, Woodhouse J et al (2005) Sound transmission through lightweight double-leaf partitions: theoretical modelling. J Sound Vib 286(4–5):817–847
39. El-Raheb M, Wagner P (2002) Effects of end cap and aspect ratio on transmission of sound across a truss-like periodic double panel. J Sound Vib 250(2):299–322
40. Lauriks W, Mees P, Allard JF (1992) The acoustic transmission through layered systems. J Sound Vib 155(1):125–132
41. Trochidis A, Kalaroutis A (1986) Sound transmission through double partitions with cavity absorption. J Sound Vib 107(2):321–327
42. Narayanan S, Shanbhag RL (1982) Sound transmission through a damped sandwich panel. J Sound Vib 80(3):315–327
43. Nilsson AC (1990) Wave propagation in and sound transmission through sandwich plates. J Sound Vib 138(1):73–94
44. Ford RD, Lord P, Walker AW (1967) Sound transmission through sandwich constructions. J Sound Vib 5(1):9–21
45. Nilsson AC (1977) Some acoustical properties of floating-floor constructions. J Acoust Soc Am 61(6):1533–1539
46. Narayanan S, Shanbhag RL (1981) Sound transmission through elastically supported sandwich panels into a rectangular enclosure. J Sound Vib 77(2):251–270
47. Huang WC, Ng CF (1998) Sound insulation improvement using honeycomb sandwich panels. Appl Acoust 53(1–3):163–177

48. Foin O, Nicolas J, Atalla N (1999) An efficient tool for predicting the structural acoustic and vibration response of sandwich plates in light or heavy fluid. Appl Acoust 57:213–242
49. Wang TA, Sokolinsky VS, Rajaram S et al (2005) Assessment of sandwich models for the prediction of sound transmission loss in unidirectional sandwich panels. Appl Acoust 66(3):245–262
50. Ng CF, Hui CK (2008) Low frequency sound insulation using stiffness control with honeycomb panels. Appl Acoust 69(4):293–301
51. Cordonnier-Cloarec P, Pauzin S, Biron D et al (1992) Contribution to the study of sound transmission and radiation of corrugated steel structures. J Sound Vib 157(3):515–530
52. Ng CF, Zheng H (1998) Sound transmission through double-leaf corrugated panel constructions. Appl Acoust 53(1–3):15–34
53. Mathur GP (1992) Sound transmission through stiffened double-panel structures lined with elastic porous materials. In: Proceedings of the 14th DGLR/AIAA aero-acoustics conference, Aachen, Germany, pp 102–105
54. Kurtze G, Watters BG (1959) New wall design for high transmission loss or high damping. J Acoust Soc Am 31(6):739–748
55. Liu T, Deng Z, Lu T (2007) Structural modeling of sandwich structures with lightweight cellular cores. Acta Mech Sin 23(5):545–559

Chapter 5
Sound Radiation, Transmission of Orthogonally Rib-Stiffened Sandwich Structures

Abstract This chapter is organized as two parts: in the first part, wave propagation in an infinite sandwich structure with two periodic sets of orthogonal rib-stiffener core is theoretically formulated when subjected to a harmonic point force excitation. The motions of the equally spaced rib-stiffeners are exactly handled by considering their tensional, bending, and torsional vibration together. As a consequence, the governing equations of the panel vibration contain the terms of the tensional forces, bending moments, and torsional moments from the corresponding motion of the rib-stiffeners. Furthermore, the inertial effects arising from the mass of the rib-stiffeners are also taken account of by introducing the inertial terms of the tensional forces, bending moments, and torsional moments. The response of the structure in wavenumber space is then determined by employing the approach of the Fourier transform and the periodic nature of the structure, which is numerically solved by truncating two infinite sets of simultaneous equations insofar as the solution converges. In terms of the response of the panel, far-field radiated sound pressure is examined with reference to that of unstiffened panel in decibel scale (dB). A number of physical interpretations of significant features are proposed, specifically including the influences of the inertial effects, the excitation position, and the periodicity spacings of the rib-stiffeners.

In the second part, an analytic model is developed to investigate the wave propagation and sound transmission characteristics of an infinite sandwich structure reinforced by two sets of orthogonal rib-stiffeners when subjected to convective fluid-loaded pressure. The rib-stiffeners are assumed to be identical and uniformly spaced, which can exert not only tensional forces and bending moments but also torsional moments on the facesheets. Inertial terms of the tensional forces, bending moments, and torsional moments are introduced to account for inertial effects arising from the mass of the rib-stiffeners. With the surrounding acoustic fluids restricted by the acoustic wave equation, fluid-structure coupling is considered by imposing velocity continuity condition at fluid-panel interfaces. By applying the Bloch theorem for periodic structures, the structural and acoustic responses are expressed in a superposition form of space harmonics for a given wavenumber.

The application of the virtual work principle for one periodic element yields two infinite sets of simultaneous algebraic coupled equations, which are numerically solved by truncating them in a finite range insofar as the solution converges. The validity and feasibility of the analytic model is qualified by comparing model predictions with existing results, in which the necessity and advantage of the exact modeling of rib-stiffener motions are also demonstrated. Specifically, the influences of inertial effects arising from the rib-stiffener mass, the periodicity spacing of rib-stiffeners, and the airborne as well as structure-borne paths on the transmission of sound across the sandwich structure are quantified, and conclusions of significant practical implications are drawn.

5.1 Sound Radiation of Sandwich Structures

5.1.1 Introduction

Wave propagation and sound radiation behaviors of periodically rib-stiffened structures are of significant interest due to their increasing applications in civil and transport engineering, e.g., as the cabin skin of aircrafts, marine ships, and express trains [1–20]. At low frequencies, a rib-stiffened structure can be approximated as an orthotropic panel when the panel flexural wave has a wavelength much greater than stiffener spacing [3, 7]. However, at high frequencies when the wavelength is comparable with stiffener spacing, the spatial periodicity of the structure should be carefully taken into account in any theoretical modeling.

There exist a multitude of analytic studies on the vibroacoustic behavior of periodically rib-stiffened structures, including beams and plates. For example, the response of a periodically supported beam subjected to spatially and temporally harmonic pressure was solved by Mead and Pujara [21] using a particular series of space harmonics. The space-harmonic method evolving from the considerations of progressive wave propagation is superior to the classical normal mode approach, since only as few as seven terms can ensure accurate convergence of the solution. Subsequently, with the emphasis placed on wave propagation characteristics, Mead and Yaman [22] developed an exact model for the harmonic response of a uniform finite beam on multiple supports. From the viewpoint of free wave propagation, Mead [23] investigated theoretically an infinite beam on regularly spaced identical supports in terms of superposed sinusoidal waves. For more details regarding wave propagation in continuous periodic structures, one may consult the review [24].

As for periodically rib-stiffened plates, a few typical works can be referred to. For instance, multimode wave propagation in a one-dimensionally (1D) stiffened plate was theoretically and experimentally investigated by Ichchou et al. [1, 2], who also examined its energy propagation features in k-space and the corresponding dispersion relations. Using the principle of superposition, Rumerman [25] proposed a general solution for the forced vibration of an infinite thin plate, periodically stiffened by identical, uniform rib-stiffeners. An approximate method was employed by Mead and Mallik [26] to estimate the sound power radiated by an infinite plate, supported elastically along parallel, equi-spaced lines, and subjected to a simple pressure field convecting uniformly over the plate. While Mead and Parthan [27] studied the propagation of flexural waves in a plate resting on an orthogonal array of equi-spaced simple line supports, Mace [28] presented a solution for the radiation of sound from a point-excited infinite fluid-loaded plate reinforced by two sets of parallel stiffeners. Several aspects related to the vibration of and sound radiation from periodically line-stiffened and fluid-loaded plates were further examined by Mace [3, 4]. The far- and near-field acoustic radiation of an infinite periodically rib-stiffened plate was obtained theoretically by Cray [29], although only the tensional forces of the rib-stiffeners were accounted for. Wang et al. [30] proposed a theoretical model of sound transmission across double-leaf partitions having periodic parallel rib-stiffeners using the space-harmonic approach: except for the torsional moments, both the tensional forces and bending moments of the rib-stiffeners were accounted for.

A few investigations also focused on the pass-/stop-band characteristics of wave propagation in periodically rib-stiffened plates. For instance, the transmission of energy in 1D periodically ribbed membrane was theoretically studied by Crighton [31] when the structure was immersed in static compressible fluid and excited by a time-harmonic line force. Later, addressing essentially the same problem, Spivack [32] gave an exact solution for general finite configurations and found that the passband response becomes increasingly sensitive to frequency as the length of rib array increases. A further investigation on the band structure of energy transmission in periodically ribbed elastic structures under fluid loading was carried out by Cooper and Crighton [33, 34] using Green's function method, from the viewpoint of spatial periodicity in the passbands and that of algebraic decay in the stop-bands, respectively.

Although the vibroacoustic behaviors of periodically rib-stiffened structures have been studied by many researchers, commonly only with the tensional forces of the rib-stiffeners considered, the influence of their bending and torsional moments as well as inertial effects remains unclear. Moreover, previous researches focused mainly on relatively simple structures, e.g., infinite periodically supported beams and 1D rib-stiffened plates. Only a noticeably few [5, 6] considered the more general two-dimensional (2D) rib-stiffened structures that are of practical importance in aeronautical and marine applications. For example, Mace examined the radiation of sound from an infinite fluid-loaded plate reinforced with two sets of orthogonal line stiffeners [5] and the vibration of a thin plate lying on point supports that form an orthogonal 2D periodic array [6]. However, in the analysis [5], it was assumed that the stiffeners only exert forces on the plate.

To address the aforementioned deficiencies, we aim to study analytically the vibration and acoustic radiation of a generic 2D periodic structure that is consisted of two infinitely large parallel plates reinforced by orthogonally extended rib-stiffeners (i.e., a sandwich panel with orthogonal rib-stiffener arrays as its core). To accurately model the motion of each rib-stiffener, its tensional, bending, and torsional vibrations are all considered. The inertial effects arising from the mass of the rib-stiffeners are also taken into account by introducing the inertial terms of their tensional forces, bending moments, and torsional moments into the governing equations of panel vibration. Fourier transform is employed to solve the resulting governing equations, leading to two sets of infinite algebraic equations, which are truncated to solve insofar as the solution converges. In terms of obtained panel responses, the radiated sound pressure at far field is numerically calculated to gain physical insight on wave propagation and sound radiation of the sandwich structure. Good agreements between model predictions with previous published results [5] validate the present analytic model and confirm the necessity of including the inertial effects and torsional moments of the rib-stiffeners in any theoretical modeling especially at high frequencies. The influences of inertial effects, excitation position, and spatial periodicity of rib-stiffeners on the vibroacoustic behavior of the sandwich are quantified with the underlying physical mechanisms explored.

Although this chapter focuses to solve and discuss a relatively specific problem, the theoretical model proposed can be readily employed to solve the similar problems of periodic structures. In particular, the theoretical model based on Fourier transform technique could be referential to mend finite element method so as to solve more generalized problems efficiently. For this point, H. Kohno et al. [35] have done an excellent job by combining the advantages of finite element method and spectral method to solve problems of wave propagation. This work should be very useful for enlightening us to extend our theoretical work for more generalized problems, for example, by combining the advantages of our theoretical model and finite element method.

5.1.2 Theoretical Modeling of Structural Dynamic Responses

Consider an infinitely large 2D sandwich structure shown schematically in Fig. 5.1, which has a lattice core in the form of orthogonal stiffeners having periodic uniform spacings in the x- and y-directions, l_x and l_y, respectively. Its geometrical dimensions are depth of orthogonal rib-stiffener core d, thickness of upper and bottom panels h_1 and h_2, and thickness of x- and y-wise stiffeners t_x and t_y. The mass densities of the x- and y-wise stiffeners are m_x and m_y, respectively. A right-handed Cartesian coordinate system (x, y, z) is established, with its x- and y-axes located on the surface of the upper panel and the positive direction of the z-axis pointing downward (Fig. 5.1).

Let a harmonic point force with amplitude q_0 be applied on the surface of the upper panel at an arbitrary location (x_0, y_0). As a result, a radially outspreading bending wave propagates from the source (x_0, y_0). The propagation of this bending

5.1 Sound Radiation of Sandwich Structures

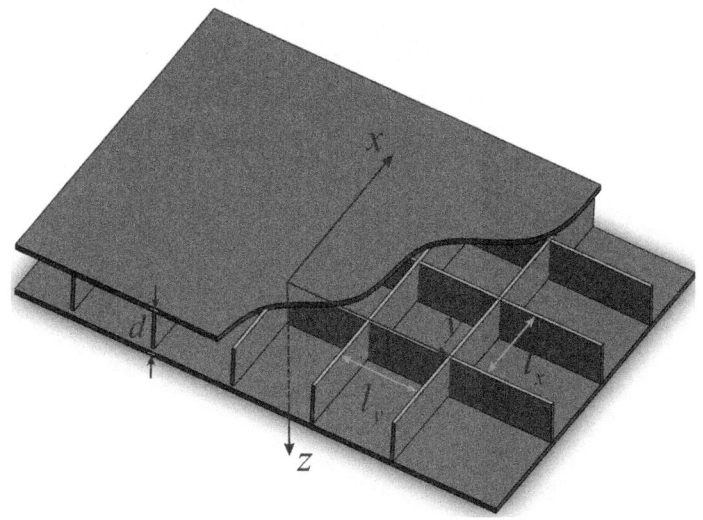

Fig. 5.1 Sandwich panel with orthogonally rib-stiffened core (With permission from Elsevier)

wave in the upper panel is affected by the attached lattice core (rib-stiffeners), which transmits the motion to the bottom panel. Both panels are modeled as a classical thin plate, following the Kirchhoff thin plate theory. As the focus is placed on the intrinsic characteristics of bending wave propagation in the structure, air-structure coupling is ignored. The theoretical formulation presented below proposes a comprehensive analytic model for bending wave propagation in the sandwich structure, accounting for not only the tensional forces, bending moments, and torsional moments of the orthogonal rib-stiffeners but also their inertial effects.

Upon point force excitation, the vibration of the upper and bottom panels can be described using two dynamic governing equations, where the influence of the rib-stiffeners exists in the form of tensional forces (general force plus inertial force), bending moments (general bending moment plus inertial bending moment), and torsional moments (general torsional moment plus inertial torsional moment). With the inertial effects of the rib-stiffeners accounted for, the resultant tensional forces, bending moments, and torsional moments acting on the upper and bottom panels per rib-stiffener are not equal, denoted here by (Q^+, M^+, M_T^+) and (Q^-, M^-, M_T^-), respectively. Figure 5.2 shows the convention employed for denoting the tensional forces as well as the bending and torsional moments between the upper panel and the x- and y-wise stiffeners. The same applies at the interface between the bottom panel and the x- and y-wise stiffeners.

Since the excitation is harmonic, the dynamic responses of the two panels should also be harmonic. For simplicity, the harmonic time term $e^{-i\omega t}$ is suppressed from the formulation below. The dynamic governing equations are thence given by

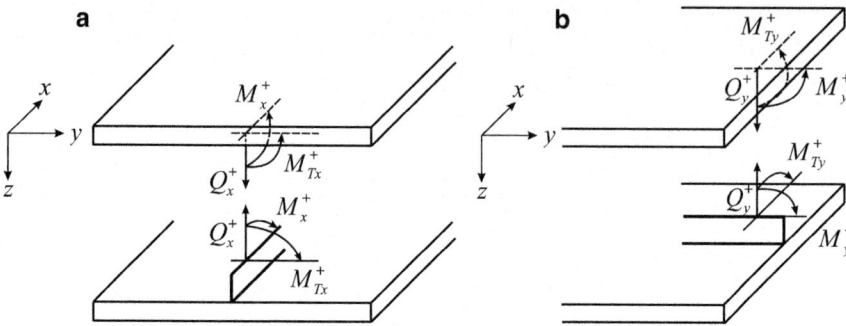

Fig. 5.2 Convention for tensional forces, bending moments, and torsional moments between upper panel and (**a**) x-wise and (**b**) y-wise stiffeners (With permission from Elsevier)

$$D_1 \nabla^4 w_1 + m_1 \frac{\partial^2 w_1}{\partial t^2} = \sum_m \left[Q_y^+ \delta(x - ml_x) + \frac{\partial}{\partial y} \{M_y^+ \delta(x - ml_x)\} \right.$$
$$\left. + \frac{\partial}{\partial x} \{M_{Ty}^+ \delta(x - ml_x)\} \right]$$
$$+ \sum_n \left[Q_x^+ \delta(y - nl_y) + \frac{\partial}{\partial x} \{M_x^+ \delta(y - nl_y)\} \right.$$
$$\left. + \frac{\partial}{\partial y} \{M_{Tx}^+ \delta(y - nl_y)\} \right]$$
$$+ q_0 \delta(x - x_0) \delta(y - y_0) \tag{5.1}$$

$$D_2 \nabla^4 w_2 + m_2 \frac{\partial^2 w_2}{\partial t^2} = -\sum_m \left[Q_y^- \delta(x - ml_x) + \frac{\partial}{\partial y} \{M_y^- \delta(x - ml_x)\} \right.$$
$$\left. + \frac{\partial}{\partial x} \{M_{Ty}^- \delta(x - ml_x)\} \right]$$
$$- \sum_n \left[Q_x^- \delta(y - nl_y) + \frac{\partial}{\partial x} \{M_x^- \delta(y - nl_y)\} \right.$$
$$\left. + \frac{\partial}{\partial y} \{M_{Tx}^- \delta(y - nl_y)\} \right] \tag{5.2}$$

where $\nabla^4 \equiv (\partial^2/\partial x^2 + \partial^2/\partial y^2)^2$; $\delta(\cdot)$ is the Dirac delta function; and (w_1, w_2), (m_1, m_2), and (D_1, D_2) are the displacement, surface mass density, and flexural rigidity of the upper and bottom panels, respectively. The material loss factor η_j ($j = 1, 2$ for upper and bottom panels, respectively) is introduced with complex Young's modulus as

5.1 Sound Radiation of Sandwich Structures

$$D_j = \frac{E_j h_j^3 (1 + i\eta_j)}{12 (1 - v_j^2)} \quad (j = 1, 2) \tag{5.3}$$

As the factual forces and moments exerting on the upper and bottom panels are not the same due to the consideration of inertial forces and moments, the terms associated with the two panels are denoted separately by superscripts + (upper) and − (bottom). Subscripts x and y are introduced to represent those terms arising from the x- and y-wise stiffeners, respectively.

Taking into account the inertial effects (due to stiffener mass) and applying both Hooke's law and Newton's second law, one obtains the tensional forces arising from the rib-stiffeners as

$$Q_x^+ = -\frac{K_x (K_x - m_x \omega^2)}{2K_x - m_x \omega^2} w_1 + \frac{K_x^2}{2K_x - m_x \omega^2} w_2 \tag{5.4}$$

$$Q_x^- = -\frac{K_x^2}{2K_x - m_x \omega^2} w_1 + \frac{K_x (K_x - m_x \omega^2)}{2K_x - m_x \omega^2} w_2 \tag{5.5}$$

$$Q_y^+ = -\frac{K_y (K_y - m_y \omega^2)}{2K_y - m_y \omega^2} w_1 + \frac{K_y^2}{2K_y - m_y \omega^2} w_2 \tag{5.6}$$

$$Q_y^- = -\frac{K_y^2}{2K_y - m_y \omega^2} w_1 + \frac{K_y (K_y - m_y \omega^2)}{2K_y - m_y \omega^2} w_2 \tag{5.7}$$

where ω is the circle frequency and (K_x, K_y) are the tensional stiffness of half the rib-stiffeners per unit length.

Similarly, the bending moments of the rib-stiffeners can be expressed as

$$M_x^+ = \frac{E_x I_x^* (E_x I_x^* - \rho_x I_x \omega^2)}{2E_x I_x^* - \rho_x I_x \omega^2} \frac{\partial^2 w_1}{\partial x^2} - \frac{E_x^2 I_x^{*2}}{2E_x I_x^* - \rho_x I_x \omega^2} \frac{\partial^2 w_2}{\partial x^2} \tag{5.8}$$

$$M_x^- = \frac{E_x^2 I_x^{*2}}{2E_x I_x^* - \rho_x I_x \omega^2} \frac{\partial^2 w_1}{\partial x^2} - \frac{E_x I_x^* (E_x I_x^* - \rho_x I_x \omega^2)}{2E_x I_x^* - \rho_x I_x \omega^2} \frac{\partial^2 w_2}{\partial x^2} \tag{5.9}$$

$$M_y^+ = \frac{E_y I_y^* (E_y I_y^* - \rho_y I_y \omega^2)}{2E_y I_y^* - \rho_y I_y \omega^2} \frac{\partial^2 w_1}{\partial y^2} - \frac{E_y^2 I_y^{*2}}{2E_y I_y^* - \rho_y I_y \omega^2} \frac{\partial^2 w_2}{\partial y^2} \tag{5.10}$$

$$M_y^- = \frac{E_y^2 I_y^{*2}}{2E_y I_y^* - \rho_y I_y \omega^2} \frac{\partial^2 w_1}{\partial y^2} - \frac{E_y I_y^* (E_y I_y^* - \rho_y I_y \omega^2)}{2E_y I_y^* - \rho_y I_y \omega^2} \frac{\partial^2 w_2}{\partial y^2} \tag{5.11}$$

where ($E_x I_x^*$, $E_y I_y^*$) are the bending stiffness of half the rib-stiffeners and (ρ_x, ρ_y) and (I_x, I_y) are the mass density and polar moment of inertia for the rib-stiffeners, with subscripts x and y indicating the direction of the stiffener.

Following similar procedures for deriving the tensional forces and bending moments, one obtains the torsional moments of the rib-stiffeners as

$$M_{Tx}^{+} = \frac{G_x J_x^* \left(G_x J_x^* - \rho_x J_x \omega^2 \right)}{2 G_x J_x^* - \rho_x J_x \omega^2} \frac{\partial^2 w_1}{\partial x \partial y} - \frac{G_x^2 J_x^{*2}}{2 G_x J_x^* - \rho_x J_x \omega^2} \frac{\partial^2 w_2}{\partial x \partial y} \quad (5.12)$$

$$M_{Tx}^{-} = \frac{G_x^2 J_x^{*2}}{2 G_x J_x^* - \rho_x J_x \omega^2} \frac{\partial^2 w_1}{\partial x \partial y} - \frac{G_x J_x^* \left(G_x J_x^* - \rho_x J_x \omega^2 \right)}{2 G_x J_x^* - \rho_x J_x \omega^2} \frac{\partial^2 w_2}{\partial x \partial y} \quad (5.13)$$

$$M_{Ty}^{+} = \frac{G_y J_y^* \left(G_y J_y^* - \rho_y J_y \omega^2 \right)}{2 G_y J_y^* - \rho_y J_y \omega^2} \frac{\partial^2 w_1}{\partial y \partial x} - \frac{G_y^2 J_y^{*2}}{2 G_y J_y^* - \rho_y J_y \omega^2} \frac{\partial^2 w_2}{\partial y \partial x} \quad (5.14)$$

$$M_{Ty}^{-} = \frac{G_y^2 J_y^{*2}}{2 G_y J_y^* - \rho_y J_y \omega^2} \frac{\partial^2 w_1}{\partial y \partial x} - \frac{G_y J_y^* \left(G_y J_y^* - \rho_y J_y \omega^2 \right)}{2 G_y J_y^* - \rho_y J_y \omega^2} \frac{\partial^2 w_2}{\partial y \partial x} \quad (5.15)$$

where $(G_x J_x^*, G_y J_y^*)$ are the torsional stiffness of half the rib-stiffeners and (J_x, J_y) are the torsional moment of inertia for the rib-stiffeners.

In the above expressions for the tensional forces, bending moments, and torsional moments of a rib-stiffener, the geometrical properties of its cross section are given by

$$K_x = \frac{E_x t_x}{d/2}, \quad K_y = \frac{E_y t_y}{d/2} \quad (5.16)$$

$$I_x^* = \frac{t_x (d/2)^3}{12}, \quad I_y^* = \frac{t_y (d/2)^3}{12}, \quad I_x = \frac{t_x d^3}{12}, \quad I_y = \frac{t_y d^3}{12} \quad (5.17)$$

$$J_x^* = \frac{t_x^3 d}{2} \left[\frac{1}{3} - \frac{64}{\pi^5} \frac{2 t_x}{d} \sum_{n=1,3,5,\ldots}^{\infty} \frac{\tanh(n \pi d / 4 t_x)}{n^5} \right] \quad (5.18)$$

$$J_y^* = \frac{t_y^3 d}{2} \left[\frac{1}{3} - \frac{64}{\pi^5} \frac{2 t_y}{d} \sum_{n=1,3,5,\ldots}^{\infty} \frac{\tanh(n \pi d / 4 t_y)}{n^5} \right] \quad (5.19)$$

$$J_x = t_x^3 d \left[\frac{1}{3} - \frac{64}{\pi^5} \frac{t_x}{d} \sum_{n=1,3,5,\ldots}^{\infty} \frac{\tanh(n \pi d / 2 t_x)}{n^5} \right] \quad (5.20)$$

$$J_y = t_y^3 d \left[\frac{1}{3} - \frac{64}{\pi^5} \frac{t_y}{d} \sum_{n=1,3,5,\ldots}^{\infty} \frac{\tanh(n \pi d / 2 t_y)}{n^5} \right] \quad (5.21)$$

where E_x and E_y are separately Young's modulus of the x- and y-wise stiffener materials.

5.1 Sound Radiation of Sandwich Structures

To simplify Eqs. (5.4), (5.5), (5.6), (5.7), (5.8), (5.9), (5.10), (5.11), (5.12), (5.13), (5.14), and (5.15), the following set of specified characteristics is introduced to replace the coefficients of general displacements:

1. Replacement of tensional force coefficients:

$$R_{Q1} = \frac{K_x(K_x - m_x\omega^2)}{2K_x - m_x\omega^2}, \quad R_{Q2} = \frac{K_x^2}{2K_x - m_x\omega^2} \quad (5.22)$$

$$R_{Q3} = \frac{K_y(K_y - m_y\omega^2)}{2K_y - m_y\omega^2}, \quad R_{Q4} = \frac{K_y^2}{2K_y - m_y\omega^2} \quad (5.23)$$

2. Replacement of bending moment coefficients:

$$R_{M1} = \frac{E_x I_x^*(E_x I_x^* - \rho_x I_x\omega^2)}{2E_x I_x^* - \rho_x I_x\omega^2}, \quad R_{M2} = \frac{E_x^2 I_x^{*2}}{2E_x I_x^* - \rho_x I_x\omega^2} \quad (5.24)$$

$$R_{M3} = \frac{E_y I_y^*(E_y I_y^* - \rho_y I_y\omega^2)}{2E_y I_y^* - \rho_y I_y\omega^2}, \quad R_{M4} = \frac{E_y^2 I_y^{*2}}{2E_y I_y^* - \rho_y I_y\omega^2} \quad (5.25)$$

3. Replacement of torsional moment coefficients:

$$R_{T1} = \frac{G_x J_x^*(G_x J_x^* - \rho_x J_x\omega^2)}{2G_x J_x^* - \rho_x J_x\omega^2}, \quad R_{T2} = \frac{G_x^2 J_x^{*2}}{2G_x J_x^* - \rho_x J_x\omega^2} \quad (5.26)$$

$$R_{T3} = \frac{G_y J_y^*(G_y J_y^* - \rho_y J_y\omega^2)}{2G_y J_y^* - \rho_y J_y\omega^2}, \quad R_{T4} = \frac{G_y^2 J_y^{*2}}{2G_y J_y^* - \rho_y J_y\omega^2} \quad (5.27)$$

Using Eqs. (5.22), (5.23), (5.24), (5.25), (5.26), and (5.27), one can simplify the expressions of the tensional forces, bending moments, and torsional moments as:

1. Tensional forces:

$$Q_x^+ = -R_{Q1}w_1 + R_{Q2}w_2, \quad Q_x^- = -R_{Q2}w_1 + R_{Q1}w_2 \quad (5.28)$$

$$Q_y^+ = -R_{Q3}w_1 + R_{Q4}w_2, \quad Q_y^- = -R_{Q4}w_1 + R_{Q3}w_2 \quad (5.29)$$

2. Bending moments:

$$M_x^+ = R_{M1}\frac{\partial^2 w_1}{\partial x^2} - R_{M2}\frac{\partial^2 w_2}{\partial x^2}, \quad M_x^- = R_{M2}\frac{\partial^2 w_1}{\partial x^2} - R_{M1}\frac{\partial^2 w_2}{\partial x^2} \quad (5.30)$$

$$M_y^+ = R_{M3}\frac{\partial^2 w_1}{\partial y^2} - R_{M4}\frac{\partial^2 w_2}{\partial y^2}, \quad M_y^- = R_{M4}\frac{\partial^2 w_1}{\partial y^2} - R_{M3}\frac{\partial^2 w_2}{\partial y^2} \quad (5.31)$$

3. Torsional moments:

$$M_{Tx}^+ = R_{T1}\frac{\partial^2 w_1}{\partial x \partial y} - R_{T2}\frac{\partial^2 w_2}{\partial x \partial y}, \quad M_{Tx}^- = R_{T2}\frac{\partial^2 w_1}{\partial x \partial y} - R_{T1}\frac{\partial^2 w_2}{\partial x \partial y} \quad (5.32)$$

$$M_{Ty}^+ = R_{T3}\frac{\partial^2 w_1}{\partial y \partial x} - R_{T4}\frac{\partial^2 w_2}{\partial y \partial x}, \quad M_{Ty}^- = R_{T4}\frac{\partial^2 w_1}{\partial y \partial x} - R_{T3}\frac{\partial^2 w_2}{\partial y \partial x} \quad (5.33)$$

5.1.3 Solutions

Fully considering the periodic nature of the present sandwich structure and applying the Poisson summation formula (Mace [5]; Rumerman [25]), one can write the wave components in the periodic structure using space-harmonic series as

$$\sum_m \delta(x - ml_x) = \frac{1}{l_x}\sum_m e^{-i(2m\pi/l_x)x} \quad (5.34)$$

$$\sum_n \delta(y - nl_y) = \frac{1}{l_y}\sum_n e^{-i(2n\pi/l_y)y} \quad (5.35)$$

The displacement of each panel is a function of coordinates (x, y) as well as the Fourier transform of its wavenumber frequency, the latter being also a function of the wavenumbers (k_x, k_y). The Fourier transform pair relating these two quantities with respect to (x, y) and (k_x, k_y) can be written as

$$w(x, y) = \int_{-\infty}^{+\infty}\int_{-\infty}^{+\infty} \tilde{w}(k_x, k_y) e^{i(k_x x + k_y y)} dk_x dk_y \quad (5.36)$$

$$\tilde{w}(k_x, k_y) = \left(\frac{1}{2\pi}\right)^2 \int_{-\infty}^{+\infty}\int_{-\infty}^{+\infty} w(x, y) e^{-i(k_x x + k_y y)} dx dy \quad (5.37)$$

Employing Eqs. (5.34) and (5.35) and then taking the Fourier transform and replacing the wavenumbers (k_x, k_y) by (α, β), respectively, one can rewrite governing Eqs. (5.1) and (5.2) as

$$\tilde{w}_1(\alpha, \beta) = \frac{1}{D_1 l_x f_1(\alpha, \beta)}\sum_m \left[\tilde{Q}_y^+(\alpha_m, \beta) + i\beta \tilde{M}_y^+(\alpha_m, \beta) + i\alpha \tilde{M}_{Ty}^+(\alpha_m, \beta)\right]$$

$$+ \frac{1}{D_1 l_y f_1(\alpha, \beta)}\sum_n \left[\tilde{Q}_x^+(\alpha, \beta_n) + i\alpha \tilde{M}_x^+(\alpha, \beta_n) + i\beta \tilde{M}_{Tx}^+(\alpha, \beta_n)\right]$$

$$+ \frac{q_0 e^{-i(\alpha x_0 + \beta y_0)}}{(2\pi)^2 D_1 f_1(\alpha, \beta)} \quad (5.38)$$

5.1 Sound Radiation of Sandwich Structures

$$\tilde{w}_2(\alpha,\beta) = \frac{-1}{D_2 l_x f_2(\alpha,\beta)} \sum_m \left[\tilde{Q}_y^-(\alpha_m,\beta) + i\beta \tilde{M}_y^-(\alpha_m,\beta) + i\alpha \tilde{M}_{Ty}^-(\alpha_m,\beta) \right]$$

$$- \frac{1}{D_2 l_y f_2(\alpha,\beta)} \sum_n \left[\tilde{Q}_x^-(\alpha,\beta_n) + i\alpha \tilde{M}_x^-(\alpha,\beta_n) + i\beta \tilde{M}_{Tx}^-(\alpha,\beta_n) \right]$$

(5.39)

where the dependence of a term on wavenumbers (α, β) is indicated using the hat sign $\sim$, meaning the corresponding Fourier transform of this term. For instance, $(\tilde{w}_1, \tilde{w}_2)$ are the Fourier transforms of (w_1, w_2). The Fourier transforms of the tensional forces, bending moments, and torsional moments are presented below:

1. Fourier transforms of tensional forces:

$$\tilde{Q}_x^+(a,\beta_n) = -R_{Q1}\tilde{w}_1(a,\beta_n) + R_{Q2}\tilde{w}_2(a,\beta_n) \quad (5.40)$$

$$\tilde{Q}_x^-(a,\beta_n) = -R_{Q2}\tilde{w}_1(a,\beta_n) + R_{Q1}\tilde{w}_2(a,\beta_n) \quad (5.41)$$

$$\tilde{Q}_y^+(a_m,\beta) = -R_{Q3}\tilde{w}_1(a_m,\beta) + R_{Q4}\tilde{w}_2(a_m,\beta) \quad (5.42)$$

$$\tilde{Q}_y^-(a_m,\beta) = -R_{Q4}\tilde{w}_1(a_m,\beta) + R_{Q3}\tilde{w}_2(a_m,\beta) \quad (5.43)$$

2. Fourier transforms of bending moments:

$$\tilde{M}_x^+(a,\beta_n) = -\alpha^2 [R_{M1}\tilde{w}_1(a,\beta_n) - R_{M2}\tilde{w}_2(a,\beta_n)] \quad (5.44)$$

$$\tilde{M}_x^-(a,\beta_n) = -\alpha^2 [R_{M2}\tilde{w}_1(a,\beta_n) - R_{M1}\tilde{w}_2(a,\beta_n)] \quad (5.45)$$

$$\tilde{M}_y^+(a_m,\beta) = -\beta^2 [R_{M3}\tilde{w}_1(a_m,\beta) - R_{M4}\tilde{w}_2(a_m,\beta)] \quad (5.46)$$

$$\tilde{M}_y^-(a_m,\beta) = -\beta^2 [R_{M4}\tilde{w}_1(a_m,\beta) - R_{M3}\tilde{w}_2(a_m,\beta)] \quad (5.47)$$

3. Fourier transforms of torsional moments:

$$\tilde{M}_{Tx}^+(a,\beta_n) = -\alpha\beta_n [R_{T1}\tilde{w}_1(a,\beta_n) - R_{T2}\tilde{w}_2(a,\beta_n)] \quad (5.48)$$

$$\tilde{M}_{Tx}^-(a,\beta_n) = -\alpha\beta_n [R_{T2}\tilde{w}_1(a,\beta_n) - R_{T1}\tilde{w}_2(a,\beta_n)] \quad (5.49)$$

$$\tilde{M}_{Ty}^+(a_m,\beta) = -\alpha_m\beta [R_{T3}\tilde{w}_1(a_m,\beta) - R_{T4}\tilde{w}_2(a_m,\beta)] \quad (5.50)$$

$$\tilde{M}_{Ty}^-(a_m,\beta) = -\alpha_m\beta [R_{T4}\tilde{w}_1(a_m,\beta) - R_{T3}\tilde{w}_2(a_m,\beta)] \quad (5.51)$$

Substitution of (5.40), (5.41), (5.42), (5.43), (5.44), (5.45), (5.46), (5.47), (5.48), (5.49), (5.50), and (5.51) into (5.38) and (5.39) yields

$$\tilde{w}_1(\alpha,\beta) = \frac{1}{D_1 l_x f_1(\alpha,\beta)} \sum_m \left[-R_{Q3} - i\beta^3 R_{M3} - i\alpha\alpha_m \beta R_{T3} \right] \tilde{w}_1(\alpha_m,\beta)$$

$$+ \frac{1}{D_1 l_y f_1(\alpha,\beta)} \sum_n \left[-R_{Q1} - i\alpha^3 R_{M1} - i\alpha\beta\beta_n R_{T1} \right] \tilde{w}_1(\alpha,\beta_n)$$

$$+ \frac{1}{D_1 l_x f_1(\alpha,\beta)} \sum_m \left[R_{Q4} + i\beta^3 R_{M4} + i\alpha\alpha_m \beta R_{T4} \right] \tilde{w}_2(\alpha_m,\beta)$$

$$+ \frac{1}{D_1 l_y f_1(\alpha,\beta)} \sum_n \left[R_{Q2} + i\alpha^3 R_{M2} + i\alpha\beta\beta_n R_{T2} \right] \tilde{w}_2(\alpha,\beta_n)$$

$$+ \frac{q_0 e^{-i(\alpha x_0 + \beta y_0)}}{(2\pi)^2 D_1 f_1(\alpha,\beta)} \tag{5.52}$$

$$\tilde{w}_2(\alpha,\beta) = \frac{-1}{D_2 l_x f_2(\alpha,\beta)} \sum_m \left[-R_{Q4} - i\beta^3 R_{M4} - i\alpha\alpha_m \beta R_{T4} \right] \tilde{w}_1(\alpha_m,\beta)$$

$$- \frac{1}{D_2 l_y f_2(\alpha,\beta)} \sum_n \left[-R_{Q2} - i\alpha^3 R_{M2} - i\alpha\beta\beta_n R_{T2} \right] \tilde{w}_1(\alpha,\beta_n)$$

$$- \frac{1}{D_2 l_x f_2(\alpha,\beta)} \sum_m \left[R_{Q3} + i\beta^3 R_{M3} + i\alpha\alpha_m \beta R_{T3} \right] \tilde{w}_2(\alpha_m,\beta)$$

$$- \frac{1}{D_2 l_y f_2(\alpha,\beta)} \sum_n \left[R_{Q1} + i\alpha^3 R_{M1} + i\alpha\beta\beta_n R_{T1} \right] \tilde{w}_2(\alpha,\beta_n) \tag{5.53}$$

where

$$f_1(\alpha,\beta) = \left(\alpha^2 + \beta^2\right)^2 - m_1 \omega^2 / D_1, \quad f_2(\alpha,\beta) = \left(\alpha^2 + \beta^2\right)^2 - m_2 \omega^2 / D_2 \tag{5.54}$$

$$\alpha_m = \alpha + 2m\pi/l_x, \quad \beta_n = \beta + 2n\pi/l_y \tag{5.55}$$

To solve Eqs. (5.52) and (5.53), one needs to replace (α, β) by (α'_m, β'_n), resulting in two sets of simultaneous algebraic equations:

5.1 Sound Radiation of Sandwich Structures

$$\tilde{w}_1\left(\alpha'_m, \beta'_n\right) + \frac{R_{Q3} + i{\beta'_n}^3 R_{M3}}{D_1 l_x f_1\left(\alpha'_m, \beta'_n\right)} \sum_m \tilde{w}_1\left(\alpha_m, \beta'_n\right) + \frac{i\alpha'_m \beta'_n R_{T3}}{D_1 l_x f_1\left(\alpha'_m, \beta'_n\right)} \sum_m \alpha_m \tilde{w}_1\left(\alpha_m, \beta'_n\right)$$

$$+ \frac{R_{Q1} + i{\alpha'_m}^3 R_{M1}}{D_1 l_y f_1\left(\alpha'_m, \beta'_n\right)} \sum_n \tilde{w}_1\left(\alpha'_m, \beta_n\right) + \frac{i\alpha'_m \beta'_n R_{T1}}{D_1 l_y f_1\left(\alpha'_m, \beta'_n\right)} \sum_n \beta_n \tilde{w}_1\left(\alpha'_m, \beta_n\right)$$

$$- \frac{R_{Q4} + i{\beta'_n}^3 R_{M4}}{D_1 l_x f_1\left(\alpha'_m, \beta'_n\right)} \sum_m \tilde{w}_2\left(\alpha_m, \beta'_n\right) - \frac{i\alpha'_m \beta'_n R_{T4}}{D_1 l_x f_1\left(\alpha'_m, \beta'_n\right)} \sum_m \alpha_m \tilde{w}_2\left(\alpha_m, \beta'_n\right)$$

$$- \frac{R_{Q2} + i{\alpha'_m}^3 R_{M2}}{D_1 l_y f_1\left(\alpha'_m, \beta'_n\right)} \sum_n \tilde{w}_2\left(\alpha'_m, \beta_n\right) - \frac{i\alpha'_m \beta'_n R_{T2}}{D_1 l_y f_1\left(\alpha'_m, \beta'_n\right)} \sum_n \beta_n \tilde{w}_2\left(\alpha'_m, \beta_n\right)$$

$$= \frac{q_0 e^{-i\left(\alpha'_m x_0 + \beta'_n y_0\right)}}{(2\pi)^2 D_1 f_1\left(\alpha'_m, \beta'_n\right)} \tag{5.56}$$

$$- \frac{R_{Q4} + i{\beta'_n}^3 R_{M4}}{D_2 l_x f_2\left(\alpha'_m, \beta'_n\right)} \sum_m \tilde{w}_1\left(\alpha_m, \beta'_n\right) - \frac{i\alpha'_m \beta'_n R_{T4}}{D_2 l_x f_2\left(\alpha'_m, \beta'_n\right)} \sum_m \alpha_m \tilde{w}_1\left(\alpha_m, \beta'_n\right)$$

$$- \frac{R_{Q2} + i{\alpha'_m}^3 R_{M2}}{D_2 l_y f_2\left(\alpha'_m, \beta'_n\right)} \sum_n \tilde{w}_1\left(\alpha'_m, \beta_n\right) - \frac{i\alpha'_m \beta'_n R_{T2}}{D_2 l_y f_2\left(\alpha'_m, \beta'_n\right)} \sum_n \beta_n \tilde{w}_1\left(\alpha'_m, \beta_n\right)$$

$$+ \tilde{w}_2\left(\alpha'_m, \beta'_n\right) + \frac{R_{Q3} + i{\beta'_n}^3 R_{M3}}{D_2 l_x f_2\left(\alpha'_m, \beta'_n\right)} \sum_m \tilde{w}_2\left(\alpha_m, \beta'_n\right)$$

$$+ \frac{i\alpha'_m \beta'_n R_{T3}}{D_2 l_x f_2\left(\alpha'_m, \beta'_n\right)} \sum_m \alpha_m \tilde{w}_2\left(\alpha_m, \beta'_n\right) + \frac{R_{Q1} + i{\alpha'_m}^3 R_{M1}}{D_2 l_y f_2\left(\alpha'_m, \beta'_n\right)} \sum_n \tilde{w}_2\left(\alpha'_m, \beta_n\right)$$

$$+ \frac{i\alpha'_m \beta'_n R_{T1}}{D_2 l_y f_2\left(\alpha'_m, \beta'_n\right)} \sum_n \beta_n \tilde{w}_2\left(\alpha'_m, \beta_n\right) = 0 \tag{5.57}$$

which contain two sets of infinite unknowns, $\tilde{w}_1\left(\alpha'_m, \beta'_n\right)$ and $\tilde{w}_2\left(\alpha'_m, \beta'_n\right)$, with $m = -\infty$ to $+\infty$ and $n = -\infty$ to $+\infty$. Insofar as the solution converges, these equations can be solved simultaneously by truncation. That is, (m, n) only take values in a finite range of $m = -\widehat{m}$ to $\widehat{m}$ and $n = -\widehat{n}$ to $\widehat{n}$ (where $\widehat{m}$ and $\widehat{n}$ both being positive integer). For brevity, the resulting simultaneous equations containing a finite number [i.e., $2MN$, where $M = 2\widehat{m} + 1$, $N = 2\widehat{n} + 1$] of unknowns can be expressed in matrix form as

$$\begin{bmatrix} T_{11} & T_{12} \\ T_{21} & T_{22} \end{bmatrix}_{2MN \times 2MN} \begin{Bmatrix} \tilde{w}_1\left(\alpha'_m, \beta'_n\right) \\ \tilde{w}_2\left(\alpha'_m, \beta'_n\right) \end{Bmatrix}_{2MN \times 1} = \begin{Bmatrix} F_{mn} \\ 0 \end{Bmatrix}_{2MN \times 1} \tag{5.58}$$

Equation (5.58) can be solved numerically to obtain the panel displacements $\tilde{w}_1(\alpha, \beta)$ and $\tilde{w}_2(\alpha, \beta)$ in their respective wavenumber space. Details of the derivation are presented in Appendix A.

5.1.4 Far-Field Radiated Sound Pressure

The radiated sound pressure is related directly to the dynamic response of the radiating panel (bottom panel in the present case). Once the dynamic response of the bottom panel $\tilde{w}_2(\alpha, \beta)$ is solved, the radiated sound pressure at far field can be obtained.

Due to the periodic nature of the sandwich, upon excitation by a harmonic point force on its upper panel, a series of space-harmonic waves are transmitting in the structure. For a given point force with wavenumbers (α_0, β_0), a flexural wave having the same wavenumbers (α_0, β_0) is excited and propagates in the face panel. It will generate the (m, n)th harmonic wavenumber components $(\alpha_0 + 2m\pi/l_x, \beta_0 + 2n\pi/l_y)$, owing to the vibration interaction of the face panel with the mth x-wise and nth y-wise stiffeners. Therefore, the face panel vibration and the radiated sound pressure both contain a series of space-harmonic wave components with wavenumbers $(\alpha_0 + 2m\pi/l_x, \beta_0 + 2n\pi/l_y)$, where $-\infty < m < +\infty$ and $-\infty < n < +\infty$.

With the origin of the spherical coordinates (r, θ, φ) located at the excitation point (x_0, y_0), the far-field sound pressure $p(r, \theta, \varphi)$ radiated from a vibrating surface with displacement $w(x, y)$ is given by Takahashi [15] and Morse and Ingard [64]:

$$p(r, \theta, \varphi) = -\frac{\rho_0 \omega^2 e^{ik_0 r}}{2\pi r} e^{i(\alpha x_0 + \beta y_0)} \int_{-\infty}^{+\infty} \int_{-\infty}^{+\infty} w(x, y) e^{-i(\alpha x + \beta y)} dx dy \quad (5.59)$$

where $k_0 = \omega/c_0$, c_0 and ρ_0 being the sound speed and air density, respectively, and the wavenumbers α and β are

$$\alpha = k_0 \cos \varphi \sin \theta, \quad \beta = k_0 \sin \varphi \sin \theta \quad (5.60)$$

Finally, with the Fourier transform of (5.37), Eq. (5.59) becomes

$$p(r, \theta, \varphi) = -2\pi \rho_0 \omega^2 \left(e^{ik_0 r}/r\right) e^{i(\alpha x_0 + \beta y_0)} \tilde{w}(\alpha, \beta) \quad (5.61)$$

With the modeling presented above describing accurately the dynamic response of an infinite orthogonally rib-stiffened sandwich structure excited by a point force and the formulation for the far-field radiated sound pressure, the on-axis (i.e., on the axis $\theta = \varphi = 0$) far-field pressure is calculated below to explore the sound radiation characteristics of the structure. Note that on the selected axis (i.e., $\theta = \varphi = 0$), the stationary phase wavenumbers $\bar{\alpha}$ and $\bar{\beta}$ are both zero.

The simultaneous algebraic equations are truncated at the $\pm \hat{m}$ and $\pm \hat{n}$ harmonic wave components in the x- and y-directions, with the frequency-dependent $\hat{m}$ and $\hat{n}$ selected as 3 and 10 at 100 Hz and 10 kHz, respectively. Numerical convergence tests have ensured that these $\hat{m}$ and $\hat{n}$ values are sufficiently large for obtaining accurate results.

5.1 Sound Radiation of Sandwich Structures

Table 5.1 Material and geometrical properties of orthogonally rib-stiffened single plate (Mace [5])

Plate				Fluid media	
D	m	η		ρ_0	c_0
2,326 N·m	39.1 kg/m	0.02		1,000 kg/m^3	1,500 m/s
Rib-stiffeners					
E	ρ	$l_x = l_y$		$t_x = t_y$	d
195 GPa	7,700 kg/m^3	0.2 m		0.00508 m	0.0508 m

For reference, the high-frequency asymptote of the far-field sound pressure radiated by an unstiffened plate (Mace [5]) is

$$p_a = \frac{\rho_0 q_0}{2\pi m} \frac{e^{ik_0 r}}{r} \tag{5.62}$$

The far-field sound pressure radiated by the present orthogonally rib-stiffened sandwich structure is then given in the form of sound pressure level (SPL) in decibel scales (dB) relative to p_a as

$$\text{SPL} = 20 \cdot \log_{10}\left(\frac{p}{p_a}\right) \tag{5.63}$$

5.1.5 Validation of Theoretical Modeling

To verify the accuracy and applicability of the present theoretical modeling on wave propagation and sound radiation behavior of an orthogonally rib-stiffened sandwich structure, results obtained using the model are compared with those of Mace [5] for sound radiation from an orthogonally rib-stiffened single plate. To facilitate the comparison, the key parameters (i.e., Young's modulus E, density ρ, and thickness h) of the bottom panel are set to negligibly small in comparison with those of the upper panel and rib-stiffeners, so that the orthogonally rib-stiffened sandwich behaves exactly like an orthogonally rib-stiffened single plate.

For the purpose of validation, the material and geometrical properties (Table 5.1) used by Mace [5] are adopted in the numerical calculations. Figures 5.3 and 5.4 present the results for two different excitation locations, $(l_x/3, l_y/3)$ and $(l_x/2, l_y/2)$. Overall, good agreement is achieved between the present results and Mace's model prediction for both excitation locations. The discrepancies at high frequencies between the two different models, however, are attributed to the fact that the inertial effects and torsional moments of the rib-stiffeners were not accounted for by Mace [5]. The reason that at high-frequency range the deviation is small in Fig. 5.3 but significant in Fig. 5.4 is because the excitation exerted at $(l_x/2, l_y/2)$ leads to stronger torsional moments of the rib-stiffeners than that exerted at $(l_x/3, l_y/3)$.

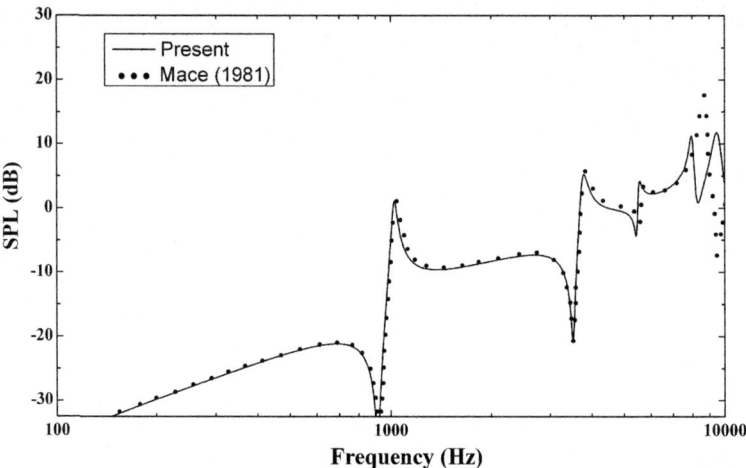

Fig. 5.3 Comparison between present model predictions and Mace [5]'s results for sound pressure level radiated by orthogonally rib-stiffened single plate excited at $(l_x/3, l_y/3)$ (With permission from Elsevier)

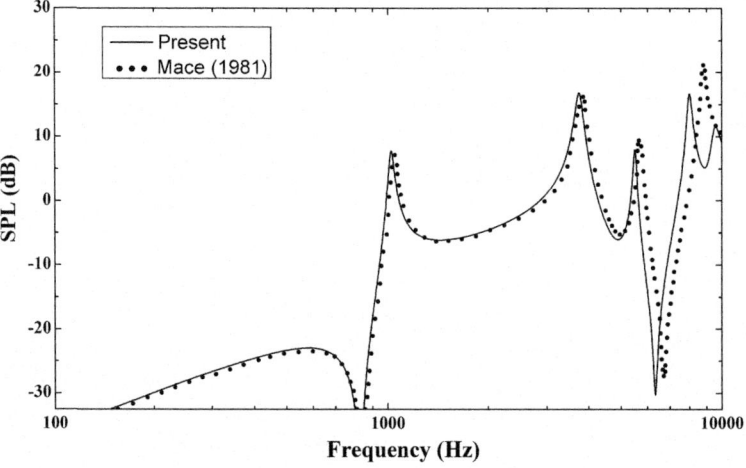

Fig. 5.4 Comparison between present model predictions and Mace [5]'s results for sound pressure level radiated by orthogonally rib-stiffened single plate excited at $(l_x/2, l_y/2)$ (With permission from Elsevier)

5.1.6 Influences of Inertial Effects Arising from Rib-Stiffener Mass

The inertial effects of the tensional forces, bending moments, and torsional moments arising from the rib-stiffeners have been accounted for by the present analytic

5.1 Sound Radiation of Sandwich Structures

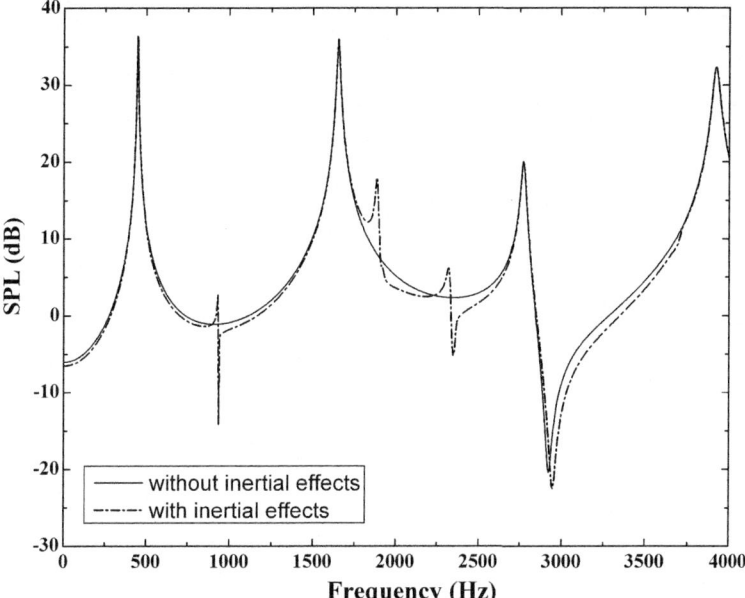

Fig. 5.5 Variation of on-axis far-field radiated sound pressure with excitation frequency: influence of inertial effects. Geometry of rib-stiffeners, $t_x = t_y = 1$ mm, $l_x = l_y = 0.2$ m; excitation location, $(x_0, y_0) = (l_x/2, l_y/2)$ (With permission from Elsevier)

model. The influence of the inertial effects is explored below by comparing the predictions obtained for orthogonally rib-stiffened sandwich structures with and without considering the inertial effects.

With the point force acting at $(l_x/2, l_y/2)$, Figs. 5.5 and 5.6 plot the on-axis far-field radiated sound pressure level as a function of excitation frequency for rib-stiffeners having square cross sections, with width $t_x = t_y = 1$ mm and $t_x = t_y = 3$ mm, respectively. It can be seen from Fig. 5.5 that the SPL curve with the inertial effects considered has a tendency similar to that without considering inertial effects, the main discrepancy being the existence of several additional peaks and dips in the former. The superposition peaks (or dips) between the inertial case and the non-inertial one are dominated by face panel vibration, which are closely related to the maximum (or minimal) sound radiation wave shapes and vibration patterns. The appearance of the additional peaks and dips controlled predominantly by the rib-stiffeners is, on the other hand, attributed to the inertial effects arising from the mass of the rib-stiffeners. By comparing Fig. 5.5 with Fig. 5.6, it is seen that the discrepancy between the inertial and non-inertial cases is enlarged when the thickness (or, equivalently, the mass) of the rib-stiffeners is increased.

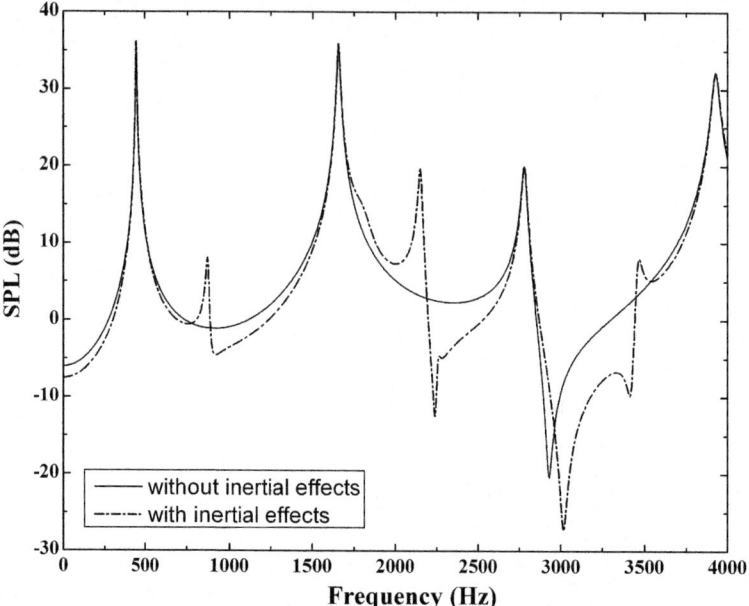

Fig. 5.6 Variation of on-axis far-field radiated sound pressure with excitation frequency: influence of inertial effects. Geometry of rib-stiffeners, $t_x = t_y = 3$ mm, $l_x = l_y = 0.2$ m; excitation location, $(x_0, y_0) = (l_x/2, l_y/2)$ (With permission from Elsevier)

5.1.7 Influence of Excitation Position

While the amplitude of any point in wave mode shape depends strongly on its position, the radiating modes excited by a point force vary with the excitation position. It is therefore expected that the on-axis far-field radiated sound pressure is significantly affected by the excitation position, which is confirmed by plotting in Fig. 5.7 the sound pressure level as a function of excitation frequency for three different excitation positions, i.e., $(x_0/l_x, y_0/l_y)$ at (0, 0), (1/4, 1/4), and (1/2, 1/2).

It is seen from Fig. 5.7 that the SPL curves of the (1/4, 1/4) and (1/2, 1/2) cases have peaks appearing at the same frequencies (e.g., 445, 1,659, 2,769, and 3,919 Hz), although there exist large discrepancies at other frequencies (e.g., 3,919 Hz in particular). In comparison, there are no evident peaks appearing in the SPL curve of the (0, 0) case at these frequencies. As these radiated sound pressure peaks are mainly controlled by the wave mode shapes and vibration patterns of the face panel, it appears that the point force excitation at (1/4, 1/4) and (1/2, 1/2) can excite the appropriate radiating mode of the face panel. In contrast, the excitation at (0, 0) is located at the joint connecting the face panel with the x- and y-wise rib-stiffeners, which excites mainly the tensional and bending motions of the x- and y-wise rib-stiffeners. Therefore, no SPL peaks appear for the (0, 0) case at radiating frequencies controlled by panel vibration.

5.1 Sound Radiation of Sandwich Structures

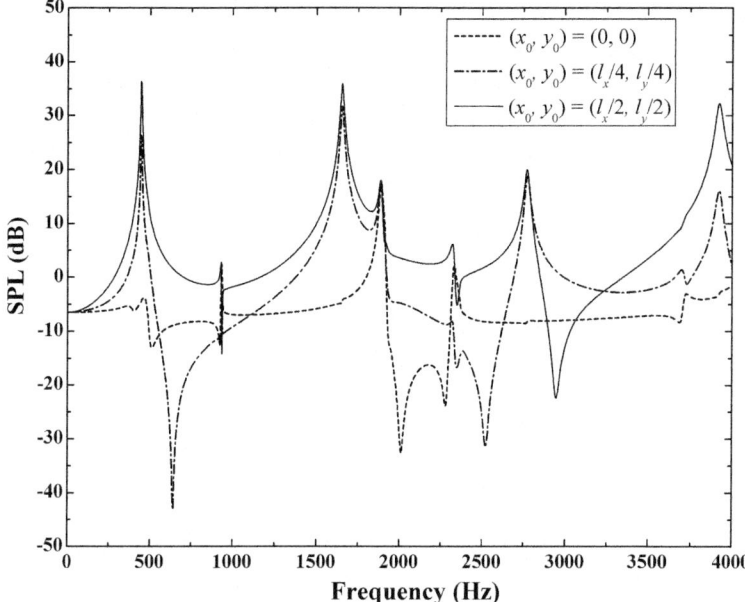

Fig. 5.7 On-axis far-field radiated sound pressure plotted as a function of excitation frequency for selected excitation positions: $(x_0/l_x, y_0/l_y)$ at $(0, 0)$, $(1/4, 1/4)$, and $(1/2, 1/2)$. Geometry of rib-stiffeners: $t_x = t_y = 1$ mm, $l_x = l_y = 0.2$ m (With permission from Elsevier)

The radiating sound pressure peaks dominated by the rib-stiffeners are well captured by the three SPL curves of Fig. 5.7 at the same frequencies (e.g., 936, 1,888, 2,329, 3,722 Hz), although some peaks may not be so evident due to complicated wave interaction at the junctions of panel, x-wise, and y-wise stiffeners.

5.1.8 Influence of Rib-Stiffener Spacings

As the periodicity spacings l_x and l_y between rib-stiffeners are key parameters describing the periodic nature of the sandwich structure (Fig. 5.1), they should influence significantly the wave propagation and sound radiation characteristics of the structure. Figure 5.8 illustrates the influence of the periodicity spacings on radiated sound pressure by plotting the SPL curve tendencies, with (l_x, l_y) selected as $(0.2, 0.2)$m, $(0.225, 0.225)$m, and $(0.25, 0.25)$m, respectively, and the point force excitation fixed at $(l_x/2, l_y/2)$.

The most attractive point about the results of Fig. 5.8 is that the magnitudes of the SPL peaks and dips (not only the panel vibration dominated but also the rib-stiffener vibration controlled) decrease as the periodicity spacings are increased. However, overall, the three SPL curves exhibit almost the same tendency although

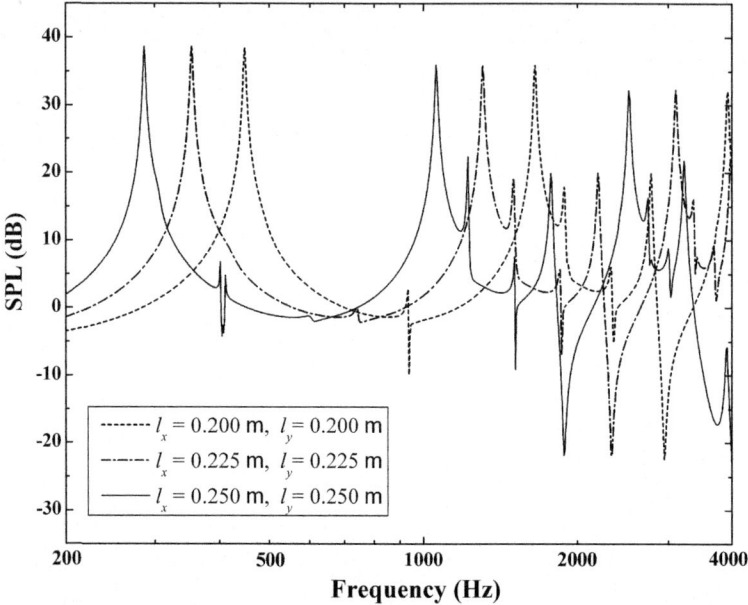

Fig. 5.8 Variation of on-axis far-field radiated sound pressure with excitation frequency: influence of periodicity spacings between rib-stiffeners. Geometry of rib-stiffeners, $t_x = t_y = 1$ mm; excitation position, $(x_0, y_0) = (l_x/2, l_y/2)$ (With permission from Elsevier)

their peaks and dips shift, which is attributed to the highly similar periodic nature of the sandwich structures.

5.1.9 Conclusions

An analytic model has been formulated to investigate the wave propagation and sound radiation behavior of a point force-excited sandwich structure having two sets of orthogonal rib-stiffeners as its core. Unlike previous researches on rib-stiffened panel without considering the inertial effects of rib-stiffeners, the vibration motion of the rib-stiffeners is accurately described by introducing their tensional forces, bending moments, and torsional moments as well as the corresponding inertial terms into the governing equations of the two face panels. The Fourier transform technique and Poisson summation formula are employed to solve the governing equations. The resulting two sets of infinite simultaneous algebraic equations are numerically solved by truncation insofar as the solution converges.

The far-field sound radiation is examined to gain physical insights of the vibroacoustic response of the sandwich structure. First, comparisons between model predictions with previous published results for orthogonally stiffened single plates validate the accuracy and feasibility of the present analytic model, which also confirm the necessity of accounting for the inertial effects and torsional moments of the rib-stiffeners in any theoretical modeling. Subsequently, the influences of the excitation position, periodicity spacings of rib-stiffeners, and inertial effects rooted in the rib-stiffener mass upon the far-field sound pressure radiated from the orthogonal sandwich structure are explored.

Since the inertial effects of the rib-stiffeners are considered in the present analytic model, a couple of additional peaks and dips on the SPL versus excitation frequency curve related to the inertial effects are well captured, which are especially evident when the mass of the rib-stiffeners is significant. Besides these rib-stiffener-controlled SPL peaks and dips, it is also found that there exist panel-controlled peaks and dips, which are related to certain wave shapes and vibration patterns possessing maximal or minimal sound radiation.

The excitation position of the point force plays a significant role in the wave propagation and sound radiation behavior of the sandwich, as different positions can excite different wave mode shapes and vibration patterns of the face panel, resulting in either panel- or rib-stiffener-controlled vibration. Therefore, different peaks and dips associated with the panel- or rib-stiffener-controlled vibration will emerge and noticeably affect the tendency of the SPL curve.

As a key parameter describing the periodic nature of the sandwich structure, rib-stiffener spacing also has a dominant role. All the SPL peaks and dips dominantly controlled by either panel vibration or rib-stiffener vibration are shifted to lower frequencies as the periodicity spacings are increased. The overall tendency of the SPL curves remains nonetheless unchanged, owing to the similar periodic nature of the sandwich structures.

5.2 Sound Transmission Through Sandwich Structures

5.2.1 Introduction

Lightweight sandwich structures consisting of two parallel plates (as the facesheets) reinforced by sets of spatially periodic rib-stiffeners (as the core) form a class of structural elements of practical importance in a wide range of engineering applications, such as aircraft fuselages and ship/submarine hulls [5–7, 16, 36–39]. Typically, these periodic rib-stiffeners construct identical and uniformly spaced sets having a repetitive structural geometry in either one or two dimensions. Since aerospace and marine vehicles are usually subjected to sound excitation and/or dynamic impact [14, 40–43], the wave propagation and acoustic characteristics of these periodic structures become increasingly important for predicting the internal and external sound pressure levels. When the wavelength of flexural wave in the periodically rib-stiffened structure is much greater than stiffener separation, the structure can be approximately regarded as an orthotropic plate. At relatively high frequencies, the wavelength is on the same order as the stiffener separation, and hence the spatially periodic rib-stiffeners should be modeled exactly to comprehend the dynamic and acoustic characteristics of the structure as a whole [3, 4, 6, 7, 39]. The aim is to formulate a physically based analytic model of desirable accuracy for the vibroacoustic response of the sandwich structure, which can in due course be employed in conjunction with optimization techniques to design more effective lightweight soundproofing structures.

A rich literature [7, 8, 16, 21, 23, 24, 38, 44–47] exists on the theoretical modeling and analysis of wave propagation and dynamic performance of periodically rib-stiffened beams and plates. Focusing on free wave propagation in periodically supported, infinite beams, Mead [23] found that a freely propagating flexural wave in such a beam must be regarded as a wave group, having components of different wavelengths, phase velocities, and directions. A mechanism of whereby slow, subsonic convected pressure fields can generate supersonic, radiating flexural waves was also elucidated. A detailed literature review was reported by Mead [24] on wave propagation in continuous periodic structures contributed by the Southampton University from 1964 to 1995. More recently, using the transfer matrix approach, Liu and Bhattacharya [48] obtained dispersion relations for elastic waves propagating in sandwich structures. Ichchou et al. [1, 2] addressed the issue of energy propagation in a ribbed plate by analyzing its response in the wavenumber space and derived dispersion relationships between the wavenumber and frequency. Wang et al. [49], Li et al. [50, 51] performed comprehensive studies on localization of elastic waves in disordered periodic structures, with the underlying mechanism of wave decay phenomenon in such structures revealed. It should be pointed out that the aforementioned contributions on wave propagation and dynamic response of structures are not meant to be exhaustive, while a few other specialized topics such as the localized waves in disorder structures, turbulent boundary layer-excited vibrations, and fluttering of aircraft wings are beyond the scope of the present research.

Existing studies on the vibroacoustic response of periodic structures may be grouped into two main categories: sound radiation under point loading [52, 53] and sound transmission due to convective fluid-loaded pressure excitation [31, 33, 34, 54–56]. Particularly, for periodically rib-stiffened structures, two approaches have been used to deal with the relevant issues. Firstly, the technique of Fourier transform was often employed [3–6, 8, 11, 15, 25, 28, 57]. For example, Rumerman [25] presented a general solution for the forced vibration of an infinite thin plate, periodically stiffened by identical, uniform ribs, with the forces and bending moments of the ribs considered via the impedances of the ribs and plate. However, the torsional moments and inertial effects of the ribs were not included in the analysis, and no numerical results were given. Mace [28] analyzed sound radiation from a point-excited infinite fluid-loaded plate reinforced by two sets of parallel stiffeners; however, the moments of the rib-stiffeners were again ignored. Considering only the forces of rib-stiffeners, Mace [5] also studied the radiation of sound from a two-dimensional (2D) plate reinforced with two sets of orthogonal line stiffeners under fluid-loaded harmonic incident pressure. Recently, from the viewpoint of vibroacoustic response in wavenumber space, Maxit [11] proposed an efficient method based on the Fourier transform technique to estimate the vibration and sound radiation from a stiffened fluid-loaded plate excited by a mechanical point force.

Secondly, it has been established that the space-harmonic approach evolving from the consideration of progressive wave propagation is also well suited for studying the vibroacoustic response of periodically rib-stiffened structures [7, 10, 21, 23, 24, 30, 58, 59]. For instance, the response of periodically supported beams to convected random loading was evaluated by Mead and Pujara [21] in terms of space-harmonic series: only as few as three terms were required to obtain a solution of acceptable accuracy in comparison with the exact solution. The same approach was adopted by Lee and Kim [7] to study the sound transmission characteristics of a thin plate reinforced by equally spaced line stiffeners, with parametric studies conducted to provide guidelines for the practical design of the system. Extending this approach to parallelly rib-stiffened sandwich structure, Wang et al. [30] developed a deterministic analytic model by coupling the acoustic and structure vibrations and then employing the virtual work principle. However, the model does not provide a complete description of the motions of the rib-stiffeners and their interaction with the face panels, as only tensional forces and bending moments are considered. A refined theoretical model of Wang et al. [30] was proposed by Legault and Atalla [58] to investigate the transmission of sound through a typical aircraft sidewall panel, i.e., sandwich structure reinforced by parallel rib-stiffeners, with fiberglass filled in the cavity: again, only the tensional forces and bending moments of the rib-stiffeners are included.

While previous researches focused mainly on relatively simple sandwich constructions and approximated the rib-stiffeners as an Euler beam or a combination of translational spring and rotational spring, an exact theoretical model concerning the vibroacoustic response of more complex structures (e.g., two-dimensional sandwich structures orthogonally reinforced by periodic rib-stiffeners) is desirable. In addition to helping exploring the underlying physical subtleties, the model should also serve as benchmark checking for approximate analytic approaches, with a small computational expense afforded compared to numerical methods such as the finite element method (FEM) and the boundary element method (BEM). With the focus placed on 2D sandwich structures reinforced orthogonally with periodic rib-stiffeners under point force excitations, Xin and Lu [39] developed such an exact model for their sound radiation characteristics using the Fourier transform technique.

Built upon the work of Ref. [39], the physical process of sound transmission through an infinite orthogonally rib-stiffened sandwich structure subjected to convective harmonic fluid-loaded pressure is analytically formulated and solved by employing the space-harmonic approach. All possible motions of the rib-stiffeners are included by introducing the tensional forces, bending moments, and torsional moments as well as the corresponding inertial terms into the governing equations of the two face panels. Furthermore, the surrounding acoustic fluids are restricted

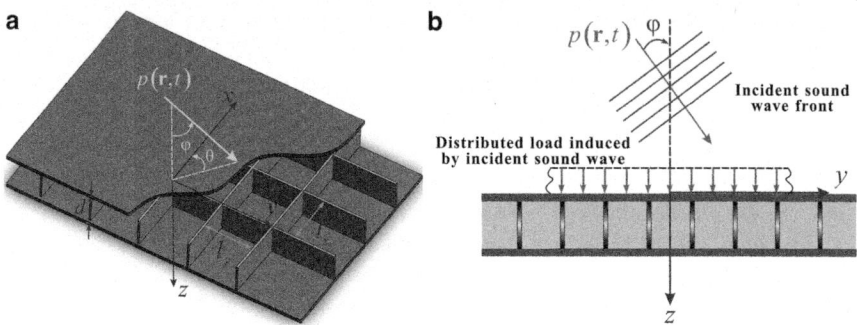

Fig. 5.9 Schematic illustration of an orthogonally rib-stiffened sandwich subjected to incident sound pressure wave: (**a**) global view; (**b**) side view of (**a**) (With permission from Elsevier)

by the acoustic wave equation, and fluid-structure coupling is incorporated by enforcing velocity continuity conditions at fluid-panel interfaces. For one periodic element, applying the principle of virtual work yields two infinite sets of simultaneous algebraic coupled equations, which are numerically solved by truncating them in a finite range insofar as the solution converges. For validation, the predictions of the present analytic model are compared with previous published results, with good overall agreement achieved. Moreover, the necessity and advantage of modeling exactly the motions of the orthogonal rib-stiffeners are also affirmed by comparing the complete model with its simplified version as well as the model of Wang et al. [30]. In the perspective of both physical understanding and practical structural design, the dependence of sound transmission of the structure upon the inertial effects arising from the rib-stiffener mass, the airborne and structure-borne paths, and the periodicity spacings of rib-stiffeners is systematically studied and conclusions of referential significance are deduced.

5.2.2 Analytic Formulation of Panel Vibration and Sound Transmission

Consider two infinite parallel face panels reinforced by two periodic sets of orthogonal rib-stiffeners having periodic uniform spacings l_x and l_y in the x- and y-directions, respectively (see Fig. 5.9). A right-handed Cartesian coordinate system (x, y, z) is established, with its x-axis and y-axis positioned separately along one pair of the orthogonal rib-stiffeners and the positive direction of the z-axis pointing downward (Fig. 5.9). The upper panel located at $z = 0$ and the bottom panel located at $z = h_1 + d$ separate the acoustic fluid in the spatial field into three parts: the upper field occupying the half-space $z < 0$, the middle field filling the space $h_1 < z < h_1 + d$ (i.e., in between the two panels and divided periodically by the rib-stiffeners), and

5.2 Sound Transmission Through Sandwich Structures

the lower field occupying the other half-space $z > h_1 + h_2 + d$. Both the upper panel (thickness h_1) and the bottom panel (thickness h_2) are modeled as Kirchhoff thin plates. Let t_x and t_y denote separately the thickness of the x- and y-wise rib-stiffeners.

An oblique plane sound wave $p(r,t)$ varying harmonically in time is incident upon the upper panel of the sandwich structure with elevation angle φ and azimuth angle θ. Consequently, a distributed load induced by the incident sound pressure wave is exerting on the panel, which in turn induces a bending wave that propagates along the panel. The bending wave in the upper panel is transmitted to the bottom panel via two paths, namely, the structure-borne path (i.e., the orthogonal rib-stiffeners) and the airborne path (i.e., the air constrained in between the two panels). The transmitted bending wave in the bottom panel radiates sound pressure wave into the semi-infinite acoustic fluid in contact with the bottom panel (see Fig. 5.9). The analytic model to be developed below not only tackles exactly with the physical process of sound transmission through the sandwich structure but also accounts for the air-structure coupling. Both the acoustic fluid constrained in between the two panels ($h_1 < z < h_1 + d$) and the semi-infinite fluids in contact with the upper panel ($z < 0$) and the bottom panel ($z > h_1 + h_2 + d$) satisfy the wave equation. Furthermore, the tensional, bending, and torsional motions of the rib-stiffeners and their corresponding inertial effects are all taken into account in the proposed model.

Given the periodic nature of the orthogonal rib-stiffened sandwich structure, the Bloch or Floquet theorem [60] is utilized here to express the panel vibration, which is well suitable to address wave propagation and vibration issues of periodic structures [61–63]. The displacements $w(x, y)$ of such a system at corresponding points in different periodic elements are related by a spatial periodic function (i.e., a bay-to-bay multiplicative factor, linking the motion of corresponding points in adjacent bays) as

$$w(x + ml_x, y + nl_y) = w(x, y) e^{-ik_x ml_x} e^{-ik_y nl_y} \quad (5.64)$$

Therefore, it is convenient to express the motion of each panel as a summation of one set of space-harmonic series. For a 2D sandwich structure stiffened by identical ribs which repeat in the x- and y-directions and excited by a harmonic plane sound wave (i.e., the convective fluid-loaded pressure) $p(x, y, z; t) = Ie^{-i(k_x x + k_y y + k_z z - \omega t)}$, the panel responses $w_j(x, y; t)$ ($j = 1, 2$ for the upper and bottom panels, respectively) can be expressed using space-harmonic expansion [5, 18, 21] as

$$w_1(x, y; t) = \sum_{m=-\infty}^{+\infty} \sum_{n=-\infty}^{+\infty} \alpha_{1,mn} e^{-i\left[(k_x + 2m\pi/l_x)x + (k_y + 2n\pi/l_y)y - \omega t\right]} \quad (5.65)$$

$$w_2(x, y; t) = \sum_{m=-\infty}^{+\infty} \sum_{n=-\infty}^{+\infty} \alpha_{2,mn} e^{-i\left[(k_x + 2m\pi/l_x)x + (k_y + 2n\pi/l_y)y - \omega t\right]} \quad (5.66)$$

where the (m, n)th harmonic wave components in the two panels have the same wavenumbers $(k_x + 2m\pi/l_x, k_y + 2n\pi/l_y)$ but different amplitudes, i.e.,

$$\alpha_{1,mn} = \frac{1}{l_x l_y} \int_0^{l_x} \int_0^{l_y} w_1(x,y;t) e^{i[(k_x+2m\pi/l_x)x+(k_y+2n\pi/l_y)y-\omega t]} dxdy \quad (5.67)$$

$$\alpha_{2,mn} = \frac{1}{l_x l_y} \int_0^{l_x} \int_0^{l_y} w_2(x,y;t) e^{i[(k_x+2m\pi/l_x)x+(k_y+2n\pi/l_y)y-\omega t]} dxdy \quad (5.68)$$

In Eqs. (5.65) and (5.66), the terms with $k_x + 2m\pi/l_x > 0$ (or $k_y + 2n\pi/l_y > 0$) stand for positive-going harmonic waves in the x-direction (or the y-direction) and those with $k_x + 2m\pi/l_x < 0$ (or $k_y + 2n\pi/l_y < 0$) denote negative-going harmonic waves in the x-direction (or the y-direction).

When sound pressure $p(\mathbf{r},t) = I e^{-i(\mathbf{k}\cdot\mathbf{r}-\omega t)}$ is incident on the upper panel, the incident sound partly reflected at the air-panel interface and the radiated sound by the vibrating panel constitute the negative-going waves in the upper semi-infinite acoustic fluid domain. The positive-going wave (i.e., the incident sound wave) and the negative-going wave (i.e., the reflected plus radiated sound waves) compose the resultant sound pressure imposed on the upper panel, which are transmitted through the sandwich structure into the semi-infinite space adjacent to the bottom panel, creating thence the transmitted sound pressure. Therefore, sound pressure in the upper semi-infinite field can be expressed as [18]

$$P_1(x,y,z;t) = I e^{-i(k_x x + k_y y + k_z z - \omega t)}$$
$$+ \sum_{m=-\infty}^{+\infty} \sum_{n=-\infty}^{+\infty} \beta_{mn} e^{-i[(k_x+2m\pi/l_x)x+(k_y+2n\pi/l_y)y-k_{z,mn}z-\omega t]}$$
$$(5.69)$$

Similarly, sound pressure in the middle field in between the two panels is expressed by space-harmonic series as

$$P_2(x,y,z;t) = \sum_{m=-\infty}^{+\infty} \sum_{n=-\infty}^{+\infty} \varepsilon_{mn} e^{-i[(k_x+2m\pi/l_x)x+(k_y+2n\pi/l_y)y+k_{z,mn}z-\omega t]}$$
$$+ \sum_{m=-\infty}^{+\infty} \sum_{n=-\infty}^{+\infty} \zeta_{mn} e^{-i[(k_x+2m\pi/l_x)x+(k_y+2n\pi/l_y)y-k_{z,mn}z-\omega t]}$$
$$(5.70)$$

The transmitted sound pressure in the bottom semi-infinite field only consists of positive-going wave:

5.2 Sound Transmission Through Sandwich Structures

$$P_3(x,y,z;t) = \sum_{m=-\infty}^{+\infty} \sum_{n=-\infty}^{+\infty} \xi_{mn} e^{-i\left[(k_x+2m\pi/l_x)x+(k_y+2n\pi/l_y)y+k_{z,mn}z-\omega t\right]} \quad (5.71)$$

In the above expressions, I is the amplitude of incident sound pressure; β_{mn} and ζ_{mn} are the (m, n)th space-harmonic amplitude of negative-going wave in the incident field and in the middle field, respectively; and ε_{mn} and ξ_{mn} are the (m, n)th space-harmonic amplitude of positive-going wave in the middle field and in the transmitted field, respectively. The wavenumber components in the x-, y-, and z-directions are determined by the elevation angle and azimuth angle of the incident sound wave as

$$k_x = k_0 \sin\varphi \cos\theta, \quad k_y = k_0 \sin\varphi \sin\theta, \quad k_z = k_0 \cos\varphi \quad (5.72)$$

where $k_{z,mn}$ is the (m, n)th space-harmonic wavenumber in the z-direction which, upon applying the Helmholtz equation, is given by

$$k_{z,mn} = \sqrt{\left(\frac{\omega}{c_0}\right)^2 - \left(k_x + \frac{2m\pi}{l_x}\right)^2 - \left(k_y + \frac{2n\pi}{l_y}\right)^2} \quad (5.73)$$

Note that when $(\omega/c_0)^2 < (k_x + 2m\pi/l_x)^2 + (k_y + 2n\pi/l_y)^2$, the pressure waves become evanescent waves [10, 30, 59] so that $k_{z,mn}$ should be taken as

$$k_{z,mn} = i\sqrt{\left(k_x + \frac{2m\pi}{l_x}\right)^2 + \left(k_y + \frac{2n\pi}{l_y}\right)^2 - \left(\frac{\omega}{c_0}\right)^2} \quad (5.74)$$

The two orthogonal sets of rib-stiffeners uniformly distributed in between the two face panels impose a strong constraint on the motions of the panels, which constitute the structure-borne path for sound transmission and wave propagation. To model accurately the vibroacoustic behavior of the sandwich, the dynamic motions of the rib-stiffeners should be carefully taken into account, which include tensional, bending, and torsional vibrations pertinent to tensional forces, bending moments, and torsional moments imposed on the connected panels. To account for the inertial effects of these motions arising from the mass of the rib-stiffeners, the resultant tensional forces, bending moments, and torsional moments acting on the upper and bottom panels are not identical, which are marked here by (Q^+, M^+, M_T^+) and (Q^-, M^-, M_T^-). Figure 5.10 illustrates the conventions used for the tensional forces, bending moments, and torsional moments at the interface between the upper panel and the x/y-wise rib-stiffeners. The same applies at the interface between the bottom panel and the x/y-wise rib-stiffeners, with (Q^+, M^+, M_T^+) replaced by (Q^-, M^-, M_T^-).

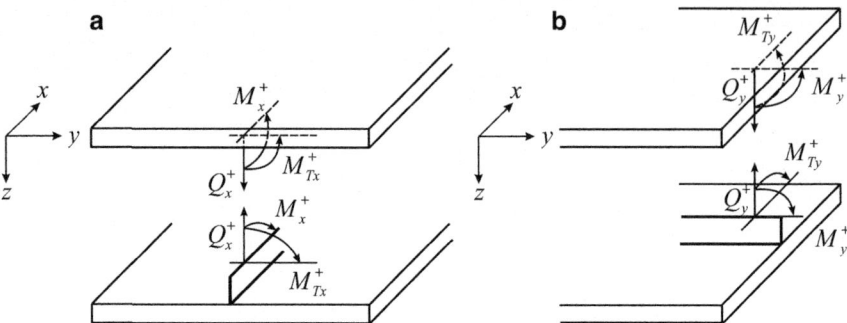

Fig. 5.10 Conventions for tensional forces, bending moments, and torsional moments between the upper panel and (a) *x*-wise rib-stiffeners and (b) *y*-wise rib-stiffeners. Similar conventions hold at the interface between the bottom panel and the *x*/*y*-wise rib-stiffeners by replacing (Q^+, M^+, M_T^+) with (Q^-, M^-, M_T^-), which is not shown here for brevity (With permission from Elsevier)

Given that the incident sound pressure wave varies harmonically in time, the dynamic responses of the two face panels are also harmonically dependent upon time. For simplicity, the harmonic time dependence $e^{-i\omega t}$ will be suppressed throughout this chapter henceforth.

The resultant pressure exerted on the upper panel is contributed by the incident sound wave, the negative-going wave on the incident side $P_1(x, y, 0)$, and the middle field pressure $P_2(x, y, h_1)$ on the other side. For the bottom panel, the net pressure is a combination of the transmitted sound pressure $P_3(x, y, h_1 + h_2 + d)$ on the transmitted side and the middle field pressure $P_2(x, y, h_1 + d)$ on the other side. Under the prescribed Cartesian coordinate system, with the tensional forces, bending moments, and torsional moments of the rib-stiffeners accounted for, the governing equations for panel vibrations are given by

$$D_1 \nabla^4 w_1 + m_1 \frac{\partial^2 w_1}{\partial t^2} = \sum_{m=-\infty}^{+\infty} \left[Q_y^+ \delta(x - ml_x) + \frac{\partial}{\partial y} \left\{ M_y^+ \delta(x - ml_x) \right\} \right.$$

$$\left. + \frac{\partial}{\partial x} \left\{ M_{Ty}^+ \delta(x - ml_x) \right\} \right]$$

$$+ \sum_{n=-\infty}^{+\infty} \left[Q_x^+ \delta(y - nl_y) + \frac{\partial}{\partial x} \left\{ M_x^+ \delta(y - nl_y) \right\} \right.$$

$$\left. + \frac{\partial}{\partial y} \left\{ M_{Tx}^+ \delta(y - nl_y) \right\} \right]$$

$$+ P_1(x, y, 0) - P_2(x, y, h_1) \qquad (5.75)$$

5.2 Sound Transmission Through Sandwich Structures

$$D_2\nabla^4 w_2 + m_2\frac{\partial^2 w_2}{\partial t^2} = -\sum_{m=-\infty}^{+\infty}\left[Q_y^-\delta(x-ml_x) + \frac{\partial}{\partial y}\{M_y^-\delta(x-ml_x)\}\right.$$
$$\left. + \frac{\partial}{\partial x}\{M_{Ty}^-\delta(x-ml_x)\}\right]$$
$$-\sum_{n=-\infty}^{+\infty}\left[Q_x^-\delta(y-nl_y) + \frac{\partial}{\partial x}\{M_x^-\delta(y-nl_y)\}\right.$$
$$\left. + \frac{\partial}{\partial y}\{M_{Tx}^-\delta(y-nl_y)\}\right]$$
$$+ P_2(x,y,h_1+d) - P_3(x,y,h_1+h_2+d)$$
(5.76)

where $\nabla^4 = (\partial^2/\partial x^2 + \partial^2/\partial y^2)^2$; (w_1, w_2), (m_1, m_2), and (D_1, D_2) are the displacements, mass density per unit area, and flexural rigidity of the upper and bottom panels, respectively; and $\delta(\cdot)$ is the Dirac delta function.

Since the inertial effects (i.e., inertial tensional forces, inertial bending moments, and inertial torsional moments) of the rib-stiffeners have been taken into account, the factual tensional forces Q, bending moments M, and torsional moments M_T imposed on the two face panels are unequal. Therefore, as shown in Fig. 5.10, superscripts $+$ and $-$ associated separately with the upper and bottom panels are introduced to differentiate this discrepancy, with subscripts x and y introduced to signify the terms arising from the x-wise and y-wise rib-stiffeners, respectively.

Taking the inertial effects of the rib-stiffeners into consideration and applying Hooke's law and Newton's second law, one can express the tensional forces of the rib-stiffeners as [39]

$$Q_x^+ = -\frac{K_x(K_x - m_x\omega^2)}{2K_x - m_x\omega^2}w_1 + \frac{K_x^2}{2K_x - m_x\omega^2}w_2 \quad (5.77)$$

$$Q_x^- = -\frac{K_x^2}{2K_x - m_x\omega^2}w_1 + \frac{K_x(K_x - m_x\omega^2)}{2K_x - m_x\omega^2}w_2 \quad (5.78)$$

$$Q_y^+ = -\frac{K_y(K_y - m_y\omega^2)}{2K_y - m_y\omega^2}w_1 + \frac{K_y^2}{2K_y - m_y\omega^2}w_2 \quad (5.79)$$

$$Q_y^- = -\frac{K_y^2}{2K_y - m_y\omega^2}w_1 + \frac{K_y(K_y - m_y\omega^2)}{2K_y - m_y\omega^2}w_2 \quad (5.80)$$

where ω is the circular frequency, (K_x, K_y) are the tensional stiffness of half the rib-stiffeners per unit length, and (m_x, m_y) are the line mass density of the x-wise and y-wise rib-stiffeners, respectively.

Similarly, the bending moments of the rib-stiffeners can be expressed as [39]

$$M_x^+ = \frac{E_x I_x^* \left(E_x I_x^* - \rho_x I_x \omega^2\right)}{2E_x I_x^* - \rho_x I_x \omega^2} \frac{\partial^2 w_1}{\partial x^2} - \frac{E_x^2 I_x^{*2}}{2E_x I_x^* - \rho_x I_x \omega^2} \frac{\partial^2 w_2}{\partial x^2} \quad (5.81)$$

$$M_x^- = \frac{E_x^2 I_x^{*2}}{2E_x I_x^* - \rho_x I_x \omega^2} \frac{\partial^2 w_1}{\partial x^2} - \frac{E_x I_x^* \left(E_x I_x^* - \rho_x I_x \omega^2\right)}{2E_x I_x^* - \rho_x I_x \omega^2} \frac{\partial^2 w_2}{\partial x^2} \quad (5.82)$$

$$M_y^+ = \frac{E_y I_y^* \left(E_y I_y^* - \rho_y I_y \omega^2\right)}{2E_y I_y^* - \rho_y I_y \omega^2} \frac{\partial^2 w_1}{\partial y^2} - \frac{E_y^2 I_y^{*2}}{2E_y I_y^* - \rho_y I_y \omega^2} \frac{\partial^2 w_2}{\partial y^2} \quad (5.83)$$

$$M_y^- = \frac{E_y^2 I_y^{*2}}{2E_y I_y^* - \rho_y I_y \omega^2} \frac{\partial^2 w_1}{\partial y^2} - \frac{E_y I_y^* \left(E_y I_y^* - \rho_y I_y \omega^2\right)}{2E_y I_y^* - \rho_y I_y \omega^2} \frac{\partial^2 w_2}{\partial y^2} \quad (5.84)$$

where $(E_x I_x^*, E_y I_y^*)$ are the bending stiffness of half the rib-stiffeners per unit length and (ρ_x, ρ_y) and (I_x, I_y) are the mass density and polar moment of inertia for the rib-stiffeners, with subscripts x and y indicating the corresponding orientations of the rib-stiffeners.

Following the same procedures, the torsional moments of the rib-stiffeners are given by [39]

$$M_{Tx}^+ = \frac{G_x J_x^* \left(G_x J_x^* - \rho_x J_x \omega^2\right)}{2G_x J_x^* - \rho_x J_x \omega^2} \frac{\partial^2 w_1}{\partial x \partial y} - \frac{G_x^2 J_x^{*2}}{2G_x J_x^* - \rho_x J_x \omega^2} \frac{\partial^2 w_2}{\partial x \partial y} \quad (5.85)$$

$$M_{Tx}^- = \frac{G_x^2 J_x^{*2}}{2G_x J_x^* - \rho_x J_x \omega^2} \frac{\partial^2 w_1}{\partial x \partial y} - \frac{G_x J_x^* \left(G_x J_x^* - \rho_x J_x \omega^2\right)}{2G_x J_x^* - \rho_x J_x \omega^2} \frac{\partial^2 w_2}{\partial x \partial y} \quad (5.86)$$

$$M_{Ty}^+ = \frac{G_y J_y^* \left(G_y J_y^* - \rho_y J_y \omega^2\right)}{2G_y J_y^* - \rho_y J_y \omega^2} \frac{\partial^2 w_1}{\partial y \partial x} - \frac{G_y^2 J_y^{*2}}{2G_y J_y^* - \rho_y J_y \omega^2} \frac{\partial^2 w_2}{\partial y \partial x} \quad (5.87)$$

$$M_{Ty}^- = \frac{G_y^2 J_y^{*2}}{2G_y J_y^* - \rho_y J_y \omega^2} \frac{\partial^2 w_1}{\partial y \partial x} - \frac{G_y J_y^* \left(G_y J_y^* - \rho_y J_y \omega^2\right)}{2G_y J_y^* - \rho_y J_y \omega^2} \frac{\partial^2 w_2}{\partial y \partial x} \quad (5.88)$$

where $(G_x J_x^*, G_y J_y^*)$ are the torsional stiffness of half the rib-stiffeners per unit length and (J_x, J_y) are the torsional moments of inertia of the rib-stiffeners.

In the above expressions for the tensional forces, bending moments, and torsional moments, the geometrical properties of rib-stiffener cross sections are given by

$$K_x = \frac{E_x t_x}{d/2}, \quad K_y = \frac{E_y t_y}{d/2} \quad (5.89)$$

$$I_x^* = \frac{t_x (d/2)^3}{12}, \quad I_y^* = \frac{t_y (d/2)^3}{12}, \quad I_x = \frac{t_x d^3}{12}, \quad I_y = \frac{t_y d^3}{12} \quad (5.90)$$

5.2 Sound Transmission Through Sandwich Structures

$$J_x^* = \frac{t_x^3 d}{2}\left[\frac{1}{3} - \frac{64}{\pi^5}\frac{2t_x}{d}\sum_{n=1,3,5,\ldots}^{\infty}\frac{\tanh\left(n\pi d/4t_x\right)}{n^5}\right] \quad (5.91)$$

$$J_y^* = \frac{t_y^3 d}{2}\left[\frac{1}{3} - \frac{64}{\pi^5}\frac{2t_y}{d}\sum_{n=1,3,5,\ldots}^{\infty}\frac{\tanh\left(n\pi d/4t_y\right)}{n^5}\right] \quad (5.92)$$

$$J_x = t_x^3 d\left[\frac{1}{3} - \frac{64}{\pi^5}\frac{t_x}{d}\sum_{n=1,3,5,\ldots}^{\infty}\frac{\tanh\left(n\pi d/2t_x\right)}{n^5}\right] \quad (5.93)$$

$$J_y = t_y^3 d\left[\frac{1}{3} - \frac{64}{\pi^5}\frac{t_y}{d}\sum_{n=1,3,5,\ldots}^{\infty}\frac{\tanh\left(n\pi d/2t_y\right)}{n^5}\right] \quad (5.94)$$

To simplify Eqs. (5.4), (5.5), (5.6), (5.7), (5.8), (5.9), (5.10), (5.11), (5.12), (5.13), (5.14), and (5.15), the following sets of specified characteristics are utilized to replace the coefficients of the general displacements:

1. Replacement of tensional force coefficients:

$$R_{Q1} = \frac{K_x\left(K_x - m_x\omega^2\right)}{2K_x - m_x\omega^2}, \quad R_{Q2} = \frac{K_x^2}{2K_x - m_x\omega^2} \quad (5.95)$$

$$R_{Q3} = \frac{K_y\left(K_y - m_y\omega^2\right)}{2K_y - m_y\omega^2}, \quad R_{Q4} = \frac{K_y^2}{2K_y - m_y\omega^2} \quad (5.96)$$

2. Replacement of bending moment coefficients:

$$R_{M1} = \frac{E_x I_x^*\left(E_x I_x^* - \rho_x I_x\omega^2\right)}{2E_x I_x^* - \rho_x I_x\omega^2}, \quad R_{M2} = \frac{E_x^2 I_x^{*2}}{2E_x I_x^* - \rho_x I_x\omega^2} \quad (5.97)$$

$$R_{M3} = \frac{E_y I_y^*\left(E_y I_y^* - \rho_y I_y\omega^2\right)}{2E_y I_y^* - \rho_y I_y\omega^2}, \quad R_{M4} = \frac{E_y^2 I_y^{*2}}{2E_y I_y^* - \rho_y I_y\omega^2} \quad (5.98)$$

3. Replacement of torsional moment coefficients:

$$R_{T1} = \frac{G_x J_x^*\left(G_x J_x^* - \rho_x J_x\omega^2\right)}{2G_x J_x^* - \rho_x J_x\omega^2}, \quad R_{T2} = \frac{G_x^2 J_x^{*2}}{2G_x J_x^* - \rho_x J_x\omega^2} \quad (5.99)$$

$$R_{T3} = \frac{G_y J_y^*\left(G_y J_y^* - \rho_y J_y\omega^2\right)}{2G_y J_y^* - \rho_y J_y\omega^2}, \quad R_{T4} = \frac{G_y^2 J_y^{*2}}{2G_y J_y^* - \rho_y J_y\omega^2} \quad (5.100)$$

Adopting Eqs. (5.22), (5.23), (5.24), (5.25), (5.26), and (5.27) and substituting Eqs. (5.65) and (5.66) into Eqs. (5.4), (5.5), (5.6), (5.7), (5.8), (5.9), (5.10), (5.11),

(5.12), (5.13), (5.14), and (5.15), we can simplify the expressions of the tensional forces, bending moments, and torsional moments as follows:

1. Tensional forces:

$$Q_x^+ = \sum_{m=-\infty}^{+\infty} \sum_{n=-\infty}^{+\infty} (-R_{Q1}\alpha_{1,mn} + R_{Q2}\alpha_{2,mn}) e^{-i[(k_x+2m\pi/l_x)x+(k_y+2n\pi/l_y)y]}$$
(5.101)

$$Q_x^- = \sum_{m=-\infty}^{+\infty} \sum_{n=-\infty}^{+\infty} (-R_{Q2}\alpha_{1,mn} + R_{Q1}\alpha_{2,mn}) e^{-i[(k_x+2m\pi/l_x)x+(k_y+2n\pi/l_y)y]}$$
(5.102)

$$Q_y^+ = \sum_{m=-\infty}^{+\infty} \sum_{n=-\infty}^{+\infty} (-R_{Q3}\alpha_{1,mn} + R_{Q4}\alpha_{2,mn}) e^{-i[(k_x+2m\pi/l_x)x+(k_y+2n\pi/l_y)y]}$$
(5.103)

$$Q_y^- = \sum_{m=-\infty}^{+\infty} \sum_{n=-\infty}^{+\infty} (-R_{Q4}\alpha_{1,mn} + R_{Q3}\alpha_{2,mn}) e^{-i[(k_x+2m\pi/l_x)x+(k_y+2n\pi/l_y)y]}$$
(5.104)

2. Bending moments:

$$M_x^+ = \sum_{m=-\infty}^{+\infty} \sum_{n=-\infty}^{+\infty} (-R_{M1}\alpha_{1,mn} + R_{M2}\alpha_{2,mn}) \alpha_m^2 e^{-i[(k_x+2m\pi/l_x)x+(k_y+2n\pi/l_y)y]}$$
(5.105)

$$M_x^- = \sum_{m=-\infty}^{+\infty} \sum_{n=-\infty}^{+\infty} (-R_{M2}\alpha_{1,mn} + R_{M1}\alpha_{2,mn}) \alpha_m^2 e^{-i[(k_x+2m\pi/l_x)x+(k_y+2n\pi/l_y)y]}$$
(5.106)

$$M_y^+ = \sum_{m=-\infty}^{+\infty} \sum_{n=-\infty}^{+\infty} (-R_{M3}\alpha_{1,mn} + R_{M4}\alpha_{2,mn}) \beta_n^2 e^{-i[(k_x+2m\pi/l_x)x+(k_y+2n\pi/l_y)y]}$$
(5.107)

$$M_y^- = \sum_{m=-\infty}^{+\infty} \sum_{n=-\infty}^{+\infty} (-R_{M4}\alpha_{1,mn} + R_{M3}\alpha_{2,mn}) \beta_n^2 e^{-i[(k_x+2m\pi/l_x)x+(k_y+2n\pi/l_y)y]}$$
(5.108)

3. Torsional moments:

$$M_{Tx}^+ = \sum_{m=-\infty}^{+\infty} \sum_{n=-\infty}^{+\infty} (-R_{T1}\alpha_{1,mn} + R_{T2}\alpha_{2,mn}) \alpha_m \beta_n e^{-i[(k_x+2m\pi/l_x)x+(k_y+2n\pi/l_y)y]}$$
(5.109)

$$M^-_{Tx} = \sum_{m=-\infty}^{+\infty}\sum_{n=-\infty}^{+\infty} (-R_{T2}\alpha_{1,mn}+R_{T1}\alpha_{2,mn})\alpha_m\beta_n e^{-i[(k_x+2m\pi/l_x)x+(k_y+2n\pi/l_y)y]}$$
(5.110)

$$M^+_{Ty} = \sum_{m=-\infty}^{+\infty}\sum_{n=-\infty}^{+\infty} (-R_{T3}\alpha_{1,mn}+R_{T4}\alpha_{2,mn})\alpha_m\beta_n e^{-i[(k_x+2m\pi/l_x)x+(k_y+2n\pi/l_y)y]}$$
(5.111)

$$M^-_{Ty} = \sum_{m=-\infty}^{+\infty}\sum_{n=-\infty}^{+\infty} (-R_{T4}\alpha_{1,mn}+R_{T3}\alpha_{2,mn})\alpha_m\beta_n e^{-i[(k_x+2m\pi/l_x)x+(k_y+2n\pi/l_y)y]}$$
(5.112)

5.2.3 The Acoustic Pressure and Continuity Condition

The acoustic pressures $P_1(x,y,z)$ in the incident field, $P_2(x,y,z)$ in the field between the two face panels, and $P_3(x,y,z)$ in the transmitted field all satisfy the wave equation [9, 12, 45]:

$$\left[\frac{\partial^2}{\partial x^2}+\frac{\partial^2}{\partial y^2}+\frac{\partial^2}{\partial z^2}+k_0^2\right]P_i = 0 \quad (i=1,2,3) \qquad (5.113)$$

where k_0 is the wavenumber of the incident sound. The momentum equation is applied to ensure the equality of panel velocity and fluid velocity on the panel surface, i.e., the continuity condition of fluid-structure coupling [14, 57]:

$$\left.\frac{\partial P_1}{\partial z}\right|_{z=0} = \omega^2\rho_0 w_1, \quad \left.\frac{\partial P_2}{\partial z}\right|_{z=h_1} = \omega^2\rho_0 w_1 \qquad (5.114)$$

$$\left.\frac{\partial P_2}{\partial z}\right|_{z=h_1+d} = \omega^2\rho_0 w_2, \quad \left.\frac{\partial P_3}{\partial z}\right|_{z=h_1+h_2+d} = \omega^2\rho_0 w_2 \qquad (5.115)$$

where ρ_0 is the ambient acoustic fluid density. Substitution of Eqs. (5.65) and (5.66) as well as Eqs. (5.69), (5.70), and (5.71) into Eqs. (5.114) and (5.115) gives rise to

$$-ik_z I e^{-i(k_x x+k_y y)} + \sum_{m=-\infty}^{+\infty}\sum_{n=-\infty}^{+\infty}(ik_{z,mn}\beta_{mn}-\omega^2\rho_0\alpha_{1,mn})$$
$$\times e^{-i[(k_x+2m\pi/l_x)x+(k_y+2n\pi/l_y)y]} = 0 \qquad (5.116)$$

$$\sum_{m=-\infty}^{+\infty}\sum_{n=-\infty}^{+\infty}\left[ik_{z,mn}\left(-\varepsilon_{mn}e^{-ik_{z,mn}h_1}+\zeta_{mn}e^{ik_{z,mn}h_1}\right)-\omega^2\rho_0\alpha_{1,mn}\right]$$
$$\times e^{-i[(k_x+2m\pi/l_x)x+(k_y+2n\pi/l_y)y]} = 0 \qquad (5.117)$$

$$\sum_{m=-\infty}^{+\infty}\sum_{n=-\infty}^{+\infty}\left[\begin{array}{c}ik_{z,mn}\left(-\varepsilon_{mn}e^{-ik_{z,mn}(h_1+d)}+\zeta_{mn}e^{ik_{z,mn}(h_1+d)}\right)\\-\omega^2\rho_0\alpha_{2,mn}\end{array}\right]$$
$$\times e^{-i[(k_x+2m\pi/l_x)x+(k_y+2n\pi/l_y)y]}=0 \qquad (5.118)$$

$$\sum_{m=-\infty}^{+\infty}\sum_{n=-\infty}^{+\infty}\left[-ik_{z,mn}\xi_{mn}e^{-ik_{z,mn}(h_1+h_2+d)}-\omega^2\rho_0\alpha_{2,mn}\right]$$
$$\times e^{-i[(k_x+2m\pi/l_x)x+(k_y+2n\pi/l_y)y]}=0 \qquad (5.119)$$

Because Eqs. (5.116), (5.117), (5.118), and (5.119) hold for all possible values of x and y, the relevant coefficients have the following relationships:

$$\beta_{00}=I+\frac{\omega^2\rho_0\alpha_{1,00}}{ik_z} \qquad (5.120)$$

$$\beta_{mn}=\frac{\omega^2\rho_0\alpha_{1,mn}}{ik_{z,mn}},\quad \text{at } m\neq 0\,||\,n\neq 0 \qquad (5.121)$$

$$\varepsilon_{mn}=\frac{\omega^2\rho_0\left[\alpha_{1,mn}e^{ik_{z,mn}(h_1+d)}-\alpha_{2,mn}e^{ik_{z,mn}h_1}\right]}{2k_{z,mn}\sin(k_{z,mn}d)} \qquad (5.122)$$

$$\zeta_{mn}=\frac{\omega^2\rho_0\left[\alpha_{1,mn}e^{-ik_{z,mn}(h_1+d)}-\alpha_{2,mn}e^{-ik_{z,mn}h_1}\right]}{2k_{z,mn}\sin(k_{z,mn}d)} \qquad (5.123)$$

$$\xi_{mn}=-\frac{\omega^2\rho_0\alpha_{2,mn}}{ik_{z,mn}}e^{ik_{z,mn}(h_1+h_2+d)} \qquad (5.124)$$

5.2.4 Solution of the Formulations with the Virtual Work Principle

As can be seen from Eqs. (5.120), (5.121), (5.122), (5.123), and (5.124), once coefficients $\alpha_{1,mn}$ and $\alpha_{2,mn}$ (i.e., modal amplitudes of the (m,n)th space-harmonic flexural waves in the upper and bottom panels, respectively) are determined, coefficients β_{mn}, ε_{mn}, ζ_{mn}, and ξ_{mn} are also determined. Coefficients $\alpha_{1,mn}$ and $\alpha_{2,mn}$ can be obtained by solving the system equations derived by applying the principle of virtual work [7, 21, 30]. In view of the spatial periodicity of the structure, it is necessary to consider only the virtual work contribution from one period of element (including the attached rib-stiffeners). As the statement of the virtual work principle,

5.2 Sound Transmission Through Sandwich Structures

the sum of the work done by all the elements in one period of the system must be zero when the system has any one of the virtual displacements:

$$\delta w_j = \delta \alpha_{j,mn} e^{-i[(k_x+2m\pi/l_x)x+(k_y+2n\pi/l_y)y]} \quad (j=1,2) \tag{5.125}$$

5.2.5 Virtual Work of Panel Elements

The equations governing the vibration responses of the two panel elements in one period of the structure are

$$D_1 \nabla^4 w_1 + m_1 \frac{\partial^2 w_1}{\partial t^2} - P_1(x,y,0) + P_2(x,y,h_1) = 0 \tag{5.126}$$

$$D_2 \nabla^4 w_2 + m_2 \frac{\partial^2 w_2}{\partial t^2} - P_2(x,y,h_1+d) + P_3(x,y,h_1+h_2+d) = 0 \tag{5.127}$$

The virtual work contributed solely by the panel elements can then be represented as

$$\delta \Pi_{p1} = \int_0^{l_x} \int_0^{l_y} \left[D_1 \nabla^4 w_1 + m_1 \frac{\partial^2 w_1}{\partial t^2} - P_1(x,y,0) + P_2(x,y,h_1) \right] \cdot \delta w_1^* dxdy \tag{5.128}$$

$$\delta \Pi_{p2} = \int_0^{l_x} \int_0^{l_y} \left[D_2 \nabla^4 w_2 + m_2 \frac{\partial^2 w_2}{\partial t^2} - P_2(x,y,h_1+d) \right.$$
$$\left. + P_3(x,y,h_1+h_2+d) \right] \cdot \delta w_2^* dxdy \tag{5.129}$$

where δw_1^* and δw_2^* denote the complex conjugate of the virtual displacement in Eq. (5.125). Together with Eqs. (5.65) and (5.66); Eqs. (5.69), (5.70), and (5.71); and Eqs. (5.120), (5.121), (5.122), (5.123), and (5.124), Eqs. (5.128) and (5.129) can be rewritten in terms of modal amplitudes $\alpha_{1,kl}$ and $\alpha_{2,kl}$ as

$$\delta \Pi_{p1} == \left\{ \left(D_1 \left[\left(k_x + \frac{2k\pi}{l_x} \right)^2 + \left(k_y + \frac{2l\pi}{l_y} \right)^2 \right]^2 - m_1 \omega^2 \right) \alpha_{1,kl} \right.$$
$$\left. - \frac{\omega^2 \rho_0 \alpha_{1,kl}}{ik_{z,kl}} + \frac{\omega^2 \rho_0 [\alpha_{1,kl} \cos(k_{z,kl} d) - \alpha_{2,kl}]}{k_{z,kl} \sin(k_{z,kl} d)} \right\} \cdot l_x l_y \cdot \delta \alpha_{1,kl}$$
$$- \int_0^{l_x} \int_0^{l_y} 2I e^{-i(k_x x + k_y y)} e^{i[(k_x+2k\pi/l_x)x+(k_y+2l\pi/l_y)y]} dxdy \cdot \delta \alpha_{1,kl}$$

$$\tag{5.130}$$

$$\delta\Pi_{p2} == \left\{ \left(D_2 \left[\left(k_x + \frac{2k\pi}{l_x}\right)^2 + \left(k_y + \frac{2l\pi}{l_y}\right)^2 \right]^2 - m_2\omega^2 \right) \alpha_{2,kl} \right.$$
$$\left. - \frac{\omega^2 \rho_0 \alpha_{2,kl}}{ik_{z,kl}} - \frac{\omega^2 \rho_0 \left[\alpha_{1,kl} - \alpha_{2,kl} \cos(k_{z,kl}d)\right]}{k_{z,kl} \sin(k_{z,kl}d)} \right\} \cdot l_x l_y \cdot \delta\alpha_{2,kl}$$

(5.131)

5.2.6 Virtual Work of x-Wise Rib-Stiffeners

The virtual work contributions from the tensional forces, bending moments, and torsional moments at the interfaces between the x-wise rib-stiffeners (aligned with $y = 0$) with the upper and bottom panels are given by

$$\delta\Pi_{x1} = -\int_0^{l_x} \left[Q_x^+(x,0) + \frac{\partial}{\partial x} M_x^+(x,0) + \frac{\partial}{\partial y} M_{Tx}^+(x,0) \right] \cdot \delta\alpha_{1,kl} e^{i(k_x + 2k\pi/l_x)x} dx$$
$$= \sum_{n=-\infty}^{+\infty} \left[R_{Q1}\alpha_{1,kn} - R_{Q2}\alpha_{2,kn} + i\left(k_x + \frac{2k\pi}{l_x}\right)^3 (-R_{M1}\alpha_{1,kn} + R_{M2}\alpha_{2,kn}) \right.$$
$$\left. + i\left(k_x + \frac{2k\pi}{l_x}\right)\left(k_y + \frac{2l\pi}{l_y}\right)\left(k_y + \frac{2n\pi}{l_y}\right)(-R_{T1}\alpha_{1,kn} + R_{T2}\alpha_{2,kn}) \right]$$
$$\cdot l_x \delta\alpha_{1,kl}$$

(5.132)

$$\delta\Pi_{x2} = \int_0^{l_x} \left[Q_x^-(x,0) + \frac{\partial}{\partial x} M_x^-(x,0) + \frac{\partial}{\partial y} M_{Tx}^-(x,0) \right] \cdot \delta\alpha_{2,kl} e^{i(k_x + 2k\pi/l_x)x} dx$$
$$= \sum_{n=-\infty}^{+\infty} \left[-R_{Q2}\alpha_{1,kn} + R_{Q1}\alpha_{2,kn} - i\left(k_x + \frac{2k\pi}{l_x}\right)^3 (-R_{M2}\alpha_{1,kn} + R_{M1}\alpha_{2,kn}) \right.$$
$$\left. - i\left(k_x + \frac{2k\pi}{l_x}\right)\left(k_y + \frac{2l\pi}{l_y}\right)\left(k_y + \frac{2n\pi}{l_y}\right)(-R_{T2}\alpha_{1,kn} + R_{T1}\alpha_{2,kn}) \right]$$
$$\cdot l_x \delta\alpha_{2,kl}$$

(5.133)

where $\frac{\partial}{\partial y} M_{Tx}^+(x, y = 0) = \frac{\partial}{\partial y} M_{Tx}^+(x, y)\big|_{y=0}$ and $\frac{\partial}{\partial y} M_{Tx}^-(x, y = 0) = \frac{\partial}{\partial y} M_{Tx}^-(x, y)\big|_{y=0}$.

5.2.7 Virtual Work of y-Wise Rib-Stiffeners

Likewise, the virtual work contributions from the tensional forces, bending moments, and torsional moments at the interfaces between the y-wise rib-stiffeners (aligned with $x = 0$) with the upper and bottom panels are

$$\delta\Pi_{y1} = -\int_0^{l_y} \left[Q_y^+(0,y) + \frac{\partial}{\partial y} M_y^+(0,y) + \frac{\partial}{\partial x} M_{Ty}^+(0,y) \right] \cdot \delta\alpha_{1,kl} e^{i(k_y + 2l\pi/l_y)y} dy$$

$$= \sum_{m=-\infty}^{+\infty} \left[R_{Q3}\alpha_{1,ml} - R_{Q4}\alpha_{2,ml} + i\left(k_y + \frac{2l\pi}{l_y}\right)^3 (-R_{M3}\alpha_{1,ml} + R_{M4}\alpha_{2,ml}) \right.$$

$$\left. + i\left(k_x + \frac{2k\pi}{l_x}\right)\left(k_x + \frac{2m\pi}{l_x}\right)\left(k_y + \frac{2l\pi}{l_y}\right)(-R_{T3}\alpha_{1,ml} + R_{T4}\alpha_{2,ml}) \right]$$

$$\cdot l_y \delta\alpha_{1,kl} \tag{5.134}$$

$$\delta\Pi_{y2} = \int_0^{l_y} \left[Q_y^-(0,y) + \frac{\partial}{\partial y} M_y^-(0,y) + \frac{\partial}{\partial x} M_{Ty}^-(0,y) \right] \cdot \delta\alpha_{2,kl} e^{i(k_y + 2l\pi/l_y)y} dy$$

$$= \sum_{m=-\infty}^{+\infty} \left[-R_{Q4}\alpha_{1,ml} + R_{Q3}\alpha_{2,ml} - i\left(k_y + \frac{2l\pi}{l_y}\right)^3 (-R_{M4}\alpha_{1,ml} + R_{M3}\alpha_{2,ml}) \right.$$

$$\left. - i\left(k_x + \frac{2k\pi}{l_x}\right)\left(k_x + \frac{2m\pi}{l_x}\right)\left(k_y + \frac{2l\pi}{l_y}\right)(-R_{T4}\alpha_{1,ml} + R_{T3}\alpha_{2,ml}) \right]$$

$$\cdot l_y \delta\alpha_{2,kl} \tag{5.135}$$

where $\frac{\partial}{\partial x} M_{Ty}^+(x=0, y) = \frac{\partial}{\partial x} M_{Ty}^+(x, y)\big|_{x=0}$ and $\frac{\partial}{\partial x} M_{Ty}^-(x=0, y) = \frac{\partial}{\partial x} M_{Ty}^-(x, y)\big|_{x=0}$.

5.2.8 Combination of Equations

Finally, the virtual work principle requires that

$$\delta\Pi_{p1} + \delta\Pi_{x1} + \delta\Pi_{y1} = 0 \tag{5.136}$$

$$\delta\Pi_{p2} + \delta\Pi_{x2} + \delta\Pi_{y2} = 0 \tag{5.137}$$

Substituting Eqs. (5.130), (5.132), and (5.134) into (5.136) and Eqs. (5.131), (5.133), and (5.135) into (5.137) and noting that the virtual displacement is arbitrary, we obtain

$$\left\{\left[D_1(\alpha_k^2+\beta_l^2)^2 - m_1\omega^2\right]\alpha_{1,kl} - \frac{\omega^2 \rho_0 \alpha_{1,kl}}{ik_{z,kl}} + \frac{\omega^2 \rho_0 \left[\alpha_{1,kl}\cos(k_{z,kl}d) - \alpha_{2,kl}\right]}{k_{z,kl}\sin(k_{z,kl}d)}\right\} \cdot l_x l_y$$

$$+ \sum_{n=-\infty}^{+\infty} \left[R_{Q1}\alpha_{1,kn} - R_{Q2}\alpha_{2,kn} + i\alpha_k \left(-R_{M1}\alpha_{1,kn} + R_{M2}\alpha_{2,kn}\right)\alpha_k^2 \right.$$

$$\left. + i\beta_l \left(-R_{T1}\alpha_{1,kn} + R_{T2}\alpha_{2,kn}\right)\alpha_k \beta_n\right] \cdot l_x$$

$$+ \sum_{m=-\infty}^{+\infty} \left[R_{Q3}\alpha_{1,ml} - R_{Q4}\alpha_{2,ml} + i\beta_l \left(-R_{M3}\alpha_{1,ml} + R_{M4}\alpha_{2,ml}\right)\beta_l^2 \right.$$

$$\left. + i\alpha_k \left(-R_{T3}\alpha_{1,ml} + R_{T4}\alpha_{2,ml}\right)\alpha_m \beta_l\right] \cdot l_y$$

$$= \begin{cases} 2Il_x l_y & \text{when } k = 0 \ \& \ l = 0 \\ 0 & \text{when } k \neq 0 \ || \ l \neq 0 \end{cases} \tag{5.138}$$

$$\left\{\left[D_2(\alpha_k^2+\beta_l^2)^2 - m_2\omega^2\right]\alpha_{2,kl} - \frac{\omega^2 \rho_0 \alpha_{2,kl}}{ik_{z,kl}} - \frac{\omega^2 \rho_0 \left[\alpha_{1,kl} - \alpha_{2,kl}\cos(k_{z,kl}d)\right]}{k_{z,kl}\sin(k_{z,kl}d)}\right\} \cdot l_x l_y$$

$$+ \sum_{n=-\infty}^{+\infty} \left[-R_{Q2}\alpha_{1,kn} + R_{Q1}\alpha_{2,kn} - i\alpha_k \left(-R_{M2}\alpha_{1,kn} + R_{M1}\alpha_{2,kn}\right)\alpha_k^2 \right.$$

$$\left. - i\beta_l \left(-R_{T2}\alpha_{1,kn} + R_{T1}\alpha_{2,kn}\right)\alpha_k \beta_n\right] \cdot l_x$$

$$+ \sum_{m=-\infty}^{+\infty} \left[-R_{Q4}\alpha_{1,ml} + R_{Q3}\alpha_{2,ml} - i\beta_l \left(-R_{M4}\alpha_{1,ml} + R_{M3}\alpha_{2,ml}\right)\beta_l^2 \right.$$

$$\left. - i\alpha_k \left(-R_{T4}\alpha_{1,ml} + R_{T3}\alpha_{2,ml}\right)\alpha_m \beta_l\right] \cdot l_y$$

$$= 0 \tag{5.139}$$

where

$$\alpha_m = k_x + \frac{2m\pi}{l_x}, \quad \beta_n = k_y + \frac{2n\pi}{l_y} \tag{5.140}$$

Note that consideration of the virtual work in any other period of the structural element would have yielded an identical set of equations.

In order to separate the variants $\alpha_{1,kl}$ and $\alpha_{2,kl}$, Eqs. (5.138) and (5.139) are rewritten as

5.2 Sound Transmission Through Sandwich Structures

$$\left[D_1(\alpha_k^2+\beta_l^2)^2 - m_1\omega^2 - \frac{\omega^2\rho_0}{ik_{z,kl}} + \frac{\omega^2\rho_0 \cos(k_{z,kl}d)}{k_{z,kl}\sin(k_{z,kl}d)} \right] \cdot l_x l_y \alpha_{1,kl} - \frac{\omega^2\rho_0}{k_{z,kl}\sin(k_{z,kl}d)} \cdot l_x l_y \alpha_{2,kl}$$

$$+ \sum_{n=-\infty}^{+\infty} \left[R_{Q1} - i\alpha_k^3 R_{M1} - i\beta_l \alpha_k \beta_n R_{T1} \right] \cdot l_x \alpha_{1,kn} + \sum_{n=-\infty}^{+\infty} \left[-R_{Q2} + i\alpha_k^3 R_{M2} + i\beta_l \alpha_k \beta_n R_{T2} \right] \cdot l_x \alpha_{2,kn}$$

$$+ \sum_{m=-\infty}^{+\infty} \left[R_{Q3} - i\beta_l^3 R_{M3} - i\alpha_k \alpha_m \beta_l R_{T3} \right] \cdot l_y \alpha_{1,ml} + \sum_{m=-\infty}^{+\infty} \left[-R_{Q4} + i\beta_l^3 R_{M4} + i\alpha_k \alpha_m \beta_l R_{T4} \right] \cdot l_y \alpha_{2,ml}$$

$$= \begin{cases} 2I l_x l_y & \text{when} \quad k=0 \,\&\, l=0 \\ 0 & \text{when} \quad k \neq 0 \,||\, l \neq 0 \end{cases} \tag{5.141}$$

$$\left[D_2(\alpha_k^2+\beta_l^2)^2 - m_2\omega^2 - \frac{\omega^2\rho_0}{ik_{z,kl}} + \frac{\omega^2\rho_0 \cos(k_{z,kl}d)}{k_{z,kl}\sin(k_{z,kl}d)} \right] \cdot l_x l_y \alpha_{2,kl} - \frac{\omega^2\rho_0}{k_{z,kl}\sin(k_{z,kl}d)} \cdot l_x l_y \alpha_{1,kl}$$

$$+ \sum_{n=-\infty}^{+\infty} \left[-R_{Q2} + i\alpha_k^3 R_{M2} + i\beta_l \alpha_k \beta_n R_{T2} \right] \cdot l_x \alpha_{1,kn} + \sum_{n=-\infty}^{+\infty} \left[R_{Q1} - i\alpha_k^3 R_{M1} - i\beta_l \alpha_k \beta_n R_{T1} \right] \cdot l_x \alpha_{2,kn}$$

$$+ \sum_{m=-\infty}^{+\infty} \left[-R_{Q4} + i\beta_l^3 R_{M4} + i\alpha_k \alpha_m \beta_l R_{T4} \right] \cdot l_y \alpha_{1,ml} + \sum_{m=-\infty}^{+\infty} \left[R_{Q3} - i\beta_l^3 R_{M3} - i\alpha_k \alpha_m \beta_l R_{T3} \right] \cdot l_y \alpha_{2,ml}$$

$$= 0 \tag{5.142}$$

where the coupling relations between the modal amplitudes of sound waves in air and those of flexural waves in panels defined in Eqs. (5.120), (5.121), (5.122), (5.123), and (5.124) have been included. Equations (5.141) and (5.142) form an infinite set of coupled algebraic simultaneous equations. The solution of a suitably restricted set of these equations allows the modal amplitudes $\alpha_{1,kl}$ and $\alpha_{2,kl}$ to be determined. Insofar as the solution converges, these equations are solved simultaneously by truncation, namely, restricting the sum index (m, n) in the finite ranges of $m = -\hat{k}$ to $\hat{k}$ and $n = -\hat{l}$ to $\hat{l}$. With laborious but straightforward algebraic manipulations, the resulting simultaneous equations containing a finite number [i.e., $2KL$, where $K = 2\hat{k} + 1$, $L = 2\hat{l} + 1$] of unknowns can be grouped into matrix form as

$$\begin{bmatrix} T_{11} & T_{12} \\ T_{21} & T_{22} \end{bmatrix}_{2KL \times 2KL} \begin{Bmatrix} \alpha_{1,kl} \\ \alpha_{2,kl} \end{Bmatrix}_{2KL \times 1} = \begin{Bmatrix} F_{kl} \\ 0 \end{Bmatrix}_{2KL \times 1} \tag{5.143}$$

Detailed derivations of Eq. (5.143) can be found in Appendix B. Once the unknowns $\alpha_{1,kl}$ and $\alpha_{2,kl}$ are determined by solving Eq. (5.143), the displacements (w_1, w_2) of the panels and the sound pressure (P_1, P_2, P_3) in the ambient acoustic fluids adjacent to the two panels are also determined, enabling the sound transmission analysis of the fluid-loaded orthogonally rib-stiffened sandwich structure.

5.2.9 Definition of Sound Transmission Loss

As can be seen from Eqs. (5.65), (5.66), (5.67), (5.68), (5.69), (5.70), and (5.71), a component of the convective fluid-loaded pressure in the form of harmonic plane sound wave with wavenumbers (k_x, k_y) would induce sets of space-harmonic wave components in the response (including sound pressure) with wavenumbers $(k_x + 2m\pi/l_x, k_y + 2n\pi/l_y)$, where (m, n) take values $(-\infty < m < +\infty, -\infty < n < +\infty)$. This implies that the groups of harmonic waves may travel in opposite directions. The appearance of the series of space-harmonic waves in the response stems from the periodic rib-stiffeners attached to the panels. For a given convective fluid-loaded pressure with wavenumbers (k_x, k_y), a bending wave having the same wavenumbers is induced which then travels in the structure. The outspreading bending wave would be polarized as a group of harmonic wave components identified by wavenumbers $(a_0 + 2m\pi/l_x, \beta_0 + 2n\pi/l_y)$, owing to the complex interaction between the bending waves in panels and the motion of the mth x-wise and nth y-wise rib-stiffeners.

Given that (k_x, k_y) are real, the wavenumber $k_{z,mn}$ of the (m, n)th harmonic wave in the z-direction may be either real or pure imaginary (see Eqs. (5.73) and (5.74)). In the case of $k_{z,mn}$ being imaginary, the (m, n)th component of the wave decays exponentially with increasing distance in the z-direction and radiates no energy. This corresponds to a subsonic wave, i.e., non-radiating wave [40], satisfying that

$$\left(k_x + \frac{2m\pi}{l_x}\right)^2 + \left(k_y + \frac{2n\pi}{l_y}\right)^2 > \left(\frac{\omega}{c_0}\right)^2 \tag{5.144}$$

Therefore, this (m, n)th component contributes only to the near field. Only when $k_{z,mn}$ is real, the (m, n)th component could contribute to the far-field sound pressure [5, 14], which pertains to a supersonic wave, i.e., radiating wave [40], satisfying that

$$\left(k_x + \frac{2m\pi}{l_x}\right)^2 + \left(k_y + \frac{2n\pi}{l_y}\right)^2 < \left(\frac{\omega}{c_0}\right)^2 \tag{5.145}$$

To facilitate the physical understanding of sound transmission, the transmission coefficient is defined here as the ratio of the transmitted sound power to the incident sound power:

$$\tau(\varphi, \theta) = \frac{\sum_{m=-\infty}^{+\infty} \sum_{n=-\infty}^{+\infty} |\xi_{mn}|^2 \mathrm{Re}\,(k_{z,mn})}{|I|^2 k_z} \tag{5.146}$$

which is dependent upon the incident angles φ and θ. Sound transmission loss (STL) is then customarily defined as the inverse of the power transmission coefficient in decibel scale [9, 12]:

5.2 Sound Transmission Through Sandwich Structures

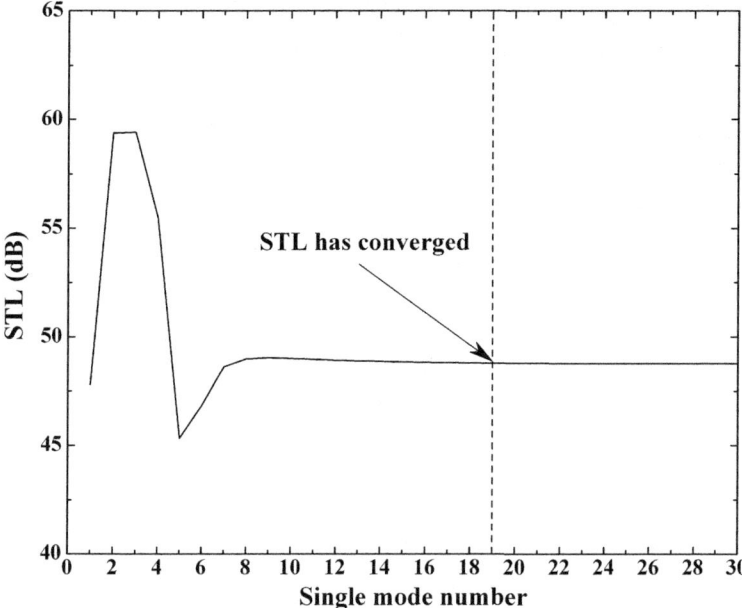

Fig. 5.11 Convergence check of double space-harmonic series solution for an infinite orthogonally rib-stiffened sandwich structure excited by a normal incident sound at 10 kHz (With permission from Elsevier)

$$\text{STL} = 10 \ \log_{10} \left(\frac{1}{\tau(\varphi,\theta)} \right) \quad (5.147)$$

Physically, STL is a measure of the effectiveness of the considered sandwich structure in isolating the transmission of convective fluid-loaded pressure.

5.2.10 Convergence Check for Space-Harmonic Series Solution

Since the analytic model is hinged on the assumed double-series solution given in Eqs. (5.65), (5.66), (5.67), (5.68), (5.69), (5.70), and (5.71), a sufficiently large number of terms have to be used to ensure the convergence and accuracy of the solution. There exists an admissible criterion (Lee and Kim [7]; Xin et al. [18]) that once the solution converges at a given frequency, it is also convergent for all frequencies lower than that. Therefore, the needed number of series terms is determined by the highest frequency of interest (i.e., 10 kHz, the frequency range below which is of concern here). Convergence check is thus performed by calculating STL value at 10 kHz, with progressively more terms used in the double-series expansion (as shown in Fig. 5.11). Once the difference between two successive STL calculations falls within a preset error band (5.01 dB selected in this work), the solution is deemed to have converged, and then the corresponding number of terms is adopted to calculate STL values at all other frequencies below 10 kHz.

In view of the symmetry of the present periodic structures in x- and y-direction, the equations are truncated as a finite set of equations with $m = -\widehat{k}$ to $\widehat{k}$ and $n = -\widehat{l}$ to $\widehat{l}$ ($\widehat{k} = \widehat{l}$ assumed) and then solved simultaneously. Figure 5.11 shows the convergence tendency of STL solution as the single-mode number $\widehat{k}$ ($=\widehat{l}$) is increased, when the sandwich structure is excited by a normal incident sound at 10 kHz. The results of Fig. 5.11 demonstrate that the solution converges when $\widehat{k} \geq 19$. In other words, it needs at least 1,521 terms (m and n both ranging from -19 to 19) to ensure solution convergence at 10 kHz. The same number (1,521 terms) is employed in subsequent STL calculations below 10 kHz, sufficient for obtaining accurate results within the error band of 5.01 dB.

5.2.11 Validation of the Analytic Model

The validity and feasibility of the proposed analytic model for sound transmission across an infinite orthogonally rib-stiffened sandwich structure subjected to convective fluid-loaded pressure is checked by comparing the model predictions and those obtained by Wang et al. [30] for sound transmission through an infinite sandwich structure with parallel rib-stiffeners as the core. While Wang et al. [30] model a single rib-stiffener as a combination of translational spring and rotational spring, the tensional, bending, and torsional vibrations of the rib-stiffener are modeled as an ensemble in the present analytic model. To make the comparison possible, the sets of orthogonal rib-stiffeners are simplified as one set of parallel rib-stiffeners. To this end, without loss of generality, the key parameters (i.e., Young's modulus E_x, density ρ_x, and thickness t_x) of the x-wise rib-stiffeners are set to zero, so that the orthogonally rib-stiffened sandwich construction is equivalent to a parallelly rib-stiffened sandwich structure. Of course, the material and geometrical properties of the structure adopted by Wang et al. [30] are fully followed in the comparison.

To highlight the necessity and advantage of the exact consideration of rib-stiffener motions in sound transmission prediction for the whole structure, results obtained using both the complete model and the simplified model are compared with the predictions of Wang et al. [30], as shown in Fig. 5.12. Here, the complete model proposed in Sects. 5.2.2–5.2.9 not only treats the motions of the rib-stiffeners as an ensemble of tensional, bending, and torsional vibrations but also considers their inertial effects. The simplified model only retains the tensional forces, inertial tensional forces, and bending moments of the y-wise rib-stiffeners in Eqs. (5.1) and (5.2), corresponding to the translational forces, inertial forces of lumped masses, and rotational forces in Wang et al.'s model, respectively.

Overall, as illustrated in Fig. 5.12, the predictions of the simplified model agree well with those obtained by Wang et al. [30]. The visible discrepancies (at relatively

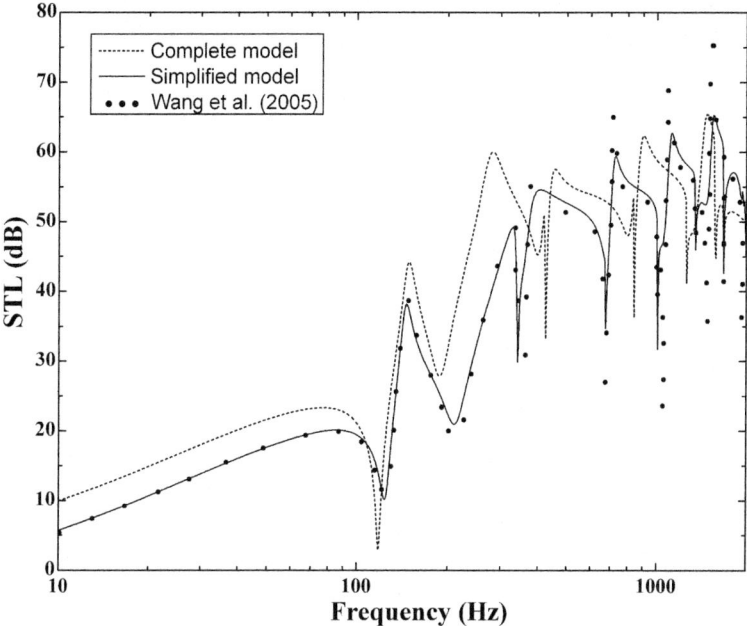

Fig. 5.12 Sound transmission loss plotted as a function of frequency for infinite sandwich structure with parallel rib-stiffeners as core subjected to oblique ($\varphi = 45°$) incident sound: comparison between predictions by the present analytic model (both complete model and simplified model) and those by Wang et al. [30] (With permission from Elsevier)

high frequencies in particular) between the two models are attributable to the fact that, for simplicity, Wang et al. approximated the lumped mass per rib-stiffener as distributed mass which was then added to the panel mass. While the STL versus frequency curves predicted by Wang et al. and the simplified model have an overall tendency of that predicted by the complete model, noticeable discrepancies are also observed in Fig. 5.12. This confirms the necessity and advantage of the present analytic formulations for modeling the structural and acoustic behaviors of rib-stiffened sandwich structures subjected to convective fluid-loaded pressure.

5.2.12 Influence of Sound Incident Angles

Since the incident azimuth angle θ plays a negligible role here, the influence of sound incident angle is mainly examined by comparing STL values calculated for three different incident elevation angles (i.e., $\varphi = 0°, 30°, 60°$) with the azimuth angle fixed at $\theta = 45°$, as shown in Fig. 5.13.

The results of Fig. 5.13 demonstrate that the incident elevation angle has a significant effect on the STL of the present sandwich structure. It is observed that

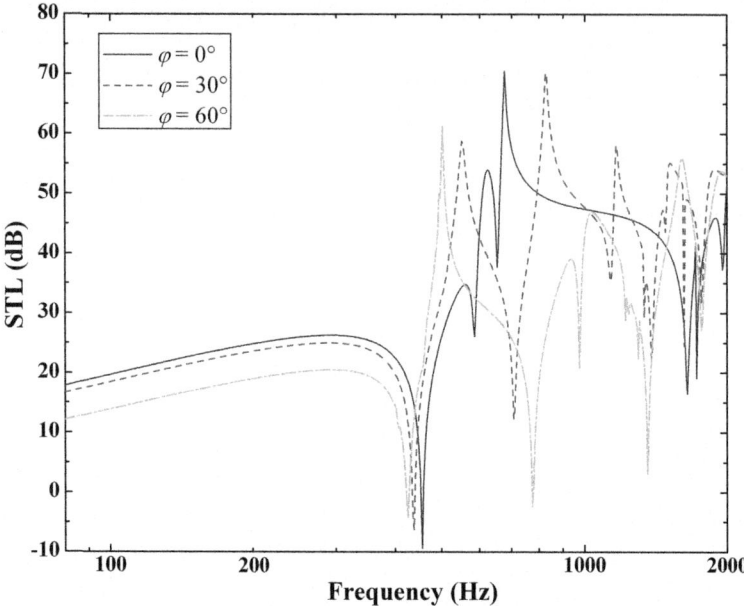

Fig. 5.13 Sound transmission loss plotted as a function of frequency for orthogonally rib-stiffened sandwich structure under sound excitation having selected incident angles (With permission from Elsevier)

the first resonance dip is shifted to a lower frequency as the elevation angle is increased, and denser resonance dips appear on the STL versus frequency curves in the oblique incident case than those in the normal incident case. Accordingly, apart from several individual peaks, the averaged STL values are smaller than that in the normal incident case, particularly so in the low-frequency range (below 400 Hz). In other words, the oblique incident sound power transmits through the structure more easily than that of the normal incident sound, due to the possibility of constructive interference between incident sound wave and structural bending waves in the former (Xin et al. [14]).

Indeed, when the trace wavelength of incident sound matches the bending wave in the face panel of the sandwich, coincidence resonance occurs in the oblique incident case but not in the normal incident case (Fahy [65]). Following Xin et al. [14], the coincidence resonance frequency may be analytically calculated as

$$f_c = \frac{c_0^2}{2\pi h \sin \varphi} \sqrt{\frac{12\rho(1-\nu^2)}{E}} \qquad (5.148)$$

The coincident resonance appearing in the oblique case is often located at high frequencies that are beyond the frequency range considered in the present study. In view of the dense resonances of the sandwich structure itself and the complex

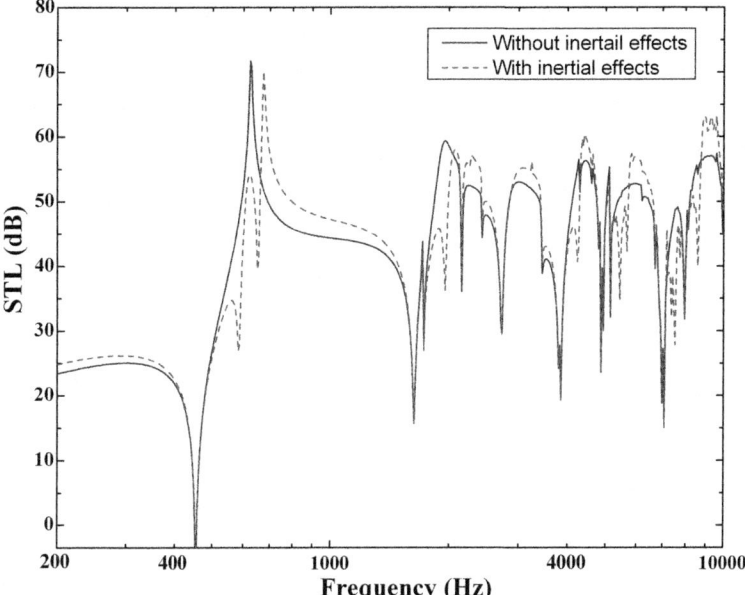

Fig. 5.14 Sound transmission loss plotted as a function of frequency for orthogonally rib-stiffened sandwich structure under normal incident sound: influence of inertial effects arising from rib-stiffener mass (With permission from Elsevier)

interaction of bending waves in the face panel and rib-stiffeners at high frequencies, the coincidence resonance dip would shift its original location, and thus it is actually impossible to identify it especially in the present complex sandwich structures.

5.2.13 Influence of Inertial Effects Arising from Rib-Stiffener Mass

The inertial effects of rib-stiffener mass should not be ignored when the rib-stiffeners are heavy. To quantify the influence of inertial effects on sound transmission characteristics, Fig. 5.14 compares the predictions obtained for an orthogonally rib-stiffened sandwich structure with and without considering the inertial effects. The inertial effects of the rib-stiffeners on sound radiation from an orthogonally rib-stiffened sandwich subjected to harmonic point force excitation have been evaluated in a companion paper [39]. As such, the influence of inertial effects on sound transmission characteristics provides additional insight into the vibroacoustic dynamics of 2D periodic sandwich structures.

It is seen from Fig. 5.14 that the STL versus frequency curve predicted with the inertial effects considered has a tendency similar to that without considering

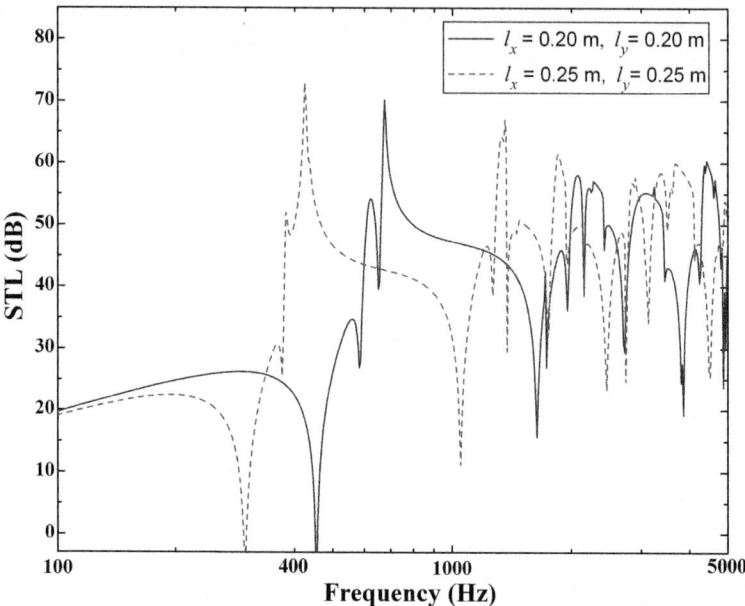

Fig. 5.15 Sound transmission loss plotted as a function of frequency for orthogonally rib-stiffened sandwich structure under normal incident sound: influence of periodicity spacings between rib-stiffeners (With permission from Elsevier)

the inertial effects, the main discrepancy being the existence of several additional peaks and dips in the former. On the one hand, the superposition peaks (or dips) between the inertial case and the non-inertial one are dominated by face panel vibration, which are closely related to the maximum (or minimal) vibration patterns. On the other hand, the appearance of the additional peaks and dips controlled predominantly by the rib-stiffeners is attributed to the inertial effects arising from the mass of the rib-stiffeners.

5.2.14 Influence of Rib-Stiffener Spacings

It is anticipated that the rib-stiffener spacings l_x and l_y (Fig. 5.9) characterizing the periodic nature of the 2D orthogonal sandwich play an important role in dictating the wave propagation and sound transmission performance of the structure. Their influence on the sound radiation behavior of the structure has been explored [39], which is further examined below in terms of sound transmission.

Figure 5.15 plots the STL as a function of frequency for two different orthogonally rib-stiffened sandwich structures, with (l_x, l_y) selected as (0.20, 0.20)m and (0.25, 0.25)m, respectively. Within the low-frequency range, it is seen from Fig. 5.15 that the characteristic curves of sound transmission corresponding to two

5.2 Sound Transmission Through Sandwich Structures

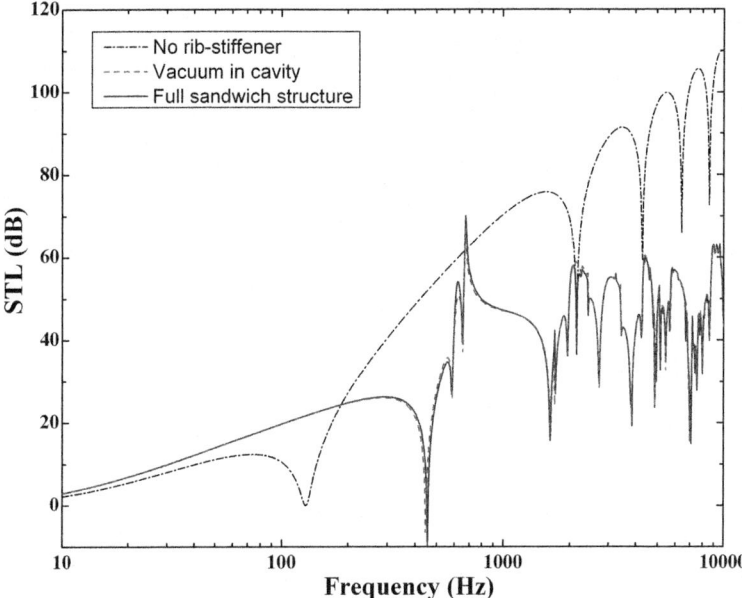

Fig. 5.16 Sound transmission loss plotted as a function of frequency for orthogonally rib-stiffened sandwich structure under normal incident sound: comparison between airborne and structure-borne transmission paths (With permission from Elsevier)

different periodicity spacings follow a similar trend, which is attributed to the fact that altering the periodicity spacings does not change the periodic nature of the sandwich markedly. However, at relatively high frequencies, visible discrepancies exist between the two cases. In addition, the STL peaks and dips are shifted to lower frequencies as the periodicity spacings increase, implying that the increment of periodicity spacings leads to noticeably reduced natural frequencies of the sandwich structure.

5.2.15 Influence of Airborne and Structure-Borne Paths

The incident sound can be transmitted via two routes from the upper panel to the bottom panel, namely, the structure-borne path (i.e., orthogonal rib-stiffeners) and the airborne path (i.e., air constrained in between the two panels). To illustrate the different roles played by the two different transmission paths, Fig. 5.16 compares the results obtained for three different cases: airborne path only (i.e., no rib-stiffeners), structure-borne path only (i.e., vacuum in cavity), and full sandwich structure.

The results of Fig. 5.16 demonstrate that, insofar as sound transmission is of concern, the case of vacuum in cavity (structure-borne path only) is nearly identical to a full sandwich structure, both significantly different from the case of no rib-stiffeners. This is understandable, as the physical process of sound transmission across the sandwich is dominated by the structure-borne path, owing to the strong constraint of structural connections (rib-stiffeners) and weak fluid-structure coupling. However, it should be pointed out that, since the transmission of sound is of concern here, the fluid-structure coupling at the incident interface (i.e., between the incident side fluid and the upper panel) and the transmitting interface (i.e., between the transmitting side fluid and the bottom panel) needs to be considered. Although the fluid-structure coupling between the air cavity and the internal interfaces of the two panels is negligibly weak compared to the constraint imposed by the rib-stiffeners, for preciseness in physics and mathematics, this coupling is included in the present analysis (which does not need much additional efforts).

In the absence of the rib-stiffeners, the first dip of the STL curve in Fig. 5.16 corresponds to the "mass-air-mass" resonance $f_a = \sqrt{\rho_0 c_0^2 (m_1 + m_2)/(m_1 m_2 d)}/2\pi$, while the following four dips (and those not shown in Fig. 5.16) are related to the standing-wave resonance $f_{s,n} = n c_0 / 2d$, where $n = 1, 2, 3, 4 \ldots$ [9, 12–14, 57]. In the case of full sandwich structure, the complex interaction of flexural wave in the panel and the rib-stiffeners creates multiple possibilities for wavenumber matching and "coincidence" [30], causing a series of resonance dips appearing in the STL curve that differ significantly from the no rib-stiffener case.

5.2.16 Conclusions

Rigorous analytic formulations are obtained with the space-harmonic approach for the structural and acoustic characteristics of an infinite orthogonally rib-stiffened sandwich structure subjected to convective fluid-loaded pressure. Unlike previous studies that focus on relatively simple structures such as rib-stiffened plates and model approximately the rib-stiffeners as Euler beams, two-dimensionally periodic sandwiches stiffened by two sets of orthogonal rib-stiffeners are considered. All possible motions of the rib-stiffeners are accurately accounted for by introducing their tensional forces, bending moments, and torsional moments as well as the corresponding inertial terms into the governing equations of the two face panels. The surrounding acoustic fluids are restricted by the acoustic wave equation, and fluid-structure coupling is included by imposing velocity continuity conditions at fluid-panel interfaces. Built upon the Bloch/Floquet theorem for periodic structures, the resulting panel motions and acoustic pressures are expressed in a superposition form of space harmonics for a given wavenumber. The application of the virtual work principle for one periodic element yields two infinite sets of simultaneous algebraic coupled equations, which are numerically solved by truncation.

To explore the physical mechanisms underlying the dynamic and acoustic performance of two-dimensionally periodic sandwich structures, the analysis is carried out from the viewpoint of sound transmission. Firstly, the validity and feasibility of the proposed analytic model is qualified by comparing the model predictions with previous published results for one-dimensionally periodic sandwich structures. The necessity and advantage of the exact modeling of rib-stiffener vibrations are also demonstrated by comparing the complete model with its simplified version and the model of Wang et al. [30]. The complete model is then used to quantify the influences of inertial effects arising from the rib-stiffener mass, the airborne and structure-borne paths, and the periodicity spacings of the rib-stiffeners on sound transmission across the sandwich structure.

Although the analytic model without considering the inertial effects of the rib-stiffeners is able to provide an overall trend of the STL versus frequency curve, the inclusion of the inertial effects in the model enables the capturing of more detailed physical features associated with the process of sound transmission, as reflected by the additional peaks and dips appearing on the STL curve.

The periodicity spacings of the rib-stiffeners play an important role in transmitting the sound across the sandwich. Two noticeable conclusions can be drawn. First, as slight alterations of the periodicity spacings do not change the periodic nature of the structure, the STL curves of different spacings exhibit similar trends. Second, increasing the periodicity spacings reduces the natural frequencies of the structure, reflected by the shifting of STL peaks and dips to lower frequencies.

For sandwich structure reinforced with rib-stiffeners, the transmission of sound via the airborne route is negligible in comparison with that transmitted via the structure-borne path, as the weak fluid-structure coupling is overwhelmed by the strong structural connections (rib-stiffeners). However, for preciseness in the viewpoints of physics and mathematics, the fluid-structure coupling present between the incident side fluid and the upper panel as well as that between the transmitting side fluid and the bottom panel needs to be considered in the analysis of sound transmission.

Appendices

Appendix A

The displacement components of the two face panels in wavenumber space are

$$\{\tilde{w}_1(\alpha'_m, \beta'_n)\} = [\tilde{w}_1(\alpha'_1, \beta'_1) \, \tilde{w}_1(\alpha'_2, \beta'_1) \cdots \tilde{w}_1(\alpha'_M, \beta'_1) \, \tilde{w}_1(\alpha'_1, \beta'_2) \, \tilde{w}_1(\alpha'_2, \beta'_2)$$
$$\times \cdots \tilde{w}_1(\alpha'_M, \beta'_2) \cdots \tilde{w}_1(\alpha'_M, \beta'_N)]^T_{MN \times 1} \qquad (5.\text{A}.1)$$

$$\{\tilde{w}_2(\alpha'_m, \beta'_n)\} = [\tilde{w}_2(\alpha'_1, \beta'_1) \, \tilde{w}_2(\alpha'_2, \beta'_1) \cdots \tilde{w}_2(\alpha'_M, \beta'_1) \, \tilde{w}_2(\alpha'_1, \beta'_2) \, \tilde{w}_2(\alpha'_2, \beta'_2)$$
$$\times \cdots \tilde{w}_2(\alpha'_M, \beta'_2) \cdots \tilde{w}_2(\alpha'_M, \beta'_N)]^T_{MN \times 1} \qquad (5.\text{A}.2)$$

The left-hand side of Eq. (5.58) represents the generalized force:

$$\{F_{mn}\} = \begin{bmatrix} F_{11} & F_{21} & \cdots & F_{M1} & F_{12} & F_{22} & \cdots & F_{M2} & \cdots & F_{MN} \end{bmatrix}^T_{MN \times 1} \quad (5.A.3)$$

where

$$F_{mn} = \frac{q_0 e^{-i(\alpha'_m x_0 + \beta'_n y_0)}}{(2\pi)^2 D_1 f_1(\alpha'_m, \beta'_n)} \quad (5.A.4)$$

$$T_{11,1} = \begin{bmatrix} 1 & & & \\ & 1 & & \\ & & \ddots & \\ & & & 1 \end{bmatrix}_{MN \times MN} \quad (5.A.5)$$

$$T_{22,1} = \begin{bmatrix} 1 & & & \\ & 1 & & \\ & & \ddots & \\ & & & 1 \end{bmatrix}_{MN \times MN} \quad (5.A.6)$$

$$\lambda^{11,2}_{Mn,j} = \begin{bmatrix} \frac{R_{Q3} + i\beta'^3_n R_{M3}}{D_j l_x f_j(\alpha'_1, \beta'_n)} & \frac{R_{Q3} + i\beta'^3_n R_{M3}}{D_j l_x f_j(\alpha'_1, \beta'_n)} & \cdots & \frac{R_{Q3} + i\beta'^3_n R_{M3}}{D_j l_x f_j(\alpha'_1, \beta'_n)} \\ \frac{R_{Q3} + i\beta'^3_n R_{M3}}{D_j l_x f_j(\alpha'_2, \beta'_n)} & \frac{R_{Q3} + i\beta'^3_n R_{M3}}{D_j l_x f_j(\alpha'_2, \beta'_n)} & \cdots & \frac{R_{Q3} + i\beta'^3_n R_{M3}}{D_j l_x f_j(\alpha'_2, \beta'_n)} \\ \vdots & \vdots & \ddots & \vdots \\ \frac{R_{Q3} + i\beta'^3_n R_{M3}}{D_j l_x f_j(\alpha'_M, \beta'_n)} & \frac{R_{Q3} + i\beta'^3_n R_{M3}}{D_j l_x f_j(\alpha'_M, \beta'_n)} & \cdots & \frac{R_{Q3} + i\beta'^3_n R_{M3}}{D_j l_x f_j(\alpha'_M, \beta'_n)} \end{bmatrix}_{M \times N} \quad (5.A.7)$$

$$T_{11,2} = \begin{bmatrix} \lambda^{11,2}_{M1,1} & & & \\ & \lambda^{11,2}_{M2,1} & & \\ & & \ddots & \\ & & & \lambda^{11,2}_{MN,1} \end{bmatrix}_{MN \times MN} \quad (5.A.8)$$

$$T_{22,2} = \begin{bmatrix} \lambda^{11,2}_{M1,2} & & & \\ & \lambda^{11,2}_{M2,2} & & \\ & & \ddots & \\ & & & \lambda^{11,2}_{MN,2} \end{bmatrix}_{MN \times MN} \quad (5.A.9)$$

$$\lambda_{Mn,j}^{11,3} = \begin{bmatrix} \dfrac{i\alpha_1\alpha'_1\beta'_n R_{T3}}{D_j l_x f_j(\alpha'_1,\beta'_n)} & \dfrac{i\alpha_2\alpha'_1\beta'_n R_{T3}}{D_j l_x f_j(\alpha'_1,\beta'_n)} & \cdots & \dfrac{i\alpha_M\alpha'_1\beta'_n R_{T3}}{D_j l_x f_j(\alpha'_1,\beta'_n)} \\ \dfrac{i\alpha_1\alpha'_2\beta'_n R_{T3}}{D_j l_x f_j(\alpha'_2,\beta'_n)} & \dfrac{i\alpha_2\alpha'_2\beta'_n R_{T3}}{D_j l_x f_j(\alpha'_2,\beta'_n)} & \cdots & \dfrac{i\alpha_M\alpha'_2\beta'_n R_{T3}}{D_j l_x f_j(\alpha'_2,\beta'_n)} \\ \vdots & \vdots & \ddots & \vdots \\ \dfrac{i\alpha_1\alpha'_M\beta'_n R_{T3}}{D_j l_x f_j(\alpha'_M,\beta'_n)} & \dfrac{i\alpha_2\alpha'_M\beta'_n R_{T3}}{D_j l_x f_j(\alpha'_M,\beta'_n)} & \cdots & \dfrac{i\alpha_M\alpha'_M\beta'_n R_{T3}}{D_j l_x f_j(\alpha'_M,\beta'_n)} \end{bmatrix}_{M \times N} \quad (5.A.10)$$

$$T_{11,3} = \begin{bmatrix} \lambda_{M1,1}^{11,3} & & & \\ & \lambda_{M2,1}^{11,3} & & \\ & & \ddots & \\ & & & \lambda_{MN,1}^{11,3} \end{bmatrix}_{MN \times MN} \quad (5.A.11)$$

$$T_{22,3} = \begin{bmatrix} \lambda_{M1,2}^{11,3} & & & \\ & \lambda_{M2,2}^{11,3} & & \\ & & \ddots & \\ & & & \lambda_{MN,2}^{11,3} \end{bmatrix}_{MN \times MN} \quad (5.A.12)$$

$$\lambda_{Mn,j}^{11,4} = \begin{bmatrix} \dfrac{R_{Q1}+i\alpha'^{3}_{1}R_{M1}}{D_j l_y f_j(\alpha'_1,\beta'_n)} & & & \\ & \dfrac{R_{Q1}+i\alpha'^{3}_{2}R_{M1}}{D_j l_y f_j(\alpha'_2,\beta'_n)} & & \\ & & \ddots & \\ & & & \dfrac{R_{Q1}+i\alpha'^{3}_{M}R_{M1}}{D_j l_y f_j(\alpha'_M,\beta'_n)} \end{bmatrix}_{M \times N} \quad (5.A.13)$$

$$T_{11,4} = \begin{bmatrix} \lambda_{M1,1}^{11,4} & \lambda_{M1,1}^{11,4} & \cdots & \lambda_{M1,1}^{11,4} \\ \lambda_{M2,1}^{11,4} & \lambda_{M2,1}^{11,4} & \cdots & \lambda_{M2,1}^{11,4} \\ \vdots & \vdots & \ddots & \vdots \\ \lambda_{MN,1}^{11,4} & \lambda_{MN,1}^{11,4} & \cdots & \lambda_{MN,1}^{11,4} \end{bmatrix}_{MN \times MN} \quad (5.A.14)$$

$$T_{22,4} = \begin{bmatrix} \lambda_{M1,2}^{11,4} & \lambda_{M1,2}^{11,4} & \cdots & \lambda_{M1,2}^{11,4} \\ \lambda_{M2,2}^{11,4} & \lambda_{M2,2}^{11,4} & \cdots & \lambda_{M2,2}^{11,4} \\ \vdots & \vdots & \ddots & \vdots \\ \lambda_{MN,2}^{11,4} & \lambda_{MN,2}^{11,4} & \cdots & \lambda_{MN,2}^{11,4} \end{bmatrix}_{MN \times MN} \quad (5.A.15)$$

$$\lambda_{Mn',jn}^{11,5} = \begin{bmatrix} \dfrac{i\alpha_1'\beta_{n'}'\beta_n R_{T1}}{D_j l_y f_j(\alpha_1',\beta_{n'}')} & & & \\ & \dfrac{i\alpha_2'\beta_{n'}'\beta_n R_{T1}}{D_j l_y f_j(\alpha_2',\beta_{n'}')} & & \\ & & \ddots & \\ & & & \dfrac{i\alpha_M'\beta_{n'}'\beta_n R_{T1}}{D_j l_y f_j(\alpha_M',\beta_{n'}')} \end{bmatrix}_{M\times N} \quad (5.A.16)$$

$$T_{11,5} = \begin{bmatrix} \lambda_{M1,11}^{11,5} & \lambda_{M1,12}^{11,5} & \cdots & \lambda_{M1,1N}^{11,5} \\ \lambda_{M2,11}^{11,5} & \lambda_{M2,12}^{11,5} & \cdots & \lambda_{M2,1N}^{11,5} \\ \vdots & \vdots & \ddots & \vdots \\ \lambda_{MN,11}^{11,5} & \lambda_{MN,12}^{11,5} & \cdots & \lambda_{MN,1N}^{11,5} \end{bmatrix}_{MN\times MN} \quad (5.A.17)$$

$$T_{22,5} = \begin{bmatrix} \lambda_{M1,21}^{11,5} & \lambda_{M1,22}^{11,5} & \cdots & \lambda_{M1,2N}^{11,5} \\ \lambda_{M2,21}^{11,5} & \lambda_{M2,22}^{11,5} & \cdots & \lambda_{M2,2N}^{11,5} \\ \vdots & \vdots & \ddots & \vdots \\ \lambda_{MN,21}^{11,5} & \lambda_{MN,22}^{11,5} & \cdots & \lambda_{MN,2N}^{11,5} \end{bmatrix}_{MN\times MN} \quad (5.A.18)$$

$$\lambda_{Mn,j}^{12,1} = -\begin{bmatrix} \dfrac{R_{Q4}+i\beta_n'^{3}R_{M4}}{D_j l_x f_j(\alpha_1',\beta_n')} & \dfrac{R_{Q4}+i\beta_n'^{3}R_{M4}}{D_j l_x f_j(\alpha_1',\beta_n')} & \cdots & \dfrac{R_{Q4}+i\beta_n'^{3}R_{M4}}{D_j l_x f_j(\alpha_1',\beta_n')} \\ \dfrac{R_{Q4}+i\beta_n'^{3}R_{M4}}{D_j l_x f_j(\alpha_2',\beta_n')} & \dfrac{R_{Q4}+i\beta_n'^{3}R_{M4}}{D_j l_x f_j(\alpha_2',\beta_n')} & \cdots & \dfrac{R_{Q4}+i\beta_n'^{3}R_{M4}}{D_j l_x f_j(\alpha_2',\beta_n')} \\ \vdots & \vdots & \ddots & \vdots \\ \dfrac{R_{Q4}+i\beta_n'^{3}R_{M4}}{D_j l_x f_j(\alpha_M',\beta_n')} & \dfrac{R_{Q4}+i\beta_n'^{3}R_{M4}}{D_j l_x f_j(\alpha_M',\beta_n')} & \cdots & \dfrac{R_{Q4}+i\beta_n'^{3}R_{M4}}{D_j l_x f_j(\alpha_M',\beta_n')} \end{bmatrix}_{M\times N} \quad (5.A.19)$$

$$T_{12,1} = \begin{bmatrix} \lambda_{M1,1}^{12,1} & & & \\ & \lambda_{M2,1}^{12,1} & & \\ & & \ddots & \\ & & & \lambda_{MN,1}^{12,1} \end{bmatrix}_{MN\times MN} \quad (5.A.20)$$

$$T_{21,1} = \begin{bmatrix} \lambda_{M1,2}^{12,1} & & & \\ & \lambda_{M2,2}^{12,1} & & \\ & & \ddots & \\ & & & \lambda_{MN,2}^{12,1} \end{bmatrix}_{MN\times MN} \quad (5.A.21)$$

$$\lambda_{Mn,j}^{12,2} = -\begin{bmatrix} \dfrac{i\alpha_1 \alpha'_1 \beta'_n R_{T4}}{D_j l_x f_j(\alpha'_1,\beta'_n)} & \dfrac{i\alpha_2 \alpha'_1 \beta'_n R_{T4}}{D_j l_x f_j(\alpha'_1,\beta'_n)} & \cdots & \dfrac{i\alpha_M \alpha'_1 \beta'_n R_{T4}}{D_j l_x f_j(\alpha'_1,\beta'_n)} \\ \dfrac{i\alpha_1 \alpha'_2 \beta'_n R_{T4}}{D_j l_x f_j(\alpha'_2,\beta'_n)} & \dfrac{i\alpha_2 \alpha'_2 \beta'_n R_{T4}}{D_j l_x f_j(\alpha'_2,\beta'_n)} & \cdots & \dfrac{i\alpha_M \alpha'_2 \beta'_n R_{T4}}{D_j l_x f_j(\alpha'_2,\beta'_n)} \\ \vdots & \vdots & \ddots & \vdots \\ \dfrac{i\alpha_1 \alpha'_M \beta'_n R_{T4}}{D_j l_x f_j(\alpha'_M,\beta'_n)} & \dfrac{i\alpha_2 \alpha'_M \beta'_n R_{T4}}{D_j l_x f_j(\alpha'_M,\beta'_n)} & \cdots & \dfrac{i\alpha_M \alpha'_M \beta'_n R_{T4}}{D_j l_x f_j(\alpha'_M,\beta'_n)} \end{bmatrix}_{M \times N} \quad (5.A.22)$$

$$T_{12,2} = \begin{bmatrix} \lambda_{M1,1}^{12,2} & & & \\ & \lambda_{M2,1}^{12,2} & & \\ & & \ddots & \\ & & & \lambda_{MN,1}^{12,2} \end{bmatrix}_{MN \times MN} \quad (5.A.23)$$

$$T_{21,2} = \begin{bmatrix} \lambda_{M1,2}^{12,2} & & & \\ & \lambda_{M2,2}^{12,2} & & \\ & & \ddots & \\ & & & \lambda_{MN,2}^{12,2} \end{bmatrix}_{MN \times MN} \quad (5.A.24)$$

$$\lambda_{Mn,j}^{12,3} = -\begin{bmatrix} \dfrac{R_{Q2}+i\alpha'^3_1 R_{M2}}{D_j l_y f_j(\alpha'_1,\beta'_n)} & & & \\ & \dfrac{R_{Q2}+i\alpha'^3_2 R_{M2}}{D_j l_y f_j(\alpha'_2,\beta'_n)} & & \\ & & \ddots & \\ & & & \dfrac{R_{Q2}+i\alpha'^3_M R_{M2}}{D_j l_y f_j(\alpha'_M,\beta'_n)} \end{bmatrix}_{M \times N} \quad (5.A.25)$$

$$T_{12,3} = \begin{bmatrix} \lambda_{M1,1}^{12,3} & \lambda_{M1,1}^{12,3} & \cdots & \lambda_{M1,1}^{12,3} \\ \lambda_{M2,1}^{12,3} & \lambda_{M2,1}^{12,3} & \cdots & \lambda_{M2,1}^{12,3} \\ \vdots & \vdots & \ddots & \vdots \\ \lambda_{MN,1}^{12,3} & \lambda_{MN,1}^{12,3} & \cdots & \lambda_{MN,1}^{12,3} \end{bmatrix}_{MN \times MN} \quad (5.A.26)$$

$$T_{21,3} = \begin{bmatrix} \lambda_{M1,2}^{12,3} & \lambda_{M1,2}^{12,3} & \cdots & \lambda_{M1,2}^{12,3} \\ \lambda_{M2,2}^{12,3} & \lambda_{M2,2}^{12,3} & \cdots & \lambda_{M2,2}^{12,3} \\ \vdots & \vdots & \ddots & \vdots \\ \lambda_{MN,2}^{12,3} & \lambda_{MN,2}^{12,3} & \cdots & \lambda_{MN,2}^{12,3} \end{bmatrix}_{MN \times MN} \quad (5.A.27)$$

$$\lambda_{Mn',jn}^{12,4} = -\begin{bmatrix} \dfrac{i\alpha'_1 \beta'_{n'} \beta_n R_{T2}}{D_j l_y f_j(\alpha'_1,\beta'_{n'})} & & & \\ & \dfrac{i\alpha'_2 \beta'_{n'} \beta_n R_{T2}}{D_j l_y f_j(\alpha'_2,\beta'_{n'})} & & \\ & & \ddots & \\ & & & \dfrac{i\alpha'_M \beta'_{n'} \beta_n R_{T2}}{D_j l_y f_j(\alpha'_M,\beta'_{n'})} \end{bmatrix}_{M \times N} \quad (5.A.28)$$

$$T_{12,4} = \begin{bmatrix} \lambda_{M1,11}^{12,4} & \lambda_{M1,12}^{12,4} & \cdots & \lambda_{M1,1N}^{12,4} \\ \lambda_{M2,11}^{12,4} & \lambda_{M2,12}^{12,4} & \cdots & \lambda_{M2,1N}^{12,4} \\ \vdots & \vdots & \ddots & \vdots \\ \lambda_{MN,11}^{12,4} & \lambda_{MN,12}^{12,4} & \cdots & \lambda_{MN,1N}^{12,4} \end{bmatrix}_{MN \times MN} \quad (5.A.29)$$

$$T_{21,4} = \begin{bmatrix} \lambda_{M1,21}^{12,4} & \lambda_{M1,22}^{12,4} & \cdots & \lambda_{M1,2N}^{12,4} \\ \lambda_{M2,21}^{12,4} & \lambda_{M2,22}^{12,4} & \cdots & \lambda_{M2,2N}^{12,4} \\ \vdots & \vdots & \ddots & \vdots \\ \lambda_{MN,21}^{12,4} & \lambda_{MN,22}^{12,4} & \cdots & \lambda_{MN,2N}^{12,4} \end{bmatrix}_{MN \times MN} \quad (5.A.30)$$

Employing the definition of the sub-matrices presented above, one obtains

$$T_{11} = T_{11,1} + T_{11,2} + T_{11,3} + T_{11,4} + T_{11,5}, \quad T_{22} = T_{22,1} + T_{22,2} + T_{22,3} + T_{22,4} + T_{22,5} \quad (5.A.31)$$

$$T_{12} = T_{12,1} + T_{12,2} + T_{12,3} + T_{12,4}, \quad T_{21} = T_{21,1} + T_{21,2} + T_{21,3} + T_{21,4} \quad (5.A.32)$$

Appendix B

The deflection coefficients of the two face panels are

$$\{\alpha_{1,kl}\} = \begin{bmatrix} \alpha_{1,11} & \alpha_{1,21} & \cdots & \alpha_{1,K1} & \alpha_{1,12} & \alpha_{1,22} & \cdots & \alpha_{1,K2} & \cdots & \alpha_{1,KL} \end{bmatrix}^T_{KL \times 1} \quad (5.B.1)$$

$$\{\alpha_{2,kl}\} = \begin{bmatrix} \alpha_{2,11} & \alpha_{2,21} & \cdots & \alpha_{2,K1} & \alpha_{2,12} & \alpha_{2,22} & \cdots & \alpha_{2,K2} & \cdots & \alpha_{2,KL} \end{bmatrix}^T_{KL \times 1} \quad (5.B.2)$$

The right-hand side of Eq. (5.143) represents the generalized force, that is,

$$\{F_{kl}\} = \begin{bmatrix} F_{11} & F_{21} & \cdots & F_{K1} & F_{12} & F_{22} & \cdots & F_{K2} & \cdots & F_{KL} \end{bmatrix}^T_{KL \times 1} \quad (5.B.3)$$

where

$$F_{kl} = \begin{cases} 2Il_xl_y & \text{at} \quad k = \frac{K+1}{2} \& l = \frac{L+1}{2} \\ 0 & \text{at} \quad k \neq \frac{K+1}{2} \parallel l \neq \frac{L+1}{2} \end{cases} \quad (5.\text{B}.4)$$

$$\lambda_{kl}^{11,1} = \left[D_1(\alpha_k^2 + \beta_l^2)^2 - m_1\omega^2 - \frac{\omega^2 \rho_0}{ik_{z,kl}} + \frac{\omega^2 \rho_0 \cos(k_{z,kl}d)}{k_{z,kl} \sin(k_{z,kl}d)} \right] \cdot l_x l_y \quad (5.\text{B}.5)$$

$$T_{11,1} = \begin{bmatrix} \lambda_{11}^{11,1} & & & & & & & \\ & \lambda_{21}^{11,1} & & & & & & \\ & & \ddots & & & & & \\ & & & \lambda_{K1}^{11,1} & & & & \\ & & & & \lambda_{12}^{11,1} & & & \\ & & & & & \lambda_{22}^{11,1} & & \\ & & & & & & \ddots & \\ & & & & & & & \lambda_{K2}^{11,1} \\ & & & & & & & & \ddots \\ & & & & & & & & & \lambda_{KL}^{11,1} \end{bmatrix}_{KL \times KL} \quad (5.\text{B}.6)$$

$$\lambda_{KL}^{11,2} = l_x \cdot \begin{bmatrix} R_{Q1} - i\alpha_1^3 R_{M1} & & & \\ & R_{Q1} - i\alpha_2^3 R_{M1} & & \\ & & \ddots & \\ & & & R_{Q1} - i\alpha_K^3 R_{M1} \end{bmatrix}_{K \times L} \quad (5.\text{B}.7)$$

$$T_{11,2} = \begin{bmatrix} \lambda_{KL}^{11,2} & \lambda_{KL}^{11,2} & \cdots & \lambda_{KL}^{11,2} \\ \lambda_{KL}^{11,2} & \lambda_{KL}^{11,2} & \cdots & \lambda_{KL}^{11,2} \\ \vdots & \vdots & \ddots & \vdots \\ \lambda_{KL}^{11,2} & \lambda_{KL}^{11,2} & \cdots & \lambda_{KL}^{11,2} \end{bmatrix}_{KL \times KL} \quad (5.\text{B}.8)$$

$$\lambda_{Kl,n}^{11,3} = l_x \cdot \begin{bmatrix} -i\alpha_1 \beta_l \beta_n R_{T1} & & & \\ & -i\alpha_2 \beta_l \beta_n R_{T1} & & \\ & & \ddots & \\ & & & -i\alpha_K \beta_l \beta_n R_{T1} \end{bmatrix}_{K \times L} \quad (5.\text{B}.9)$$

$$T_{11,3} = \begin{bmatrix} \lambda_{K1,1}^{11,3} & \lambda_{K1,2}^{11,3} & \cdots & \lambda_{K1,L}^{11,3} \\ \lambda_{K2,1}^{11,3} & \lambda_{K2,2}^{11,3} & \cdots & \lambda_{K2,L}^{11,3} \\ \vdots & \vdots & \ddots & \vdots \\ \lambda_{KL,1}^{11,3} & \lambda_{KL,2}^{11,3} & \cdots & \lambda_{KL,L}^{11,3} \end{bmatrix}_{KL \times KL} \quad (5.\text{B}.10)$$

$$\lambda_{Kl}^{11,4} = l_y \cdot \begin{bmatrix} R_{Q3} - i\beta_l^3 R_{M3} & R_{Q3} - i\beta_l^3 R_{M3} & \cdots & R_{Q3} - i\beta_l^3 R_{M3} \\ R_{Q3} - i\beta_l^3 R_{M3} & R_{Q3} - i\beta_l^3 R_{M3} & \cdots & R_{Q3} - i\beta_l^3 R_{M3} \\ \vdots & \vdots & \ddots & \vdots \\ R_{Q3} - i\beta_l^3 R_{M3} & R_{Q3} - i\beta_l^3 R_{M3} & \cdots & R_{Q3} - i\beta_l^3 R_{M3} \end{bmatrix}_{K \times L}$$
(5.B.11)

$$T_{11,4} = \begin{bmatrix} \lambda_{K1}^{11,4} & & & \\ & \lambda_{K2}^{11,4} & & \\ & & \ddots & \\ & & & \lambda_{KL}^{11,4} \end{bmatrix}_{KL \times KL}$$
(5.B.12)

$$\lambda_{Kl}^{11,5} = l_y \cdot \begin{bmatrix} -i\alpha_1 \beta_l \alpha_1 R_{T3} & -i\alpha_1 \beta_l \alpha_2 R_{T3} & \cdots & -i\alpha_1 \beta_l \alpha_K R_{T3} \\ -i\alpha_2 \beta_l \alpha_1 R_{T3} & -i\alpha_2 \beta_l \alpha_2 R_{T3} & \cdots & -i\alpha_2 \beta_l \alpha_K R_{T3} \\ \vdots & \vdots & \ddots & \vdots \\ -i\alpha_K \beta_l \alpha_1 R_{T3} & -i\alpha_K \beta_l \alpha_2 R_{T3} & \cdots & -i\alpha_K \beta_l \alpha_K R_{T3} \end{bmatrix}_{K \times L}$$
(5.B.13)

$$T_{11,5} = \begin{bmatrix} \lambda_{K1}^{11,5} & & & \\ & \lambda_{K2}^{11,5} & & \\ & & \ddots & \\ & & & \lambda_{KL}^{11,5} \end{bmatrix}_{KL \times KL}$$
(5.B.14)

$$\lambda_{kl}^{12,1} = -\frac{\omega^2 \rho_0}{k_{z,kl} \sin(k_{z,kl} d)} \cdot l_x l_y$$
(5.B.15)

$$T_{12,1} = \begin{bmatrix} \lambda_{11}^{12,1} & & & & & & & \\ & \lambda_{21}^{12,1} & & & & & & \\ & & \ddots & & & & & \\ & & & \lambda_{K1}^{12,1} & & & & \\ & & & & \lambda_{12}^{12,1} & & & \\ & & & & & \lambda_{22}^{12,1} & & \\ & & & & & & \ddots & \\ & & & & & & & \lambda_{K2}^{12,1} \\ & & & & & & & & \ddots \\ & & & & & & & & & \lambda_{KL}^{12,1} \end{bmatrix}_{KL \times KL}$$
(5.B.16)

$$\lambda_{KL}^{12,2} = l_x \cdot \begin{bmatrix} -R_{Q2} + i\alpha_1^3 R_{M2} & & & \\ & -R_{Q2} + i\alpha_2^3 R_{M2} & & \\ & & \ddots & \\ & & & -R_{Q2} + i\alpha_K^3 R_{M2} \end{bmatrix}_{K \times L} \tag{5.B.17}$$

$$T_{12,2} = \begin{bmatrix} \lambda_{KL}^{12,2} & \lambda_{KL}^{12,2} & \cdots & \lambda_{KL}^{12,2} \\ \lambda_{KL}^{12,2} & \lambda_{KL}^{12,2} & \cdots & \lambda_{KL}^{12,2} \\ \vdots & \vdots & \ddots & \vdots \\ \lambda_{KL}^{12,2} & \lambda_{KL}^{12,2} & \cdots & \lambda_{KL}^{12,2} \end{bmatrix}_{KL \times KL} \tag{5.B.18}$$

$$\lambda_{Kl,n}^{12,3} = l_x \cdot \begin{bmatrix} i\alpha_1 \beta_l \beta_n R_{T2} & & & \\ & i\alpha_2 \beta_l \beta_n R_{T2} & & \\ & & \ddots & \\ & & & i\alpha_K \beta_l \beta_n R_{T2} \end{bmatrix}_{K \times L} \tag{5.B.19}$$

$$T_{12,3} = \begin{bmatrix} \lambda_{K1,1}^{12,3} & \lambda_{K1,2}^{12,3} & \cdots & \lambda_{K1,L}^{12,3} \\ \lambda_{K2,1}^{12,3} & \lambda_{K2,2}^{12,3} & \cdots & \lambda_{K2,L}^{12,3} \\ \vdots & \vdots & \ddots & \vdots \\ \lambda_{KL,1}^{12,3} & \lambda_{KL,2}^{12,3} & \cdots & \lambda_{KL,L}^{12,3} \end{bmatrix}_{KL \times KL} \tag{5.B.20}$$

$$\lambda_{Kl}^{12,4} = l_y \cdot \begin{bmatrix} -R_{Q4} + i\beta_l^3 R_{M4} & -R_{Q4} + i\beta_l^3 R_{M4} & \cdots & -R_{Q4} + i\beta_l^3 R_{M4} \\ -R_{Q4} + i\beta_l^3 R_{M4} & -R_{Q4} + i\beta_l^3 R_{M4} & \cdots & -R_{Q4} + i\beta_l^3 R_{M4} \\ \vdots & \vdots & \ddots & \vdots \\ -R_{Q4} + i\beta_l^3 R_{M4} & -R_{Q4} + i\beta_l^3 R_{M4} & \cdots & -R_{Q4} + i\beta_l^3 R_{M4} \end{bmatrix}_{K \times L} \tag{5.B.21}$$

$$T_{12,4} = \begin{bmatrix} \lambda_{K1}^{12,4} & & & \\ & \lambda_{K2}^{12,4} & & \\ & & \ddots & \\ & & & \lambda_{KL}^{12,4} \end{bmatrix}_{KL \times KL} \tag{5.B.22}$$

$$\lambda_{Kl}^{12,5} = l_y \cdot \begin{bmatrix} i\alpha_1 \beta_l \alpha_1 R_{T4} & i\alpha_1 \beta_l \alpha_2 R_{T4} & \cdots & i\alpha_1 \beta_l \alpha_K R_{T4} \\ i\alpha_2 \beta_l \alpha_1 R_{T4} & i\alpha_2 \beta_l \alpha_2 R_{T4} & \cdots & i\alpha_2 \beta_l \alpha_K R_{T4} \\ \vdots & \vdots & \ddots & \vdots \\ i\alpha_K \beta_l \alpha_1 R_{T4} & i\alpha_K \beta_l \alpha_2 R_{T4} & \cdots & i\alpha_K \beta_l \alpha_K R_{T4} \end{bmatrix}_{K \times L} \tag{5.B.23}$$

$$T_{12,5} = \begin{bmatrix} \lambda_{K1}^{12,5} & & & \\ & \lambda_{K2}^{12,5} & & \\ & & \ddots & \\ & & & \lambda_{KL}^{12,5} \end{bmatrix}_{KL \times KL} \tag{5.B.24}$$

$$\lambda_{kl}^{21,1} = -\frac{\omega^2 \rho_0}{k_{z,kl} \sin(k_{z,kl} d)} \cdot l_x l_y \tag{5.B.25}$$

$$T_{21,1} = \begin{bmatrix} \lambda_{11}^{21,1} & & & & & & & \\ & \lambda_{21}^{21,1} & & & & & & \\ & & \ddots & & & & & \\ & & & \lambda_{K1}^{21,1} & & & & \\ & & & & \lambda_{12}^{21,1} & & & \\ & & & & & \lambda_{22}^{21,1} & & \\ & & & & & & \ddots & \\ & & & & & & & \lambda_{K2}^{21,1} \\ & & & & & & & & \ddots \\ & & & & & & & & & \lambda_{KL}^{21,1} \end{bmatrix}_{KL \times KL} \tag{5.B.26}$$

$$\lambda_{KL}^{21,2} = l_x \cdot \begin{bmatrix} -R_{Q2} + i\alpha_1^3 R_{M2} & & & \\ & -R_{Q2} + i\alpha_2^3 R_{M2} & & \\ & & \ddots & \\ & & & -R_{Q2} + i\alpha_K^3 R_{M2} \end{bmatrix}_{K \times L} \tag{5.B.27}$$

$$T_{21,2} = \begin{bmatrix} \lambda_{KL}^{21,2} & \lambda_{KL}^{21,2} & \cdots & \lambda_{KL}^{21,2} \\ \lambda_{KL}^{21,2} & \lambda_{KL}^{21,2} & \cdots & \lambda_{KL}^{21,2} \\ \vdots & \vdots & \ddots & \vdots \\ \lambda_{KL}^{21,2} & \lambda_{KL}^{21,2} & \cdots & \lambda_{KL}^{21,2} \end{bmatrix}_{KL \times KL} \tag{5.B.28}$$

$$\lambda_{Kl,n}^{21,3} = l_x \cdot \begin{bmatrix} i\alpha_1 \beta_l \beta_n R_{T2} & & & \\ & i\alpha_2 \beta_l \beta_n R_{T2} & & \\ & & \ddots & \\ & & & i\alpha_K \beta_l \beta_n R_{T2} \end{bmatrix}_{K \times L} \tag{5.B.29}$$

$$T_{21,3} = \begin{bmatrix} \lambda_{K1,1}^{21,3} & \lambda_{K1,2}^{21,3} & \cdots & \lambda_{K1,L}^{21,3} \\ \lambda_{K2,1}^{21,3} & \lambda_{K2,2}^{21,3} & \cdots & \lambda_{K2,L}^{21,3} \\ \vdots & \vdots & \ddots & \vdots \\ \lambda_{KL,1}^{21,3} & \lambda_{KL,2}^{21,3} & \cdots & \lambda_{KL,L}^{21,3} \end{bmatrix}_{KL \times KL} \quad (5.B.30)$$

$$\lambda_{Kl}^{21,4} = l_y \cdot \begin{bmatrix} -R_{Q4} + i\beta_l^3 R_{M4} & -R_{Q4} + i\beta_l^3 R_{M4} & \cdots & -R_{Q4} + i\beta_l^3 R_{M4} \\ -R_{Q4} + i\beta_l^3 R_{M4} & -R_{Q4} + i\beta_l^3 R_{M4} & \cdots & -R_{Q4} + i\beta_l^3 R_{M4} \\ \vdots & \vdots & \ddots & \vdots \\ -R_{Q4} + i\beta_l^3 R_{M4} & -R_{Q4} + i\beta_l^3 R_{M4} & \cdots & -R_{Q4} + i\beta_l^3 R_{M4} \end{bmatrix}_{K \times L} \quad (5.B.31)$$

$$T_{21,4} = \begin{bmatrix} \lambda_{K1}^{21,4} & & & \\ & \lambda_{K2}^{21,4} & & \\ & & \ddots & \\ & & & \lambda_{KL}^{21,4} \end{bmatrix}_{KL \times KL} \quad (5.B.32)$$

$$\lambda_{Kl}^{21,5} = l_y \cdot \begin{bmatrix} i\alpha_1\beta_l\alpha_1 R_{T4} & i\alpha_1\beta_l\alpha_2 R_{T4} & \cdots & i\alpha_1\beta_l\alpha_K R_{T4} \\ i\alpha_2\beta_l\alpha_1 R_{T4} & i\alpha_2\beta_l\alpha_2 R_{T4} & \cdots & i\alpha_2\beta_l\alpha_K R_{T4} \\ \vdots & \vdots & \ddots & \vdots \\ i\alpha_K\beta_l\alpha_1 R_{T4} & i\alpha_K\beta_l\alpha_2 R_{T4} & \cdots & i\alpha_K\beta_l\alpha_K R_{T4} \end{bmatrix}_{K \times L} \quad (5.B.33)$$

$$T_{21,5} = \begin{bmatrix} \lambda_{K1}^{21,5} & & & \\ & \lambda_{K2}^{21,5} & & \\ & & \ddots & \\ & & & \lambda_{KL}^{21,5} \end{bmatrix}_{KL \times KL} \quad (5.B.34)$$

$$\lambda_{kl}^{22,1} = \left[D_2(\alpha_k^2 + \beta_l^2)^2 - m_2\omega^2 - \frac{\omega^2 \rho_0}{ik_{z,kl}} + \frac{\omega^2 \rho_0 \cos(k_{z,kl}d)}{k_{z,kl}\sin(k_{z,kl}d)} \right] \cdot l_x l_y \quad (5.B.35)$$

$$T_{22,1} = \begin{bmatrix} \lambda_{11}^{22,1} & & & & & & & \\ & \lambda_{21}^{22,1} & & & & & & \\ & & \ddots & & & & & \\ & & & \lambda_{K1}^{22,1} & & & & \\ & & & & \lambda_{12}^{22,1} & & & \\ & & & & & \lambda_{22}^{22,1} & & \\ & & & & & & \ddots & \\ & & & & & & & \lambda_{K2}^{22,1} \\ & & & & & & & & \ddots \\ & & & & & & & & & \lambda_{KL}^{22,1} \end{bmatrix}_{KL \times KL} \quad (5.B.36)$$

$$\lambda_{KL}^{22,2} = l_x \cdot \begin{bmatrix} R_{Q1} - i\alpha_1^3 R_{M1} & & & \\ & R_{Q1} - i\alpha_2^3 R_{M1} & & \\ & & \ddots & \\ & & & R_{Q1} - i\alpha_K^3 R_{M1} \end{bmatrix}_{K \times L} \quad (5.B.37)$$

$$T_{22,2} = \begin{bmatrix} \lambda_{KL}^{22,2} & \lambda_{KL}^{22,2} & \cdots & \lambda_{KL}^{22,2} \\ \lambda_{KL}^{22,2} & \lambda_{KL}^{22,2} & \cdots & \lambda_{KL}^{22,2} \\ \vdots & \vdots & \ddots & \vdots \\ \lambda_{KL}^{22,2} & \lambda_{KL}^{22,2} & \cdots & \lambda_{KL}^{22,2} \end{bmatrix}_{KL \times KL} \quad (5.B.38)$$

$$\lambda_{Kl,n}^{22,3} = l_x \cdot \begin{bmatrix} -i\alpha_1 \beta_l \beta_n R_{T1} & & & \\ & -i\alpha_2 \beta_l \beta_n R_{T1} & & \\ & & \ddots & \\ & & & -i\alpha_K \beta_l \beta_n R_{T1} \end{bmatrix}_{K \times L} \quad (5.B.39)$$

$$T_{22,3} = \begin{bmatrix} \lambda_{K1,1}^{22,3} & \lambda_{K1,2}^{22,3} & \cdots & \lambda_{K1,L}^{22,3} \\ \lambda_{K2,1}^{22,3} & \lambda_{K2,2}^{22,3} & \cdots & \lambda_{K2,L}^{22,3} \\ \vdots & \vdots & \ddots & \vdots \\ \lambda_{KL,1}^{22,3} & \lambda_{KL,2}^{22,3} & \cdots & \lambda_{KL,L}^{22,3} \end{bmatrix}_{KL \times KL} \quad (5.B.40)$$

$$\lambda_{Kl}^{22,4} = l_y \cdot \begin{bmatrix} R_{Q3} - i\beta_l^3 R_{M3} & R_{Q3} - i\beta_l^3 R_{M3} & \cdots & R_{Q3} - i\beta_l^3 R_{M3} \\ R_{Q3} - i\beta_l^3 R_{M3} & R_{Q3} - i\beta_l^3 R_{M3} & \cdots & R_{Q3} - i\beta_l^3 R_{M3} \\ \vdots & \vdots & \ddots & \vdots \\ R_{Q3} - i\beta_l^3 R_{M3} & R_{Q3} - i\beta_l^3 R_{M3} & \cdots & R_{Q3} - i\beta_l^3 R_{M3} \end{bmatrix}_{K \times L} \quad (5.B.41)$$

$$T_{22,4} = \begin{bmatrix} \lambda_{K1}^{22,4} & & & \\ & \lambda_{K2}^{22,4} & & \\ & & \ddots & \\ & & & \lambda_{KL}^{22,4} \end{bmatrix}_{KL \times KL} \quad (5.B.42)$$

$$\lambda_{Kl}^{22,5} = l_y \cdot \begin{bmatrix} -i\alpha_1 \beta_l \alpha_1 R_{T3} & -i\alpha_1 \beta_l \alpha_2 R_{T3} & \cdots & -i\alpha_1 \beta_l \alpha_K R_{T3} \\ -i\alpha_2 \beta_l \alpha_1 R_{T3} & -i\alpha_2 \beta_l \alpha_2 R_{T3} & \cdots & -i\alpha_2 \beta_l \alpha_K R_{T3} \\ \vdots & \vdots & \ddots & \vdots \\ -i\alpha_K \beta_l \alpha_1 R_{T3} & -i\alpha_K \beta_l \alpha_2 R_{T3} & \cdots & -i\alpha_K \beta_l \alpha_K R_{T3} \end{bmatrix}_{K \times L}$$
(5.B.43)

$$T_{22,5} = \begin{bmatrix} \lambda_{K1}^{22,5} & & & \\ & \lambda_{K2}^{22,5} & & \\ & & \ddots & \\ & & & \lambda_{KL}^{22,5} \end{bmatrix}_{KL \times KL} \quad (5.B.44)$$

Using the definition of the sub-matrices presented above, one obtains

$$T_{11} = T_{11,1} + T_{11,2} + T_{11,3} + T_{11,4} + T_{11,5}, \quad T_{22} = T_{22,1} + T_{22,2} + T_{22,3} + T_{22,4} + T_{22,5} \quad (5.B.45)$$

$$T_{12} = T_{12,1} + T_{12,2} + T_{12,3} + T_{12,4} + T_{12,5}, \quad T_{21} = T_{21,1} + T_{21,2} + T_{21,3} + T_{21,4} + T_{21,5} \quad (5.B.46)$$

References

1. Ichchou MN, Berthaut J, Collet M (2008) Multi-mode wave propagation in ribbed plates. Part I: Wavenumber-space characteristics. Int J Solids Struct 45(5):1179–1195
2. Ichchou MN, Berthaut J, Collet M (2008) Multi-mode wave propagation in ribbed plates. Part II: Predictions and comparisons. Int J Solids Struct 45(5):1196–1216
3. Mace BR (1980) Periodically stiffened fluid-loaded plates, I: Response to convected harmonic pressure and free wave propagation. J Sound Vib 73(4):473–486
4. Mace BR (1980) Periodically stiffened fluid-loaded plates, II: Response to line and point forces. J Sound Vib 73(4):487–504
5. Mace BR (1981) Sound radiation from fluid loaded orthogonally stiffened plates. J Sound Vib 79(3):439–452
6. Mace BR (1996) The vibration of plates on two-dimensionally periodic point supports. J Sound Vib 192(3):629–643
7. Lee JH, Kim J (2002) Analysis of sound transmission through periodically stiffened panels by space-harmonic expansion method. J Sound Vib 251(2):349–366

8. Yin XW, Gu XJ, Cui HF et al (2007) Acoustic radiation from a laminated composite plate reinforced by doubly periodic parallel stiffeners. J Sound Vib 306(3–5):877–889
9. Xin FX, Lu TJ, Chen CQ (2008) Vibroacoustic behavior of clamp mounted double-panel partition with enclosure air cavity. J Acoust Soc Am 124(6):3604–3612
10. Xin FX, Lu TJ, Chen C (2008) Sound transmission through lightweight all-metallic sandwich panels with corrugated cores. Multi-Funct Mater Struct Parts 1 and 2 47–50:57–60
11. Maxit L (2009) Wavenumber space and physical space responses of a periodically ribbed plate to a point drive: a discrete approach. Appl Acoust 70(4):563–578
12. Xin FX, Lu TJ (2009) Analytical and experimental investigation on transmission loss of clamped double panels: implication of boundary effects. J Acoust Soc Am 125(3):1506–1517
13. Xin FX, Lu TJ, Chen CQ (2009) Dynamic response and acoustic radiation of double-leaf metallic panel partition under sound excitation. Comput Mater Sci 46(3):728–732
14. Xin FX, Lu TJ, Chen CQ (2009) External mean flow influence on noise transmission through double-leaf aeroelastic plates. AIAA J 47(8):1939–1951
15. Takahashi D (1983) Sound radiation from periodically connected double-plate structures. J Sound Vib 90(4):541–557
16. Fahy FJ, Lindqvist E (1976) Wave propagation in damped, stiffened structures characteristic of ship construction. J Sound Vib 45(1):115–138
17. Mead DJ (1990) Plates with regular stiffening in acoustic media: vibration and radiation. J Acoust Soc Am 88(1):391–401
18. Xin FX, Lu TJ, Chen CQ (2010) Sound transmission through simply supported finite double-panel partitions with enclosed air cavity. J Vib Acoust 132(1):011008: 011001–011011
19. Xin FX, Lu TJ (2010) Sound radiation of orthogonally rib-stiffened sandwich structures with cavity absorption. Compos Sci Technol 70(15):2198–2206
20. Xin FX, Lu TJ (2010) Analytical modeling of sound transmission across finite aeroelastic panels in convected fluids. J Acoust Soc Am 128(3):1097–1107
21. Mead DJ, Pujara KK (1971) Space-harmonic analysis of periodically supported beams: response to convected random loading. J Sound Vib 14(4):525–532
22. Mead DJ, Yaman Y (1990) The harmonic response of uniform beams on multiple linear supports: a flexural wave analysis. J Sound Vib 141(3):465–484
23. Mead DJ (1970) Free wave propagation in periodically supported, infinite beams. J Sound Vib 11(2):181–197
24. Mead DJ (1996) Wave propagation in continuous periodic structures: research contributions from Southampton, 1964–1995. J Sound Vib 190(3):495–524
25. Rumerman ML (1975) Vibration and wave propagation in ribbed plates. J Acoust Soc Am 57(2):370–373
26. Mead DJ, Mallik AK (1978) An approximate theory for the sound radiated from a periodic line-supported plate. J Sound Vib 61(3):315–326
27. Mead DJ, Parthan S (1979) Free wave propagation in two-dimensional periodic plates. J Sound Vib 64(3):325–348
28. Mace BR (1980) Sound radiation from a plate reinforced by two sets of parallel stiffeners. J Sound Vib 71(3):435–441
29. Cray BA (1994) Acoustic radiation from periodic and sectionally aperiodic rib-stiffened plates. J Acoust Soc Am 95(1):256–264
30. Wang J, Lu TJ, Woodhouse J et al (2005) Sound transmission through lightweight double-leaf partitions: theoretical modelling. J Sound Vib 286(4–5):817–847
31. Crighton DG (1807) Transmission of energy down periodically ribbed elastic structures under fluid loading. Proc R Soc Lond A Math Phys Sci 1984(394):405–436
32. Spivack M (1895) Wave propagation in finite periodically ribbed structures with fluid loading. Proc R Soc Lond A Math Phys Sci 1991(435):615–634
33. Cooper AJ, Crighton DG (1979) Transmission of energy down periodically ribbed elastic structures under fluid loading: spatial periodicity in the pass bands. Proc R Soc Lond A Math Phys Sci 1998(454):2893–2909

References

34. Cooper AJ, Crighton DG (1973) Transmission of energy down periodically ribbed elastic structures under fluid loading: algebraic decay in the stop bands. Proc R Soc Lond A Math Phys Sci 1998(454):1337–1355
35. Kohno H, Bathe K-J, Wright JC (2010) A finite element procedure for multiscale wave equations with application to plasma waves. Comput Struct 88(1–2):87–94
36. Langley RS, Bardell NS, Ruivo HM (1997) The response of two-dimensional periodic structures to harmonic point loading: a theoretical and experimental study of a beam grillage. J Sound Vib 207(4):521–535
37. Dozio L, Ricciardi M (2009) Free vibration analysis of ribbed plates by a combined analytical-numerical method. J Sound Vib 319(1–2):681–697
38. Mead DJ, Yaman Y (1991) The harmonic response of rectangular sandwich plates with multiple stiffening: a flexural wave analysis. J Sound Vib 145(3):409–428
39. Xin FX, Lu TJ (2011) Analytical modeling of wave propagation in orthogonally rib-stiffened sandwich structures: sound radiation. Comput Struct 89(5–6):507–516
40. Graham WR (1995) High-frequency vibration and acoustic radiation of fluid-loaded plates. Philos Trans R Soc Phys Eng Sci (1990–1995) 352(1698):1–43
41. Lucey AD (1998) The excitation of waves on a flexible panel in a uniform flow. Philos Trans R Soc A Math Phys Eng Sci 356(1749):2999–3039
42. Hambric SA, Hwang YF, Bonness WK (2004) Vibrations of plates with clamped and free edges excited by low-speed turbulent boundary layer flow. J Fluids Struct 19(1):93–110
43. Maury C, Gardonio P, Elliott SJ (2001) Active control of the flow-induced noise transmitted through a panel. AIAA J 39(10):1860–1867
44. Ruzzene M (2004) Vibration and sound radiation of sandwich beams with honeycomb truss core. J Sound Vib 277(4–5):741–763
45. Spadoni A, Ruzzene M (2006) Structural and acoustic behavior of chiral truss-core beams. J Vib Acoust 128(5):616–626
46. Rao UN, Mallik AK (1977) Response of finite periodic beams to convected loading–An approximate method. J Sound Vib 55(3):395–403
47. Mead DJ, Mallik AK (1976) An approximate method of predicting the response of periodically supported beams subjected to random convected loading. J Sound Vib 47(4):457–471
48. Liu L, Bhattacharya K (2009) Wave propagation in a sandwich structure. Int J Solids Struct 46(17):3290–3300
49. Wang YZ, Li FM, Huang WH et al (2008) The propagation and localization of Rayleigh waves in disordered piezoelectric phononic crystals. J Mech Phys Solids 56(4):1578–1590
50. Li FM, Wang YS, Hu C et al (2005) Localization of elastic waves in periodic rib-stiffened rectangular plates under axial compressive load. J Sound Vib 281(1–2):261–273
51. Li F-M, Wang Y-S (2005) Study on wave localization in disordered periodic layered piezoelectric composite structures. Int J Solids Struct 42(24–25):6457–6474
52. Leppington FG, Broadbent EG, Heron KH (1984) Acoustic radiation from rectangular panels with constrained edges. Proc R Soc Lond A Math Phys Sci (1934–1990) 393(1804):67–84
53. Leppington FG, Broadbent EG, Heron KH et al (1986) Resonant and non-resonant acoustic properties of elastic panels. I. The radiation problem. Proc R Soc Lond A Math Phys Sci (1934–1990) 406(1831):139–171
54. Leppington FG, Heron KH, Broadbent EG et al (1987) Resonant and non-resonant acoustic properties of elastic panels. II. The transmission problem. Proc R Soc Lond A Math Phys Sci (1934–1990) 412(1843):309–337
55. Leppington FG (1870) The transmission of randomly incident sound through an elastic panel. Proc R Soc Lond A Math Phys Sci 1989(426):153–165
56. Leppington FG, Heron KH, Broadbent EG (2019) Resonant and non-resonant transmission of random noise through complex plates. Proc R Soc Lond A Math Phys Eng Sci 2002(458):683–704
57. Lin G-F, Garrelick JM (1977) Sound transmission through periodically framed parallel plates. J Acoust Soc Am 61(4):1014–1018

58. Legault J, Atalla N (2009) Numerical and experimental investigation of the effect of structural links on the sound transmission of a lightweight double panel structure. J Sound Vib 324(3–5):712–732
59. Xin FX, Lu TJ (2011) Effects of core topology on sound insulation performance of lightweight all-metallic sandwich panels. Mater Manuf Processes 26(9):1213–1221
60. Brillouin L (1953) Wave propagation in periodic structures. Dover, New York
61. Wang YZ, Li FM, Kishimoto K et al (2009) Elastic wave band gaps in magnetoelectroelastic phononic crystals. Wave Motion 46(1):47–56
62. Wang YZ, Li FM, Wang YS et al (2009) Tuning of band gaps for a two-dimensional piezoelectric phononic crystal with a rectangular lattice. Acta Mech Sin 25(1):65–71
63. Wang Y-Z, Li F-M, Kishimoto K et al (2010) Band gaps of elastic waves in three-dimensional piezoelectric phononic crystals with initial stress. Eur J Mech A/Solids 29(2):182–189
64. Morse PM, Ingard KU (1968) Theoretical acoustics. McGraw-Hill, New York
65. Fahy F (1985) Sound and structural vibration: radiation, transmission and response. Academic, London

Chapter 6
Sound Propagation in Rib-Stiffened Sandwich Structures with Cavity Absorption

Abstract This chapter is organized as two parts: in the first part, a comprehensive theoretical model is developed for the radiation of sound from an infinite orthogonally rib-stiffened sandwich structure filled with fibrous sound absorptive material in the partitioned cavity, when excited by a time-harmonic point force. The vibrations of the rib-stiffeners are accounted for by considering all possible motions. Built upon the concepts of dynamic density and bulk modulus, both frequency dependent, an equivalent fluid model is employed to characterize the absorption of sound in the fibrous material. Given the periodicity of the sandwich structure, Fourier transform technique is employed to solve the series of panel vibration equations and acoustic equations. In the absence of fibrous sound absorptive material, the model can be favorably degraded to the case of an infinite rib-stiffened structure with air or vacuum cavity. Validation of the model is performed by comparing the present model predictions with previously published data, with excellent agreements achieved. The influences of air-structure coupling effect and cavity-filling fibrous material on the sound radiation are systematically examined. The physical features associated with sound penetration across these sandwich structures are interpreted by considering the combined effects of fiberglass stiffness and damping, the balance of which is significantly affected by stiffener separation. The proposed model provides a convenient and efficient tool for the factual engineering design of this kind of sandwich structures.

In the second part, the transmission loss of sound through infinite orthogonally rib-stiffened double-panel structures having cavity-filling fibrous sound absorptive materials is theoretically investigated. The propagation of sound across the fibrous material is characterized using an equivalent fluid model, and the motions of the rib-stiffeners are described by including all possible vibrations, i.e., tensional displacements, bending, and torsional rotations. The effects of fluid-structure coupling are accounted for by enforcing velocity continuity conditions at fluid-panel interfaces. By fully taking advantage of the periodic nature of the double panel, the space-harmonic approach and virtual work principle are applied to solve the sets of resultant governing equations, which are eventually truncated as a finite system of simultaneous algebraic equations and numerically solved insofar as the solution converges. To validate the proposed model, a comparison between the present model predictions and existing numerical and experimental results for a simplified version of the double-panel structure is carried out, with overall agreement achieved. The model is subsequently employed to explore the influence of the fluid-structure coupling between fluid in the cavity and the two panels on sound transmission across the orthogonally rib-stiffened double-panel structure. Obtained results demonstrate that this fluid-structure coupling affects significantly sound transmission loss (STL) at low frequencies and cannot be ignored when the rib-stiffeners are sparsely distributed. As a highlight of this research, an integrated optimal algorithm toward lightweight, high-stiffness, and superior sound insulation capability is proposed, based on which a preliminary optimal design of the double-panel structure is performed.

6.1 Sound Radiation of Absorptive Sandwich Structures

6.1.1 Introduction

With the increasing use of periodically rib-stiffened composite sandwich structures as the cabin skin of aircrafts, marine ships, express trains, etc. [1–5], great efforts have been made in the pursuit of efficient theoretical methods for predicting the vibration and acoustic behaviors of these lightweight structures, so as to design optimized configurations competent for practical low-noise requirements.

Active control algorithms with sensors and actuators have been developed to reduce structure vibration and sound radiation [6], which however inevitably brings the penalty of increasing system complexity and financial costs. Alternatively, passive measures such as inserting fibrous sound absorptive materials in the partitioned cavity of sandwich structures may be a preferable choice to achieve a compromise between noise reduction efficiency and financial cost. For instance, the fuselages of commercial aircrafts are commonly made of periodically rib-stiffened composite structures filled with fiberglass to enhance thermal and sound insulation [6–9]. This provides strong impetus for the development of effective theoretical models to predict the sound radiation characteristics of periodically rib-stiffened sandwich structures filled with sound absorptive materials.

There exist numerous theoretical models for the vibroacoustic behaviors of periodic rib-stiffened structures, which may be grouped into two main categories: one is based on the Fourier transform method [2–4, 10–13], which is able to handle both sound radiation and sound transmission problems, and the other is built upon the space-harmonic approach [14–18], which is suited particularly for sound transmission problems. Mace [2] employed the Fourier transform method to solve the problem of sound radiation from a fluid-loaded infinite plate reinforced

6.1 Sound Radiation of Absorptive Sandwich Structures

by two sets of parallel stiffeners when excited by a point force; for simplification, only the tensional force of the rib-stiffeners was considered. Subsequently, Mace [3] proposed a theoretical model for the radiation of sound from an infinite fluid-loaded plate when the plate is reinforced with two sets of orthogonal line stiffeners; again, only the tensional force of the rib-stiffeners was accounted for. Similarly, by only taking account of the normal force interaction between panel and rib-stiffeners, Yin et al. [4] presented a simplified theoretical model for acoustic radiation from a point-driven, fluid-loaded infinite laminated composite plate reinforced by periodic parallel rib-stiffeners.

As an essentially equivalent method, the space-harmonic approach evolved from progressive wave propagation was initiated by Mead and Pujara [14] when they studied the acoustic response of periodic stiffened beams subjected to a spatial and temporal harmonic pressure. It was demonstrated that as few as three terms of space harmonics could lead to solutions of acceptable accuracy. By combining the space-harmonic approach and virtual energy method, Lee and Kim [15] analyzed the sound transmission characteristics of a thin plate stiffened by equally spaced line stiffeners. By modeling the rib-stiffeners as a combination of translational springs and rotational springs, Wang et al. [16] proposed an analytic model for sound transmission loss across double-leaf partitions stiffened with periodically placed studs. Recently, Xin and Lu [18] developed a comprehensive analytic model for sound transmission through orthogonally rib-stiffened sandwich structures: all possible motions of the rib-stiffeners were accounted for by introducing the tensional forces, bending moments, and torsional moments as well as the corresponding inertial terms into the governing equations of the two face panels.

None of the abovementioned investigations dealt with sound radiation and/or sound transmission issues of composite sandwich structures filled with porous sound absorptive materials. As far as the sound radiation/transmission problems of double partitions with cavity absorption are of concern, a number of theoretical [12, 13], numerical [19, 20], and experimental [21] studies do exist. However, all of these studies *did not* consider the effects of structural rib-connections between two face panels, which may be far away from the factual engineering structures. To address this deficiency, a comprehensive theoretical model is developed here for the radiation of sound from an infinite orthogonally rib-stiffened sandwich structure filled with fibrous sound absorptive material in the partitioned cavity when excited by a time-harmonic point force. The equivalent forces and moments (both bending and torsional) imposed on the two face panels by the rib-stiffeners are accounted for by considering all possible motions of the rib-stiffeners. By employing the well-known equivalent fluid model [12, 22], wave propagation in the fibrous sound absorptive material can be accurately described. Both viscous drag forces and thermal exchanges between air and solid fibers are accounted for by introducing frequency-dependent dynamic density and bulk modulus. Taking advantage of the periodic property of the composite sandwich structure, the Fourier transform technique is adopted to solve both the structural and acoustic governing equations. In limiting cases, the developed model can be favorably degraded to deal

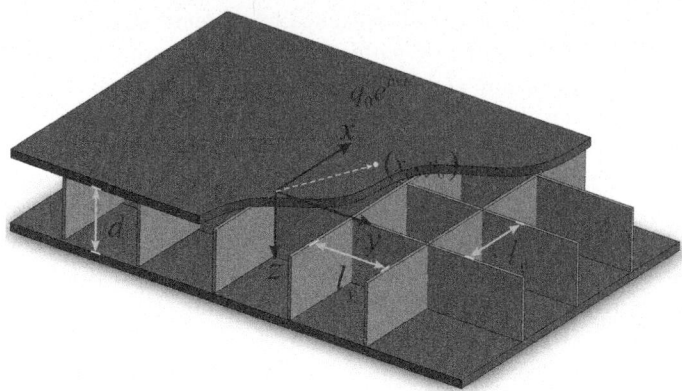

Fig. 6.1 Schematic illustration of orthogonally rib-stiffened sandwich structure (three different kinds: vacuum cavity, air cavity, and fiberglass-filled cavity) excited by time-harmonic point force at (x_0, y_0) (With permission from Elsevier)

with sound radiation issues of sandwich structures with vacuum or air cavities. Therefore, model validation is carried out by comparing the present predictions for simplified sandwich structures with those available in the open literature. To explore the influence of fibrous sound absorptive materials on sound radiation of orthogonally rib-stiffened composite structures, numerical results are presented, with relevant physical features interpreted in detail. Conclusions drawn from the present theoretical study may provide fundamental principles for factual engineering design of rib-stiffened composite structures filled with fibrous sound absorptive materials.

6.1.2 Structural Dynamic Responses to Time-Harmonic Point Force

Consider an infinite sandwich structure as shown schematically in Fig. 6.1, which is reinforced by two periodic sets of orthogonal rib-stiffeners having periodic uniform separations l_x and l_y in the x- and y-directions, respectively. A right-handed Cartesian coordinate system (x, y, z) is established, with its x-axis and y-axis positioned separately along one pair of the orthogonal rib-stiffeners, the positive direction of the z-axis pointing downward (Fig. 6.1). Three different kinds of sandwich structures will be considered in the proceeding sections, namely, the gap between the two parallel face panels and portioned by the orthogonal lattice cores is in vacuum, air filled, or filled with fibrous sound absorptive material (e.g., fiberglass), respectively. A theoretical model will be formulated for the complex structure (i.e., orthogonally rib-stiffened sandwich structure filled with fibrous sound absorptive material), which can be degraded to deal with the other two sandwiches.

6.1 Sound Radiation of Absorptive Sandwich Structures

Assume that a time-harmonic point force $q_0 e^{i\omega t}$ acts on the surface of the upper panel at location (x_0, y_0); see Fig. 6.1. Consequently, a radially outspreading bending wave propagates in the upper panel from the source (x_0, y_0). The vibration of the upper panel is transmitted to the bottom panel via the orthogonal rib-stiffeners and sound absorbing material (or air cavity). Subsequently, the bottom panel vibrates and radiates sound pressure waves.

The dynamic responses of the sandwich structure are time harmonic as the excitation is in the form of $q_0 e^{i\omega t}$. For simplicity, the harmonic time term $e^{i\omega t}$ is suppressed henceforth. With the equivalent forces and moments of the lattice core and the pressure in the fibrous sound absorptive material (or air cavity) accounted for, the equations governing panel vibrations are given by

$$D_1 \nabla^4 w_1 + m_1 \frac{\partial^2 w_1}{\partial t^2}$$

$$= \sum_{m=-\infty}^{+\infty} \left[Q_y^+ \delta(x - ml_x) + \frac{\partial}{\partial y} \{M_y^+ \delta(x - ml_x)\} + \frac{\partial}{\partial x} \{M_{Ty}^+ \delta(x - ml_x)\} \right]$$

$$+ \sum_{n=-\infty}^{+\infty} \left[Q_x^+ \delta(y - nl_y) + \frac{\partial}{\partial x} \{M_x^+ \delta(y - nl_y)\} + \frac{\partial}{\partial y} \{M_{Tx}^+ \delta(y - nl_y)\} \right]$$

$$+ q_0 \delta(x - x_0) \delta(y - y_0) - p_{\text{cav}}(x, y, h_1) \tag{6.1}$$

$$D_2 \nabla^4 w_2 + m_2 \frac{\partial^2 w_2}{\partial t^2}$$

$$= -\sum_{m=-\infty}^{+\infty} \left[Q_y^- \delta(x - ml_x) + \frac{\partial}{\partial y} \{M_y^- \delta(x - ml_x)\} + \frac{\partial}{\partial x} \{M_{Ty}^- \delta(x - ml_x)\} \right]$$

$$- \sum_{n=-\infty}^{+\infty} \left[Q_x^- \delta(y - nl_y) + \frac{\partial}{\partial x} \{M_x^- \delta(y - nl_y)\} + \frac{\partial}{\partial y} \{M_{Tx}^- \delta(y - nl_y)\} \right]$$

$$+ p_{\text{cav}}(x, y, h_1 + d) \tag{6.2}$$

where $\nabla^4 \equiv (\partial^2/\partial x^2 + \partial^2/\partial y^2)^2$; $\delta(\cdot)$ is the Dirac delta function; and (w_1, w_2), (m_1, m_2), and (D_1, D_2) are the displacement, surface mass density, and flexural rigidity of the upper panel and bottom panel, respectively. The material loss factor η_j ($j = 1, 2$ for upper panel and bottom panel, respectively) is introduced with the complex Young's modulus $D_j = E_j h_j^3 (1 + i\eta_j)/12(1 - \nu_j^2)$ (where $j = 1, 2$).

In the above equations, (Q^+, M^+, M_T^+) and (Q^-, M^-, M_T^-) denote separately the tensional forces, bending moments, and torsional moments of the lattice core acting on the upper panel and bottom panel, as shown in Fig. 6.2 (more details given in Appendix A). As the corresponding inertial terms are also considered, the forces and moments exerting on the upper and bottom panels are not quite the same and are thus differentiated using superscripts + (upper) and − (bottom).

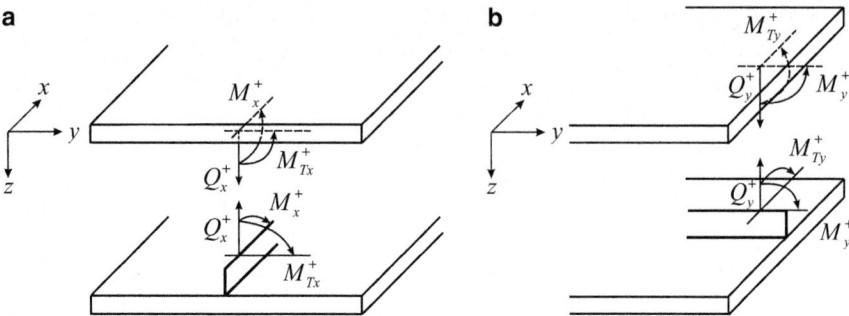

Fig. 6.2 Conventions for tensional forces, bending moments, and torsional moments between upper panel and (**a**) *x*-wise and (**b**) *y*-wise stiffeners, which also hold at the interface between bottom panel and *x*- and *y*-wise stiffeners (With permission from Elsevier)

Given the 2D (two-dimensional) periodic nature of the sandwich structure as shown in Fig. 6.1, applying the Poisson summation formula [3, 23], the wave components in the structure can be expressed by using space-harmonic series as

$$\sum_{m=-\infty}^{+\infty} \delta(x - ml_x) = \frac{1}{l_x} \sum_{m=-\infty}^{+\infty} e^{-i(2m\pi/l_x)x}, \quad \sum_{n=-\infty}^{+\infty} \delta(y - nl_y) = \frac{1}{l_y} \sum_{n=-\infty}^{+\infty} e^{-i(2n\pi/l_y)y}$$
(6.3)

The Fourier transform pair of a function with respect to (x, y) and (α, β) can be defined as

$$w(x, y) = \int_{-\infty}^{+\infty} \int_{-\infty}^{+\infty} \tilde{w}(\alpha, \beta) \, e^{i(\alpha x + \beta y)} d\alpha d\beta$$
(6.4)

$$\tilde{w}(\alpha, \beta) = \left(\frac{1}{2\pi}\right)^2 \int_{-\infty}^{+\infty} \int_{-\infty}^{+\infty} w(x, y) \, e^{-i(\alpha x + \beta y)} dx dy$$
(6.5)

Applying the Poisson summation formula and then taking the Fourier transform of Eqs. (6.4) and (6.5), one gets

$$\left[D_1(\alpha^2 + \beta^2)^2 - m_1\omega^2\right]\tilde{w}_1(\alpha, \beta)$$

$$= \frac{1}{l_x}\sum_m \left[\tilde{Q}_y^+(\alpha_m, \beta) + i\beta\tilde{M}_y^+(\alpha_m, \beta) + i\alpha\tilde{M}_{Ty}^+(\alpha_m, \beta)\right]$$

$$+ \frac{1}{l_y}\sum_n \left[\tilde{Q}_x^+(\alpha, \beta_n) + i\alpha\tilde{M}_x^+(\alpha, \beta_n) + i\beta\tilde{M}_{Tx}^+(\alpha, \beta_n)\right]$$

$$+ \frac{q_0}{(2\pi)^2}e^{-i(\alpha x_0 + \beta y_0)} - \tilde{p}_{\text{cav}}(\alpha, \beta, h_1)$$
(6.6)

6.1 Sound Radiation of Absorptive Sandwich Structures

$$\left[D_2(\alpha^2 + \beta^2)^2 - m_2\omega^2\right]\tilde{w}_2(\alpha, \beta)$$
$$= -\frac{1}{l_x}\sum_m \left[\tilde{Q}_y^-(\alpha_m, \beta) + i\beta \tilde{M}_y^-(\alpha_m, \beta) + i\alpha \tilde{M}_{Ty}^-(\alpha_m, \beta)\right]$$
$$- \frac{1}{l_y}\sum_n \left[\tilde{Q}_x^-(\alpha, \beta_n) + i\alpha \tilde{M}_x^-(\alpha, \beta_n) + i\beta \tilde{M}_{Tx}^-(\alpha, \beta_n)\right]$$
$$+ \tilde{p}_{\text{cav}}(\alpha, \beta, h_1 + d) \tag{6.7}$$

where $\alpha_m = \alpha + 2m\pi/l_x$, $\beta_n = \beta + 2n\pi/l_y$, and the dependence of a term on the wavenumbers (α, β) is indicated using the hat sign ~, meaning the corresponding Fourier transform of this term. Expressions for the Fourier transform of the tensional forces, bending moments, and torsional moments are listed in Appendix A.

6.1.3 The Acoustic Pressure and Fluid-Structure Coupling

The absorption of sound absorption by porous materials mainly arises from viscous drag forces and thermal exchange loss when sound penetrates through the material [19, 24–26]. There exist numerous theoretical models to address these issues, while different models may be specialized to deal with different types of porous materials. For instance, Lu et al. proposed a model for high-porosity cellular metallic foams with open cells [24, 25, 27, 28] and another model for semi-open metal foams [26]. As for fibrous materials considered here, there are two main classes of models [19]. The first one models the fibrous material as an equivalent fluid with effective density and bulk modulus [22, 29, 30]: under the assumption of the solid fibers being a rigid skeleton, only one compression wave propagates in the air-saturated medium, which thereby is governed by the Helmholtz equation. The other one employs the more rigorous theory of Biot [31, 32] with the elasticity of the skeleton taken into account, the solution of which often seeks help from the finite element method (FEM) and suffers from huge computational expenses.

In view of the complexity of the proposed structure vibration model and the primary focus of the present study on sound radiation of the sandwich structure as a whole, the well-developed empirical expressions (i.e., equivalent fluid model) of Allard and Champoux [22] are adopted to model the acoustic pressure in fibrous absorption materials such as glass/rock wools widely used in noise absorption engineering. In terms of scholar description, these may be defined as Newtonian fluid-saturated rigid frame fibrous materials, with the frame fibers randomly distributed. Although Allard and Champoux [22] called their empirical equations as the equivalent fluid model, this model is in fact based on Johnson et al.'s two-phase theory [33]. It accurately accounted for the viscous forces between fluid and solid and the physical transposition in the process of sound propagation, by adopting two variables – the dynamic density $\rho(\omega)$ and the dynamic bulk modulus $K(\omega)$ –

assuming that the fibrous material is isotropic. The equivalent fluid model has been demonstrated to be capable of providing accurate predictions of sound wave propagation across fibrous sound absorptive materials, over a wide frequency range, and hence has been widely acknowledged by the acoustic community [19, 24–26]. To be more precise, the equivalent fluid model is valid for most glass/rock wools for f/R smaller than 1.0 kg^{-1} m^3, where f is the frequency and R is the flow resistivity of the fibrous material [22]. Generally, the flow resistivity R of typical glass/rock wools is approximately 20,000 Nm/s^4, and hence the equivalent fluid model works well for frequencies below 20 kHz.

According to the equivalent fluid model, wave propagation in fibrous sound absorptive material (e.g., fiberglass or mineral wool) is governed by [12, 13, 22]

$$\left(\partial^2/\partial x^2 + \partial^2/\partial y^2 + \partial^2/\partial z^2\right) p_{\text{cav}} + k_{\text{cav}}^2 p_{\text{cav}} = 0 \qquad (6.8)$$

where p_{cav} is the sound pressure in the fibrous material and k_{cav} is the corresponding complex wavenumber, which is related to the dynamic density $\rho(\omega)$ and dynamic bulk modulus $K(\omega)$ of the fibrous material by

$$k_{\text{cav}} = 2\pi f \sqrt{\rho(\omega)/K(\omega)} \qquad (6.9)$$

In accordance with the complex physical phenomena taking place in the fibrous material, such as thermal exchanges between air and fibers showing a significant transition with increasing frequency (i.e., isothermal process at low frequency turning to adiabatic process at high frequency) [22], the equivalent density and bulk modulus are both dynamic. In other words, the dynamic density and dynamic bulk modulus are frequency dependent, given by [22]

$$\rho(\omega) = \rho_0 \left[1 + (R/\rho_0 f) G_1 (\rho_0 f/R)/i2\pi\right] \qquad (6.10)$$

$$K(\omega) = \gamma_s P_0 \left(\gamma_s - \frac{\gamma_s - 1}{1 + (1/i8\pi N_{pr})(\rho_0 f/R)^{-1} G_2(\rho_0 f/R)}\right)^{-1} \qquad (6.11)$$

where $G_1(\rho_0 f/R) = \sqrt{1 + i\pi(\rho_0 f/R)}$, $G_2(\rho_0 f/R) = G_1[(\rho_0 f/R)4N_{pr}]$, R is the static flow resistivity of the fibrous material, γ_s and ρ_0 are separately the specific heat ratio (i.e., $\gamma_s = c_p/c_v$, c_p and c_v being the specific heat per unit mass of the air at constant pressure and constant volume) and density of air, P_0 is the air equilibrium pressure, and N_{pr} is the Prandtl number. As a further understanding of physical meanings, the dynamic density $\rho(\omega)$ contains the inertial and viscous forces per unit volume of air in fibrous material, while the dynamic bulk modulus $K(\omega)$ gives the relationship between the averaged molecular displacement of air and the averaged variation of pressure. As a conclusion of Lu et al.'s model, it is found that the viscous drag forces operating at the fiber surface govern the complex density $\rho(\omega)$ and the

6.1 Sound Radiation of Absorptive Sandwich Structures

thermal forces control the complex bulk modulus $K(\omega)$. As seen in Eqs. (6.10) and (6.11), these two quantities are strongly dependent on the term $\rho_0 f/R$, reflecting the inherent dynamic property of sound absorbing process and flow resistance being the fundamental origin of sound absorption.

Generally, in contrast with the facesheet to stiffener interaction and the facesheet to fibrous-material (or air) interaction, the stiffener to fibrous-material (or air) interaction is negligible. It is easy to understand that the direct structural connection between the facesheets and the rib-stiffeners is far stronger than the stiffener to fibrous-material (or air) interaction. As for the facesheet to fibrous-material interaction, note that the stiffener separations l_x and l_y are generally much larger than the stiffener height d, implying that the contact surface area between the facesheet and fibrous material is much larger than that between the stiffener and fibrous material. Therefore, while the vibration of the facesheet is affected significantly by the fibrous material in contact, it has negligible influence on the motion of the short stiffeners. As a result, the fluid-structure coupling here only needs to consider the facesheet to fibrous-material interaction. To ensure the equality of panel velocity and fluid velocity on the panel surface, the momentum equation (i.e., continuity condition of fluid-structure coupling [7, 10, 34]) is applied:

$$\left[\frac{\partial p_{cav}}{\partial z}\right]_{z=h_1} = \rho_{cav}\omega^2 w_1, \quad \left[\frac{\partial p_{cav}}{\partial z}\right]_{z=h_1+d} = \rho_{cav}\omega^2 w_2 \quad (6.12)$$

where the complex density ρ_{cav} of the fibrous material is related to the complex wavenumber k_{cav} and porosity σ as [12]

$$\frac{k_{cav}^2}{k_0^2} = \frac{\gamma_s \sigma \rho_{cav}}{\rho_0} \quad (6.13)$$

Applying the Fourier transform to Eqs. (6.8) and (6.12), one obtains

$$\tilde{p}_{cav}(\alpha, \beta, z) = -\frac{\rho_{cav}\omega^2}{\gamma(\alpha, \beta)\left(e^{\gamma d} - e^{-\gamma d}\right)} \left\{ \begin{array}{l} \left[\tilde{w}_1(\alpha, \beta) e^{\gamma(h_1+d)} - \tilde{w}_2(\alpha, \beta) e^{\gamma h_1}\right] e^{-\gamma z} \\ + \left[\tilde{w}_1(\alpha, \beta) e^{-\gamma(h_1+d)} - \tilde{w}_2(\alpha, \beta) e^{-\gamma h_1}\right] e^{\gamma z} \end{array} \right\} \quad (6.14)$$

where $\gamma^2 = \alpha^2 + \beta^2 - k_{cav}^2$. More specifically, the pressures acting on the upper and bottom panels are given by

$$\tilde{p}_{cav}(\alpha, \beta, h_1) = -\frac{\rho_{cav}\omega^2 \left[\tilde{w}_1(\alpha, \beta) \cosh(\gamma d) - \tilde{w}_2(\alpha, \beta)\right]}{\gamma(\alpha, \beta) \sinh(\gamma d)} \quad (6.15)$$

$$\tilde{p}_{cav}(\alpha, \beta, h_1+d) = -\frac{\rho_{cav}\omega^2 \left[\tilde{w}_1(\alpha, \beta) - \tilde{w}_2(\alpha, \beta) \cosh(\gamma d)\right]}{\gamma(\alpha, \beta) \sinh(\gamma d)} \quad (6.16)$$

Substituting Eqs. (6.15) and (6.16) into Eqs. (6.6) and (6.7), respectively, one can rewrite the governing equations as

$$\left[D_1(\alpha^2+\beta^2)^2 - m_1\omega^2 - \frac{\rho_{cav}\omega^2 \cosh(\gamma d)}{\gamma(\alpha,\beta)\sinh(\gamma d)}\right]\tilde{w}_1(\alpha,\beta) + \frac{\rho_{cav}\omega^2}{\gamma(\alpha,\beta)\sinh(\gamma d)}\tilde{w}_2(\alpha,\beta)$$

$$= \frac{1}{l_x}\sum_m \left[-R_{Q3} - i\beta^3 R_{M3} - i\alpha\alpha_m\beta R_{T3}\right]\tilde{w}_1(\alpha_m,\beta)$$

$$+ \frac{1}{l_y}\sum_n \left[-R_{Q1} - i\alpha^3 R_{M1} - i\alpha\beta\beta_n R_{T1}\right]\tilde{w}_1(\alpha,\beta_n)$$

$$+ \frac{1}{l_x}\sum_m \left[R_{Q4} + i\beta^3 R_{M4} + i\alpha\alpha_m\beta R_{T4}\right]\tilde{w}_2(\alpha_m,\beta)$$

$$+ \frac{1}{l_y}\sum_n \left[R_{Q2} + i\alpha^3 R_{M2} + i\alpha\beta\beta_n R_{T2}\right]\tilde{w}_2(\alpha,\beta_n)$$

$$+ \frac{q_0}{(2\pi)^2}e^{-i(\alpha x_0+\beta y_0)} \tag{6.17}$$

$$\left[D_2(\alpha^2+\beta^2)^2 - m_2\omega^2 - \frac{\rho_{cav}\omega^2 \cosh(\gamma d)}{\gamma(\alpha,\beta)\sinh(\gamma d)}\right]\tilde{w}_2(\alpha,\beta) + \frac{\rho_{cav}\omega^2}{\gamma(\alpha,\beta)\sinh(\gamma d)}\tilde{w}_1(\alpha,\beta)$$

$$= -\frac{1}{l_x}\sum_m \left[-R_{Q4} - i\beta^3 R_{M4} - i\alpha\alpha_m\beta R_{T4}\right]\tilde{w}_1(\alpha_m,\beta)$$

$$- \frac{1}{l_y}\sum_n \left[-R_{Q2} - i\alpha^3 R_{M2} - i\alpha\beta\beta_n R_{T2}\right]\tilde{w}_1(\alpha,\beta_n)$$

$$- \frac{1}{l_x}\sum_m \left[R_{Q3} + i\beta^3 R_{M3} + i\alpha\alpha_m\beta R_{T3}\right]\tilde{w}_2(\alpha_m,\beta)$$

$$- \frac{1}{l_y}\sum_n \left[R_{Q1} + i\alpha^3 R_{M1} + i\alpha\beta\beta_n R_{T1}\right]\tilde{w}_2(\alpha,\beta_n) \tag{6.18}$$

To simplify the derivation procedures, the following definitions are introduced:

$$f_1(\alpha,\beta) = (\alpha^2+\beta^2)^2 - \frac{m_1\omega^2}{D_1} - \frac{\rho_{cav}\omega^2 \cosh(\gamma d)}{D_1\gamma(\alpha,\beta)\sinh(\gamma d)} \tag{6.19}$$

$$f_2(\alpha,\beta) = (\alpha^2+\beta^2)^2 - \frac{m_2\omega^2}{D_2} - \frac{\rho_{cav}\omega^2 \cosh(\gamma d)}{D_2\gamma(\alpha,\beta)\sinh(\gamma d)} \tag{6.20}$$

As seen in Eqs. (6.17) and (6.18), the panel displacements $\tilde{w}_1(\alpha,\beta)$ and $\tilde{w}_2(\alpha,\beta)$ to be solved are not independent but have coupling terms $\tilde{w}_1(\alpha_m,\beta)$, $\tilde{w}_1(\alpha,\beta_n)$, $\tilde{w}_2(\alpha_m,\beta)$, and $\tilde{w}_2(\alpha,\beta_n)$ in the corresponding sum formula. To solve

6.1 Sound Radiation of Absorptive Sandwich Structures

these unknowns, one needs to replace (α, β) by (α'_m, β'_n), leading to two sets of simultaneous algebraic equations:

$$D_1 f_1\left(\alpha'_m, \beta'_n\right) \tilde{w}_1\left(\alpha'_m, \beta'_n\right) + \frac{\rho_{\text{cav}} \omega^2}{\gamma\left(\alpha'_m, \beta'_n\right) \sinh\left[\gamma\left(\alpha'_m, \beta'_n\right) \cdot d\right]} \tilde{w}_2\left(\alpha'_m, \beta'_n\right)$$

$$+ \frac{1}{l_x}\left[\left(R_{Q3} + i\beta'^3_n R_{M3}\right) \sum_m \tilde{w}_1\left(\alpha_m, \beta'_n\right) + i\alpha'_m \beta'_n R_{T3} \sum_m \alpha_m \tilde{w}_1\left(\alpha_m, \beta'_n\right)\right]$$

$$+ \frac{1}{l_y}\left[\left(R_{Q1} + i\alpha'^3_m R_{M1}\right) \sum_n \tilde{w}_1\left(\alpha'_m, \beta_n\right) + i\alpha'_m \beta'_n R_{T1} \sum_n \beta_n \tilde{w}_1\left(\alpha'_m, \beta_n\right)\right]$$

$$- \frac{1}{l_x}\left[\left(R_{Q4} + i\beta'^3_n R_{M4}\right) \sum_m \tilde{w}_2\left(\alpha_m, \beta'_n\right) + i\alpha'_m \beta'_n R_{T4} \sum_m \alpha_m \tilde{w}_2\left(\alpha_m, \beta'_n\right)\right]$$

$$- \frac{1}{l_y}\left[\left(R_{Q2} + i\alpha'^3_m R_{M2}\right) \sum_n \tilde{w}_2\left(\alpha'_m, \beta_n\right) + i\alpha'_m \beta'_n R_{T2} \sum_n \beta_n \tilde{w}_2\left(\alpha'_m, \beta_n\right)\right]$$

$$= \frac{q_0}{(2\pi)^2} e^{-i\left(\alpha'_m x_0 + \beta'_n y_0\right)} \quad (6.21)$$

$$\frac{\rho_{\text{cav}} \omega^2}{\gamma\left(\alpha'_m, \beta'_n\right) \sinh\left[\gamma\left(\alpha'_m, \beta'_n\right) \cdot d\right]} \tilde{w}_1\left(\alpha'_m, \beta'_n\right) + D_2 f_2\left(\alpha'_m, \beta'_n\right) \tilde{w}_2\left(\alpha'_m, \beta'_n\right)$$

$$- \frac{1}{l_x}\left[\left(R_{Q4} + i\beta'^3_n R_{M4}\right) \sum_m \tilde{w}_1\left(\alpha_m, \beta'_n\right) + i\alpha'_m \beta'_n R_{T4} \sum_m \alpha_m \tilde{w}_1\left(\alpha_m, \beta'_n\right)\right]$$

$$- \frac{1}{l_y}\left[\left(R_{Q2} + i\alpha'^3_m R_{M2}\right) \sum_n \tilde{w}_1\left(\alpha'_m, \beta_n\right) + i\alpha'_m \beta'_n R_{T2} \sum_n \beta_n \tilde{w}_1\left(\alpha'_m, \beta_n\right)\right]$$

$$+ \frac{1}{l_x}\left[\left(R_{Q3} + i\beta'^3_n R_{M3}\right) \sum_m \tilde{w}_2\left(\alpha_m, \beta'_n\right) + i\alpha'_m \beta'_n R_{T3} \sum_m \alpha_m \tilde{w}_2\left(\alpha_m, \beta'_n\right)\right]$$

$$+ \frac{1}{l_y}\left[\left(R_{Q1} + i\alpha'^3_m R_{M1}\right) \sum_n \tilde{w}_2\left(\alpha'_m, \beta_n\right) + i\alpha'_m \beta'_n R_{T1} \sum_n \beta_n \tilde{w}_2\left(\alpha'_m, \beta_n\right)\right]$$

$$= 0 \quad (6.22)$$

which contain two sets of infinite unknowns: $\tilde{w}_1\left(\alpha'_m, \beta'_n\right)$ and $\tilde{w}_2\left(\alpha'_m, \beta'_n\right)$, with $m = -\infty$ to $+\infty$ and $n = -\infty$ to $+\infty$. Insofar as the solution converges, these equations can be truncated to retain one set of finite unknowns $\tilde{w}_1\left(\alpha'_m, \beta'_n\right)$ and $\tilde{w}_2\left(\alpha'_m, \beta'_n\right)$, with $m = -\hat{m}$ to $\hat{m}$ and $n = -\hat{n}$ to $\hat{n}$ (both $\hat{m}$ and $\hat{n}$ are positive integers), and hence can be numerically solved.

6.1.4 Far-Field Sound-Radiated Pressure

Owing to the fluid-structure interaction of the vibrating panel (bottom panel in the present case) and its surrounding fluid, sound pressure will be radiated from the fluid-structure interface into the far field. Therefore, once the response of the bottom panel $\tilde{w}_2(\alpha, \beta)$ is numerically solved, the radiated sound pressure at the far field can be obtained by employing the established sound radiation theory.

With the origin of the spherical coordinates (r, θ, φ) located at the excitation point (x_0, y_0), the far-field sound pressure $p(r, \theta, \varphi)$ radiated from a vibrating surface with displacement $w(x, y)$ is given by [30]

$$p(r, \theta, \varphi) = -\rho_0 \omega^2 \frac{e^{ik_0 r}}{2\pi r} e^{i(\alpha x_0 + \beta y_0)} \int_{-\infty}^{+\infty} \int_{-\infty}^{+\infty} w(x, y) e^{-i(\alpha x + \beta y)} dx dy \quad (6.23)$$

where $k_0 = \omega/c_0$, c_0 and ρ_0 are separately the sound speed and air density, and the wavenumbers α and β are

$$\alpha = k_0 \cos \varphi \sin \theta, \quad \beta = k_0 \sin \varphi \sin \theta \quad (6.24)$$

By adopting the Fourier transform of Eq. (6.5), Eq. (6.23) becomes

$$p(r, \theta, \varphi) = -2\pi \rho_0 \omega^2 \left(\frac{e^{ik_0 r}}{r}\right) e^{i(\alpha x_0 + \beta y_0)} \tilde{w}(\alpha, \beta) \quad (6.25)$$

For reference, the high-frequency asymptote of far-field sound pressure radiated by an *unstiffened* plate [3] is introduced as

$$p_{\text{asy}} = \frac{\rho_0 q_0 e^{ik_0 r}}{2\pi m r} \quad (6.26)$$

The far-field sound pressure radiated by the present orthogonally rib-stiffened sandwich structure with cavity absorption is then given in the form of sound pressure level (SPL) in decibel scales (dB) with respect to p_{asy} as

$$\text{SPL} = 10 \cdot \log_{10}(p/p_{\text{asy}})^2 \quad (6.27)$$

6.1.5 Convergence Check for Numerical Solution

As previously mentioned, the infinite simultaneous algebraic equations are truncated so that one only needs to solve a finite system of equations containing a finite number of unknowns. More specifically, only $M = 2\hat{m} + 1$ and $N = 2\hat{n} + 1$ unknowns are retained, associated separately with subscripts m and n, leading to

6.1 Sound Radiation of Absorptive Sandwich Structures

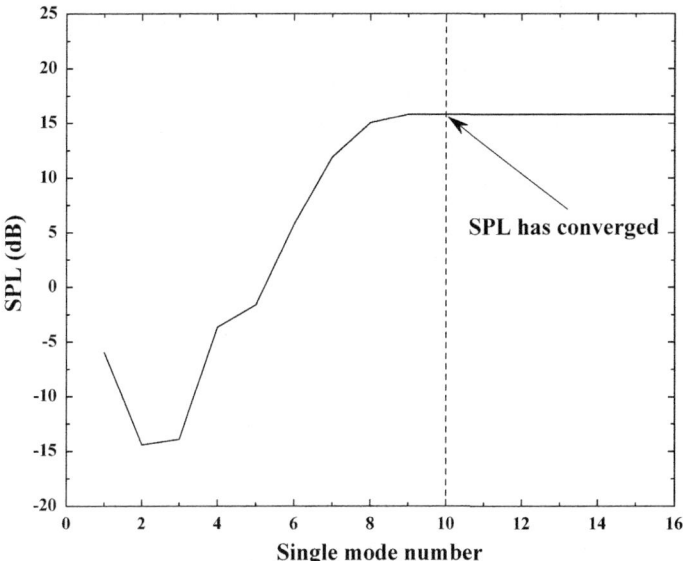

Fig. 6.3 Convergence check of numerical solution for sound radiation of an infinite orthogonally rib-stiffened sandwich structure with stiffener separations $(l_x, l_y) = (0.20 \text{ m}, 0.20 \text{ m})$ when excited by a harmonic point force at 10 kHz (With permission from Elsevier)

the same number of harmonic wave components in the x- and y-directions. Insofar as a sufficiently large number of terms are retained, the finite system is capable of ensuring the convergence and accuracy of the solution. The well-acknowledged criterion [1, 15] is employed, which assumes that once the solution converges at a given frequency, it converges for all lower frequencies. Therefore, the required number of unknowns is determined by the highest frequency of interest (10 kHz in the present study). To check the convergence of the solution, a numerical test is carried out by calculating the SPL at 10 kHz, with increasingly more terms used in Eqs. (6.21) and (6.22), as shown in Fig. 6.3. It can be seen from Fig. 6.3 that when $\hat{m}$ and $\hat{n}$ both have a value of 10, solution convergence is ensured at 10 kHz. Consequently, following the abovementioned criterion, the values of $\hat{m}$ and $\hat{n}$ are both taken as 10 (i.e., retaining 441 unknowns in the finite system) for all frequencies below 10 kHz, which is sufficient to ensure the convergence and accuracy of the solution.

6.1.6 Validation of Theoretical Modeling

To check the validity of the proposed model, the model (simplified version) is used to calculate the sound pressure level radiated from an orthogonally rib-stiffened single panel, and the predictions are compared in Fig. 6.4 with those obtained

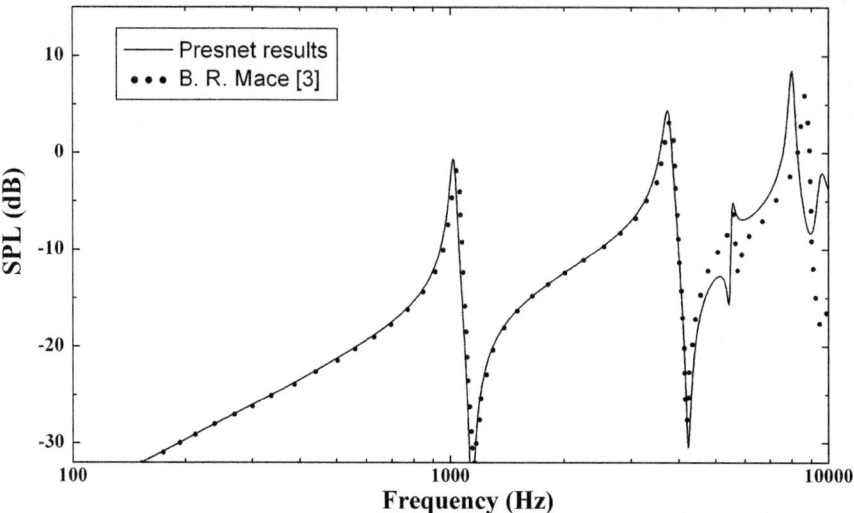

Fig. 6.4 Comparison between present model predictions and those by Mace [3] for orthogonally rib-stiffened single panel excited by time-harmonic point force at location (0, 0) (With permission from Elsevier)

by Mace [3]. To degrade the present model for sandwich structures to cover rib-stiffened single panels, negligibly small values are assigned to the prime parameters (i.e., Young's modulus E, density ρ, and thickness h) of one face panel of the sandwich, while the remaining system parameters are identical to those used by Mace [3].

It can be seen from Fig. 6.4 that overall the present predictions agree excellently well with those of Mace: only slight deviations exist beyond 5,000 Hz. These discrepancies in the high-frequency range are expected, which can be attributed to the difference in vibration modeling of the rib-stiffeners between the present model and Mace's theory. The rib-stiffeners were modeled as Euler beams in Mace's theory [3], meaning that only the bending moments and the inertial effect of the tensional forces of the rib-stiffeners are considered. In contrast, the present model accounts for all possible motions of the rib-stiffeners, including tensional forces, bending moments, and torsional moments as well as their inertial effects. Therefore, insofar as the dynamic responses and sound radiation of rib-stiffened plates are of concern, the present model provides a more precise theoretical tool than the beam-based theory of Mace. The discrepancies between the two theories in the high-frequency range of Fig. 6.4 just demonstrate the necessity of accurately modeling the motion of the rib-stiffeners.

To further check the accuracy of the present model for the double-panel case, the model is degraded to reproduce Takahashi's results [11] for rib-stiffened double-panel structures, as shown in Fig. 6.5. The relevant geometrical dimensions and material property parameters are identical as those of Takahashi. Again, the model

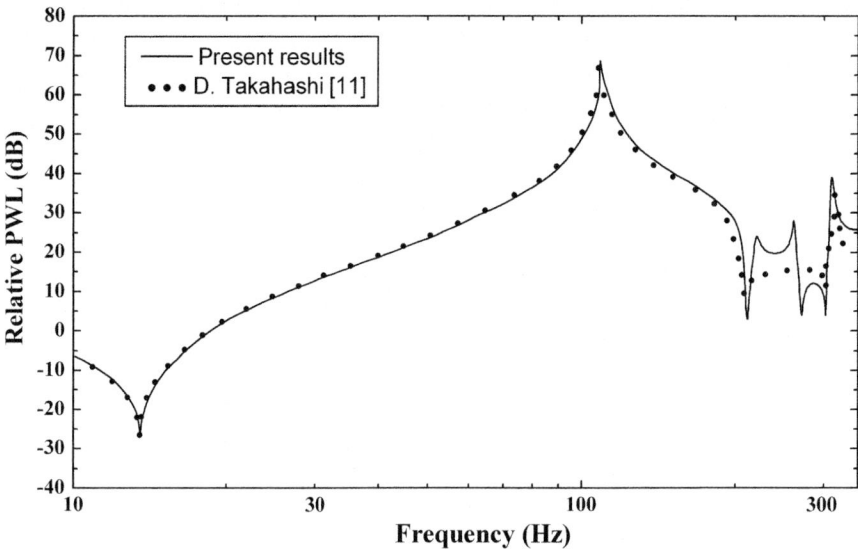

Fig. 6.5 Comparison between model predictions and theoretical results of Takahashi [11] for rib-stiffened double-panel structure excited by time-harmonic point force at location $(l_x/2, l_y/2)$ (With permission from Elsevier)

predictions fit well with Takahashi's theoretical results, with only slight divergences appearing at relatively high frequencies. These divergences are attributed to the additional consideration of inertial effects corresponding to the bending moments and torsional moments in the present model, which Takahashi did not take into account.

To a large extent, the comparisons made above may be regarded as acceptable validations for the proposed theoretical model, because all the theoretical formulations have been involved in the numerical calculation. In particular, if a theoretical model can be degraded to obtain the same results for simplified cases, its accuracy and feasibility would be better than the case when it can only give results similar to those obtained with its counterpart models.

6.1.7 Influence of Air-Structure Coupling Effect

Together with the equivalent fluid model for fibrous sound absorptive materials, the present model is able to characterize the sound radiation characteristics of lightweight lattice-cored sandwich structures filled with fibrous materials, such as fiberglass considered here. Note also that the model can be degraded to describe sandwich structures with either air cavity (i.e., air-structure coupling effect included) or vacuum cavity (i.e., fluid-structure coupling effect ignored).

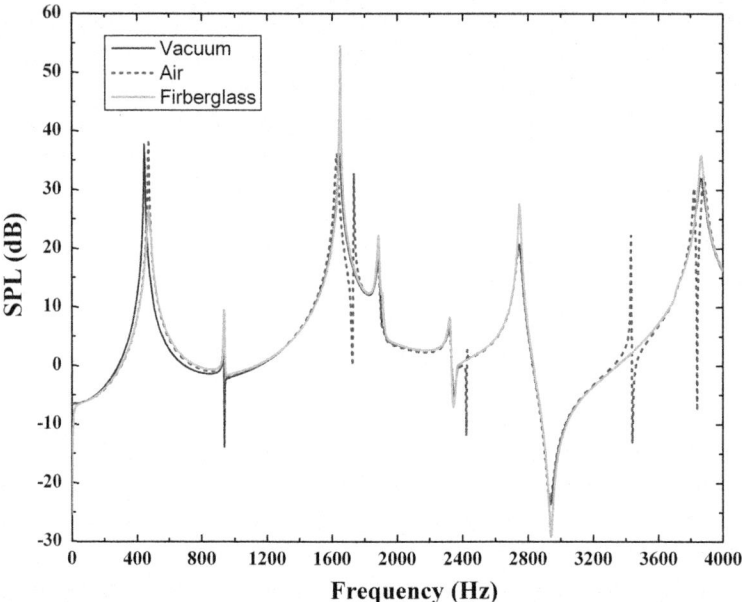

Fig. 6.6 Sound pressure levels radiated by different orthogonally rib-stiffened sandwich structures plotted as functions of frequency for stiffener separations $(l_x, l_y) = (0.20 \text{ m}, 0.20 \text{ m})$ (With permission from Elsevier)

Therefore, comparisons among the three different kinds of sandwiches under time-harmonic point force excitation can be performed to assess the influences of air-structure coupling effect and fibrous filling material on sound radiation.

To better evaluate the influences of air-structure coupling effect and fibrous material, the location of point force acting on the face panel is selected at the center of one lattice cell, i.e., $(l_x/2, l_y/2)$, away from the conjunction between the face panel and rib-stiffeners. The predicted sound pressure level (SPL) radiated by the three different sandwich structures is plotted in Figs. 6.6, 6.7, and 6.8 as a function of frequency for $(l_x, l_y) = (0.20 \text{ m}, 0.20 \text{ m})$, $(0.35 \text{ m}, 0.35 \text{ m})$, and $(0.50 \text{ m}, 0.50 \text{ m})$, respectively. For each pair of stiffener spacing selected, three kinds of sandwich configurations are compared: (1) vacuum cavity, (2) air cavity, and (3) cavity filled with fiberglass.

At first glance, it can be seen from Figs. 6.6, 6.7, and 6.8 that the air cavity case shows several additional peaks and dips on the SPL versus frequency curve. This is caused by air cavity interacting with the face panels through air-structure coupling. Besides these additional peaks and dips, it is also observed that the air-structure coupling effect plays an increasingly significant role in structure sound radiation with increasing rib-stiffener separation. This is reflected by the enlarged deviations between the two curves associated separately with the vacuum case and the air cavity case as the rib-stiffener separation is increased. In particular, when

6.1 Sound Radiation of Absorptive Sandwich Structures 305

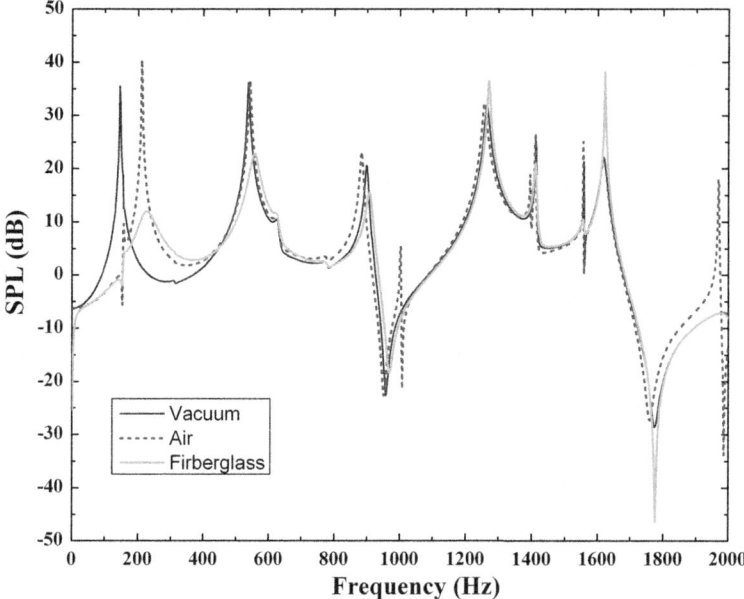

Fig. 6.7 Sound pressure levels radiated by different orthogonally rib-stiffened sandwich structures plotted as functions of frequency for stiffener separations ($l_x = 0.35$ m, $l_y = 0.35$ m) (With permission from Elsevier)

the separation is relatively large, air-structure coupling exerts a visible effect on the location of maximum sound radiation especially in low-frequency range. The air-structure coupling is in effect by means of pumping effect, that is, the air cavity partitioned by the face panels and rib-stiffeners has timely changing pressure as its volume alters with the dynamic displacements of these two face panels, often imposing a converse force on the panels. In the case of rib-stiffener separation being relatively large, a considerable area of the panels is exposed to the impinging of air cavity pressure. It is thence understandable that the air-structure coupling effect may not be ignored when the rib-stiffeners are sparsely distributed.

6.1.8 Influence of Fibrous Sound Absorptive Filling Material

In contrast to the air cavity case, the fiberglass case exhibits almost the same trends as the vacuum one, especially when the stiffener separation is relatively small, although the discrepancies between the two cases increase as the stiffener separation is increased. Note that the air-structure coupling effect is not present in the vacuum case, while it is eliminated in the fiberglass case (the presence of fiberglass in the cavity significantly changes the behavior of the cavity). This is also the reason

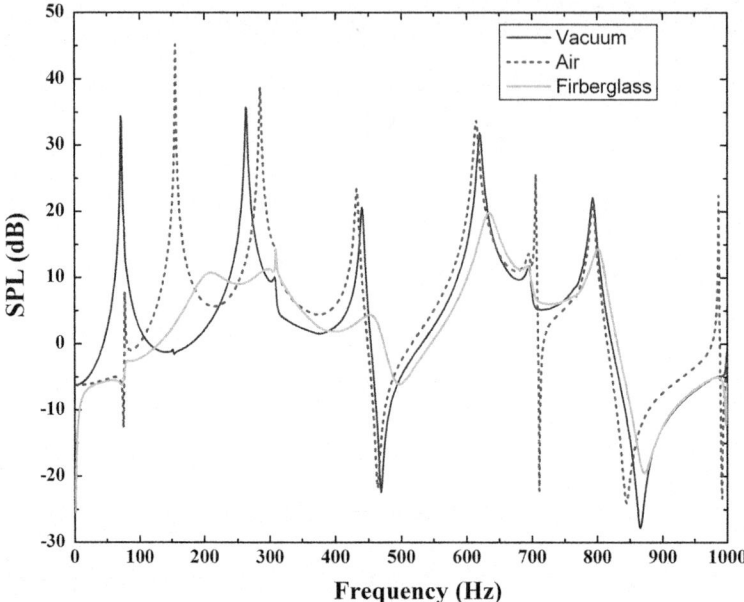

Fig. 6.8 Sound pressure levels radiated by different orthogonally rib-stiffened sandwich structures plotted as functions of frequency for stiffener separations ($l_x = 0.50$ m, $l_y = 0.50$ m) (With permission from Elsevier)

why the fiberglass case exhibits almost the same trend as the vacuum one: the discrepancies between the two cases enlarging with increasing stiffener separation actually reflect the combined effect of fiberglass stiffness and damping on structure responses.

It is understandable that the stiffness of the cavity-filling fiberglass reinforces the structural connection between the two face panels, enabling more vibration energies transmitted from the upper panel to the bottom one and thus causing larger sound radiation pressure levels. Conversely, fiberglass can dissipate acoustic energy via viscous drag forces and thermal exchange between the air and fibers and hence decreases sound radiation. In addition, both the stiffness and damping of the fiberglass material are frequency dependent [17, 19, 22]. Consequently, the fact that the discrepancies between the vacuum and fiberglass cases increase with increasing stiffener separation can be well explained.

The periodically distributed rib-stiffeners with relatively narrow separations restrict the deformation of fiberglass in between, offering therefore the fiberglass a larger stiffness than that inserted between those stiffeners having wider separations. That the fiberglass case exhibits the same trend as the vacuum one when the separation is small (e.g., $l_x = 0.20$ m and $l_y = 0.20$ m, as shown in Fig. 6.6) may be attributed to the balance of the converse effects of fiberglass stiffness and damping on sound radiation. More specifically, while damping is dominant at low frequencies, causing decreased sound radiation in the first peak, stiffness dominates at high frequencies, resulting in increased sound radiation in the following peaks (Figs. 6.7 and 6.8). As mentioned above, the stiffness of fiberglass decreases with increasing stiffener separation. Therefore, as the separation is increased, the frequency range dominated by stiffness (i.e., stiffness-controlled region) is shifted to higher frequencies and that dominated by damping (i.e., damping-controlled region) is widened. Correspondingly, in Fig. 6.7, the first three sound radiation peaks of the fiberglass case are lower than those of the vacuum one, and all the sound radiation peaks of the fiberglass case are significantly lower than those of the vacuum one in Fig. 6.8.

It may thence be deduced that the fiberglass-filled cavity affects structural radiation through the combined effects of fiberglass stiffness and damping (both being frequency dependent), the balance of which is significantly influenced by stiffener separation. It is therefore possible to optimize the stiffener separation and fiberglass porosity (both indirectly related to the stiffness and damping of fiberglass) to reduce structure sound radiation to an acceptable level required in specific cases.

6.1.9 Conclusions

The sound radiation characteristics of an infinite orthogonally rib-stiffened sandwich structure having cavity-filling fibrous sound absorptive material have been formulated by a comprehensive theoretical model when the structure is excited by a time-harmonic point force. The novelty of this work is to provide a general theoretical framework to address sound radiation issues of sandwich structures filled with fibrous sound absorptive materials, which can be degraded to deal with relatively simple structures. In the theoretical model, the vibration behaviors of the rib-stiffeners are accounted for by including all possible forces and moments exerted on the face panels by the rib-stiffeners in the governing equations. The propagation of sound in the fibrous material is modeled by adopting an equivalent fluid model with frequency-dependent dynamic density and bulk modulus, with viscous drag force and thermal exchanges between air and fibers taken into account. The technique of Fourier transform is applied to solve the governing equations, resulting in an infinite set of simultaneous algebraic equations, which can be truncated and numerically solved.

Numerical calculations are subsequently carried out to explore the influences of air-structure coupling effect and fibrous sound absorptive materials on structure sound radiation. The model is validated by comparing the present model predictions with previously published data, with excellent agreement achieved especially at low frequencies. Nevertheless, slight deviations emerge at high frequencies, which just demonstrate the superiority of the present model.

Special attention is then focused on the effects of air-structure coupling and fibrous sound absorptive materials on sound penetration. This is explored by comparing three different sandwich structures: partitioned cavity in vacuum, filled with air, and filled with fiberglass. Interesting physical features emerging from the comparison are well interpreted by considering the combined effects of fiberglass stiffness and damping as well as the influence of rib-stiffener separation. It is found that the air-structure coupling effect induces additional peaks and dips in the SPL versus frequency curves, which plays an increasingly significant effect on structure sound radiation as the stiffener separation is increased. In particular, it is concluded that the fiberglass-filled cavity exerts its impact on wave penetration (finally on structural radiation) through the combined effects of fiberglass stiffness and damping (both frequency dependent), the balance of which is significantly affected by stiffener separation. This may provide a convenient and efficient tool to optimize the porosity, cell size, and other topological parameters of fiberglass (indirectly altering its stiffness and damping) in conjunction with stiffener separation to reduce the structure vibration and sound radiation to an acceptable level required in specific situations.

As a future research forecast, the theoretical model for sandwich composite structures considered here (i.e., square lattice-cored sandwich structures filled with fibrous materials) can be further extended to study the acoustic performance of sandwich composite structures having laminated composites as skins, since these structures have been increasingly applied in aerospace and astronautic fields.

6.2 Sound Transmission Through Absorptive Sandwich Structure

6.2.1 Introduction

Applications of lightweight, periodically rib-stiffened structures are increasingly found in mechanical, aeronautics, aerospace, and marine industries [2, 3, 8–10, 14, 15, 35]. When these structures are applied as hulls or fuselages, external dynamic loadings (e.g., dynamic impact, sound wave impingement, and turbulent boundary layer excitation) are often encountered. The dynamic responses, sound radiation/transmission, and other relevant issues of structures have therefore been put forward and attracted much attention.

Numerous researchers have studied the sound radiation and transmission problems of periodic rib-stiffened structures [2–4, 14–18, 35, 36]. It has been established that the rib-stiffeners play a significant role in the vibroacoustic behavior of the whole structure, especially when the bending wavelength is comparable with the periodic spacing of the stiffeners [15, 35]. Consequently, the equivalent forces and moments of the stiffeners should be carefully taken into account in theoretical modeling. Two different theoretical approaches have been used to address the issue.

The first one is the Fourier transform technique. Lin and Garrelick [10] employed this technique to study the transmission of sound through two infinite parallel plates connected by identical periodically spaced frames. Subsequently, a range of sound radiation problems associated with different structures, such as infinite single plates attached with identical rib-stiffeners [35], with two different kinds of stiffeners (e.g., bulkheads and intermediate frames) [2], or with orthogonally distributed stiffeners [3], were considered using the same method, although only the equivalent forces of the stiffeners were accounted for. As an extension of Mace's work [35], Yin et al.[4] theoretically analyzed the acoustic radiation from a point-driven laminated composite plate reinforced by doubly periodic parallel stiffeners, wherein the plate was modeled using the classical composite plate theory.

The other is the space-harmonic method, which is essentially equivalent to the Fourier transform technique. This approach was introduced by Mead and Pujara [14, 37] to describe structural responses and acoustic pressures in terms of space-harmonic series. Based upon Mead and Pujara's works, Lee and Kim [15] developed an analytic method to study the sound transmission characteristics of a thin plate stiffened by equally spaced line stiffeners, with the resulting governing equations solved by utilizing the virtual work principle. Wang et al. [16] extended this approach to lightweight double-leaf partitions stiffened with periodically distributed studs and explored the underlying sound transmission mechanisms by incorporating the dispersion relation of the structure. Following the schemes of Wang et al., Legault and Atalla [17, 36] analyzed the effect of structural links on sound transmission across periodically rib-stiffened double-panel structures.

Existing studies on the vibroacoustic behavior of periodically rib-stiffened structures are often limited to one-dimensional (1D) systems. Two-dimensional (2D) orthogonally rib-stiffened double-panel structures received much less attention [3, 18, 38] and have not been adequately addressed, not to mention the even more complicated scenario when the cavity of the double panel is filled with fibrous sound absorptive materials. With focus placed upon aircraft sidewalls made of rib-stiffened structures having cavity-filling fiberglass, this research proposes a relatively comprehensive theoretical model for sound transmission through orthogonally rib-stiffened double-panel structures with cavity absorption. The effect of fibrous sound absorptive filling materials on sound transmission is accounted for with an equivalent fluid model. As a highlight, an integrated optimal algorithm toward lightweight, high-stiffness, and superior sound insulation capability is presented. A relatively rough optimal design regarding key structural geometry ratios is performed, and general optimal principles are presented. The optimal scheme suggests that the integrated optimization of double-panel structures involving various physical attributes is feasible.

6.2.2 Analytic Formulation of Panel Vibration and Sound Transmission

With reference to Fig. 6.9 where (x, y, z) denote the Cartesian coordinates, consider two parallel infinite Kirchhoff thin plates lying separately in the planes of $z = 0$

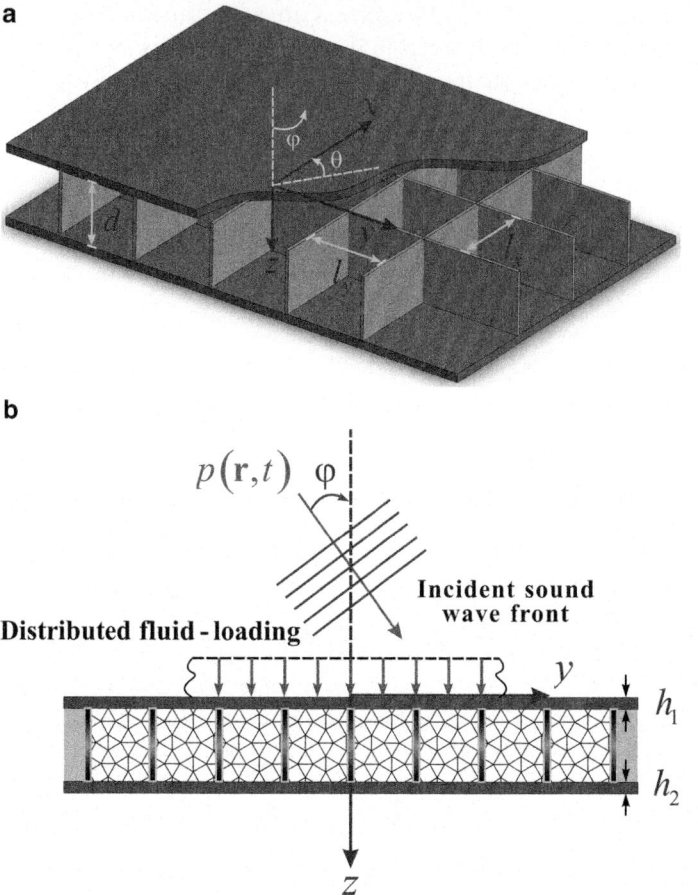

Fig. 6.9 Schematic illustration of sound pressure wave incident on an orthogonally rib-stiffened sandwich structure filled with, but not necessarily limited to, fibrous sound absorptive materials: (**a**) global view; (**b**) side view of (**a**) (With permission from Acoustical Society of America)

and $z = h_1 + d$ and connected with periodically distributed rib-stiffeners along two orthogonal lines $x = ml_x$ and $y = nl_y$ (m and n both being positive or negative integer). Let d denote the thickness of the rib-stiffeners (or cavity height), h_1 denote the thickness of the upper plate, and h_2 denote that of the bottom plate; see Fig. 6.9. The cavities in between the two faceplates and partitioned by the rib-stiffeners are filled with fibrous sound absorptive materials (see Fig. 6.9b). The upper plate located at $z = 0$ is loaded by a time-harmonic acoustic fluid $p(\mathbf{r}, t)$ with elevation angle φ and azimuth angle θ:

$$p(\mathbf{r}, t) = I e^{-i(k_x x + k_y y + k_z z - \omega t)} \tag{6.28}$$

6.2 Sound Transmission Through Absorptive Sandwich Structure

The wavenumber components in the x-, y-, and z-directions are determined by the elevation angle and azimuth angle of the incident acoustic loading as

$$k_x = k_0 \sin\varphi \cos\theta, \quad k_y = k_0 \sin\varphi \sin\theta, \quad k_z = k_0 \cos\varphi \quad (6.29)$$

where $k_0 = \omega/c_0$ is the acoustic wavenumber in air, ω being the angular frequency and c_0 the sound speed in air.

As a result of the acoustic loading, a distributed load impinges on the upper plate, inducing vibration of the upper plate which is then transmitted to the bottom plate via both structure- and fluid-borne paths. The fluid-structure interaction between the bottom plate and the nearby acoustic medium causes the radiation of sound.

As shown in Fig. 6.9, the acoustic field is divided into three main parts by the sandwich structure: upper field occupying the half-space $z < 0$, middle field filling the space $h_1 < z < h_1 + d$ (i.e., in between the two face panels and partitioned periodically by the rib-stiffeners), and bottom field occupying the other half-space $z > h_1 + h_2 + d$. The corresponding acoustic pressure in the incident field $p_i(\mathbf{r}, t)$ should satisfy the scalar Helmholtz equation

$$\left(\partial^2/\partial x^2 + \partial^2/\partial y^2 + \partial^2/\partial z^2\right) p_i + k_0^2 p_i = 0, \quad z < 0 \quad (6.30)$$

It is assumed that the cavities of the sandwich structure are filled with fibrous sound absorptive materials. As is well known, the absorption of sound by a porous absorptive material mainly arises from viscous drag forces and thermal exchange loss when sound penetrates through it [19, 24–26]. With the help of a well-developed equivalent fluid model [22, 39] for such materials, the absorbent effect induced by viscous drag force and thermal exchange between air and solid fibers is accounted for by introducing a complex wavenumber k_{cav} and a complex density ρ_{cav}. Both k_{cav} and ρ_{cav} are frequency dependent in accordance with thermal exchange transition with increasing frequency [22] (i.e., isothermal process at low frequency turning to adiabatic process at high frequency). The complex wavenumber may be expressed as $ik_{cav} = \Gamma = \alpha + i\beta$, wherein Γ is the wave propagation constant, α is the attenuation, and β is the phase constant. The corresponding acoustic pressure $p_{cav}(\mathbf{r}, t)$ in the fibrous sound absorptive material (i.e., in between the two faceplates) obeys the equation [12, 13, 17]

$$\left(\partial^2/\partial x^2 + \partial^2/\partial y^2 + \partial^2/\partial z^2\right) p_{cav} + k_{cav}^2 p_{cav} = 0, \quad h_1 < z < h_1 + d \quad (6.31)$$

where k_{cav} is closely related to the dynamic density $\rho(\omega)$ and dynamic bulk modulus $K(\omega)$ of the fibrous sound absorptive materials:

$$k_{cav} = 2\pi f \sqrt{\rho(\omega)/K(\omega)} \quad (6.32)$$

The dynamic density and the dynamic bulk modulus are given by [22]

$$\rho(\omega) = \rho_0 \left[1 + \frac{1}{i2\pi}\left(\frac{R}{\rho_0 f}\right) G_1\left(\frac{\rho_0 f}{R}\right)\right] \quad (6.33)$$

$$K(\omega) = \gamma_s P_0 \left(\gamma_s - \frac{\gamma_s - 1}{1 + (1/i8\pi N_{pr})(\rho_0 f/R)^{-1} G_2(\rho_0 f/R)}\right)^{-1} \quad (6.34)$$

where $G_1(\rho_0 f/R) = \sqrt{1 + i\pi(\rho_0 f/R)}$, $G_2(\rho_0 f/R) = G_1[(\rho_0 f/R)4N_{pr}]$, R is the flow resistivity, γ_s is the specific heat ratio, P_0 is the air equilibrium pressure, and N_{pr} is the Prandtl number.

Finally, in the transmitted field, the acoustic pressure $p_t(r, t)$ is also a solution of the scalar Helmholtz equation

$$(\partial^2/\partial x^2 + \partial^2/\partial y^2 + \partial^2/\partial z^2) p_t + k_0^2 p_t = 0, \quad z > h_1 + h_2 + d \quad (6.35)$$

Assuming that the fibrous material is in perfect contact with the two plates, one can use the momentum equation to ensure the equality of plate velocity and fluid velocity at the fluid-plate interface, i.e., the continuity condition of fluid-structure coupling [7, 10]:

$$\left[\frac{\partial p_i}{\partial z}\right]_{z=0} = \rho_0 \omega^2 w_1, \quad \left[\frac{\partial p_{cav}}{\partial z}\right]_{z=h_1} = \rho_{cav} \omega^2 w_1 \quad (6.36)$$

$$\left[\frac{\partial p_{cav}}{\partial z}\right]_{z=h_1+d} = \rho_{cav} \omega^2 w_2, \quad \left[\frac{\partial p_t}{\partial z}\right]_{z=h_1+h_2+d} = \rho_0 \omega^2 w_2 \quad (6.37)$$

The complex density ρ_{cav} of the fibrous material appearing in Eqs. (6.36) and (6.37) is related to the complex wavenumber k_{cav} by [12]

$$\frac{k_{cav}^2}{k_0^2} = \frac{\gamma_s \sigma \rho_{cav}}{\rho_0} \quad (6.38)$$

where γ_s is the ratio of specific heats, σ is the porosity of the fibrous material, and ρ_0 is the air density.

The sandwich structure is driven by the difference of acoustic pressure between the two sides of each faceplate. The resultant pressure imposed on the upper panel is the pressure difference between $p_i(x, y, 0; t)$ in the incident side and $p_{cav}(x, y, h_1; t)$ in the fibrous material. Similarly, the bottom panel bears the net pressure that is a subtraction of $p_{cav}(x, y, h_1 + d; t)$ in the fibrous material and $p_t(x, y, h_1 + h_2 + d; t)$ in the transmitted side. Meanwhile, with the structural constraints of the orthogonal rib-stiffeners on the faceplates duly accounted for, the vibration of the plates is governed by

6.2 Sound Transmission Through Absorptive Sandwich Structure

$$D_1 \nabla^4 w_1 + m_1 \frac{\partial^2 w_1}{\partial t^2} = \sum_{m=-\infty}^{+\infty} \left[Q_y^+ \delta(x - ml_x) + \frac{\partial}{\partial y} \left\{ M_y^+ \delta(x - ml_x) \right\} \right.$$
$$\left. + \frac{\partial}{\partial x} \left\{ M_{Ty}^+ \delta(x - ml_x) \right\} \right]$$
$$+ \sum_{n=-\infty}^{+\infty} \left[Q_x^+ \delta(y - nl_y) + \frac{\partial}{\partial x} \left\{ M_x^+ \delta(y - nl_y) \right\} + \frac{\partial}{\partial y} \left\{ M_{Tx}^+ \delta(y - nl_y) \right\} \right]$$
$$+ p_i(x, y, 0) - p_{cav}(x, y, h_1) \tag{6.39}$$

$$D_2 \nabla^4 w_2 + m_2 \frac{\partial^2 w_2}{\partial t^2} = -\sum_{m=-\infty}^{+\infty} \left[Q_y^- \delta(x - ml_x) + \frac{\partial}{\partial y} \left\{ M_y^- \delta(x - ml_x) \right\} \right.$$
$$\left. + \frac{\partial}{\partial x} \left\{ M_{Ty}^- \delta(x - ml_x) \right\} \right]$$
$$- \sum_{n=-\infty}^{+\infty} \left[Q_x^- \delta(y - nl_y) + \frac{\partial}{\partial x} \left\{ M_x^- \delta(y - nl_y) \right\} + \frac{\partial}{\partial y} \left\{ M_{Tx}^- \delta(y - nl_y) \right\} \right]$$
$$+ p_{cav}(x, y, h_1 + d) - p_t(x, y, h_1 + h_2 + d) \tag{6.40}$$

where $\nabla^4 = (\partial^2/\partial x^2 + \partial^2/\partial y^2)^2$; (w_1, w_2), (m_1, m_2), and (D_1, D_2) are the displacements, surface mass density, and bending stiffness of the upper and bottom panel, respectively; and $\delta(\cdot)$ stands for the Dirac delta function.

Due to the consideration of inertial effects, the resultant tensional forces, bending moments, and torsional moments exerted on the upper and bottom plates are not identical, denoted here separately as (Q^+, M^+, M_T^+) and (Q^-, M^-, M_T^-). Superscripts $+$ and $-$ denote separately the upper and bottom plates, while subscripts x and y signify the terms arising from the x- and y-wise rib-stiffeners, respectively. An illustration of the present conventions for tensional forces, bending moments, and torsional moments is given in Fig. 6.10. Detailed derivations of these quantities can be found in Appendix B.

6.2.3 Application of the Periodicity of Structures

Taking advantage of the periodic property of the orthogonally rib-stiffened sandwich structures considered here, one can simplify the theoretical formulations presented above to obtain analytic solutions of the problem. As mentioned in an earlier work [18], following the key conclusion of Bloch or Floquet's theorem [40] for wave

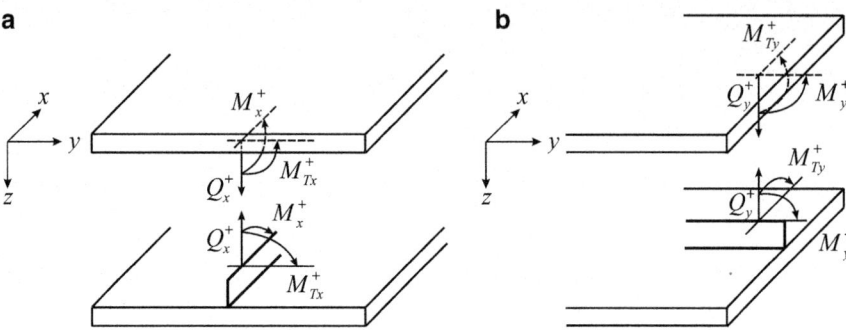

Fig. 6.10 Conventions for tensional forces, bending moments, and torsional moments at the interface between upper plate and (**a**) *x*-wise stiffeners and (**b**) *y*-wise stiffeners. The same applies at the interface between bottom plate and *x*-/*y*-wise stiffeners (With permission from Acoustical Society of America)

propagation in periodic structures, the displacements $w(x, y)$ of such a system at the corresponding points in different periodic elements are related by the periodicity condition as

$$w(x + ml_x, y + nl_y) = w(x, y) e^{-ik_x ml_x} e^{-ik_y nl_y}, \quad (m, n \text{ being integers}) \tag{6.41}$$

The space-harmonic expansion series can thence be favorably applied to express the panel displacement $w_j(x, y; t)$ as [1, 3, 14, 18]

$$w_j(x, y; t) = \sum_{m=-\infty}^{+\infty} \sum_{n=-\infty}^{+\infty} \alpha_{1,mn} e^{-i[(k_x + 2m\pi/l_x)x + (k_y + 2n\pi/l_y)y - \omega t]} \quad (j = 1, 2) \tag{6.42}$$

where $j = 1$ for the upper panel, $j = 2$ for the bottom panel, and the (m, n)th harmonic wave has wavenumber components $(k_x + 2m\pi/l_x, k_y + 2n\pi/l_y)$, illustrating its propagation direction in the structure, and

$$\alpha_{j,mn} = \frac{1}{l_x l_y} \int_0^{l_x} \int_0^{l_y} w_1(x, y; t) e^{i[(k_x + 2m\pi/l_x)x + (k_y + 2n\pi/l_y)y - \omega t]} dx dy \quad (j = 1, 2) \tag{6.43}$$

Due to sound pressure $p(\mathbf{r}, t) = I e^{-i(\mathbf{k} \cdot \mathbf{r} - \omega t)}$ incident on the sandwich, the set of sound pressures can be expressed as [1, 18]

$$p_i(x, y, z; t) = I e^{-i(k_x x + k_y y + k_z z - \omega t)}$$
$$+ \sum_{m=-\infty}^{+\infty} \sum_{n=-\infty}^{+\infty} \beta_{mn} e^{-i[(k_x + 2m\pi/l_x)x + (k_y + 2n\pi/l_y)y - k_{z,mn}z - \omega t]} \tag{6.44}$$

6.2 Sound Transmission Through Absorptive Sandwich Structure

$$p_{\text{cav}}(x, y, z; t) = \sum_{m=-\infty}^{+\infty} \sum_{n=-\infty}^{+\infty} \varepsilon_{mn} e^{-i\left[(k_x+2m\pi/l_x)x+(k_y+2n\pi/l_y)y+k_{z,\text{cav},mn}z-\omega t\right]}$$

$$+ \sum_{m=-\infty}^{+\infty} \sum_{n=-\infty}^{+\infty} \zeta_{mn} e^{-i\left[(k_x+2m\pi/l_x)x+(k_y+2n\pi/l_y)y-k_{z,\text{cav},mn}z-\omega t\right]}$$

(6.45)

$$p_t(x, y, z; t) = \sum_{m=-\infty}^{+\infty} \sum_{n=-\infty}^{+\infty} \xi_{mn} e^{-i\left[(k_x+2m\pi/l_x)x+(k_y+2n\pi/l_y)y+k_{z,mn}z-\omega t\right]} \quad (6.46)$$

In the above expressions, I is the amplitude of incident sound pressure; β_{mn} and ζ_{mn} are the (m, n)th space-harmonic amplitude of negative-going wave in the incident field and in the middle field, respectively; and ε_{mn} and ξ_{mn} are the (m, n)th space-harmonic amplitude of positive-going wave in the middle filed and in the transmitted field, respectively. Furthermore, $k_{z,mn}$ and $k_{z,\text{cav},mn}$ are the (m, n)th space-harmonic wavenumbers in the z-direction (related separately to wave propagation in air and fibrous absorptive material) which, upon applying the Helmholtz equation, are given by [16, 17, 36]

$$k_{z,mn} = \sqrt{k_0^2 - \alpha_m^2 - \beta_n^2}, \quad k_{z,\text{cav},mn} = \sqrt{k_{\text{cav}}^2 - \alpha_m^2 - \beta_n^2} \quad (6.47)$$

where the (m, n)th harmonic wavenumber components in the x- and y-directions are expressed as

$$\alpha_m = k_x + 2m\pi/l_x, \quad \beta_n = k_y + 2n\pi/l_y \quad (6.48)$$

Given that the z-direction wavenumber component is determined by Eq. (6.48), two different modes of sound propagation in the transmitting field can be distinguished [18, 41]: (1) non-radiating wave (i.e., subsonic wave) when $\alpha_m^2 + \beta_n^2 > k_0^2$ and (2) radiating wave (i.e., supersonic wave) when $\alpha_m^2 + \beta_n^2 < k_0^2$. While the (m, n)th sound wave component contributes only to the near field in the first case, it is able to contribute to the far field in the second case.

Substitution of Eqs. (6.42) and Eqs. (6.44), (6.45), and (6.46) into Eqs. (6.36) and (6.37) yields

$$-ik_z I e^{-i(k_x x+k_y y)} + \sum_{m=-\infty}^{+\infty} \sum_{n=-\infty}^{+\infty} \left(ik_{z,mn}\beta_{mn} - \rho_0\omega^2 \alpha_{1,mn}\right) e^{-i(\alpha_m x+\beta_n y)} = 0$$

(6.49)

$$\sum_{m=-\infty}^{+\infty} \sum_{n=-\infty}^{+\infty} \left[\begin{array}{c} ik_{z,\text{cav},mn}\left(-\varepsilon_{mn} e^{-ik_{z,\text{cav},mn}h_1} + \zeta_{mn} e^{ik_{z,\text{cav},mn}h_1}\right) \\ - \rho_{\text{cav}}\omega^2 \alpha_{1,mn} \end{array}\right] e^{-i(\alpha_m x+\beta_n y)} = 0$$

(6.50)

$$\sum_{m=-\infty}^{+\infty}\sum_{n=-\infty}^{+\infty}\left[ik_{z,\text{cav},mn}\left(-\varepsilon_{mn}e^{-ik_{z,\text{cav},mn}(h_1+d)} + \zeta_{mn}e^{ik_{z,\text{cav},mn}(h_1+d)}\right) - \rho_{\text{cav}}\omega^2\alpha_{2,mn}\right]$$
$$e^{-i(\alpha_m x+\beta_n y)} = 0 \tag{6.51}$$

$$\sum_{m=-\infty}^{+\infty}\sum_{n=-\infty}^{+\infty}\left[-ik_{z,mn}\xi_{mn}e^{-ik_{z,mn}(h_1+h_2+d)} - \rho_0\omega^2\alpha_{2,mn}\right]e^{-i(\alpha_m x+\beta_n y)} = 0 \tag{6.52}$$

Because Eqs. (6.49), (6.50), (6.51), and (6.52) hold for all possible values of x and y, it can be shown that the relevant coefficients have the following relationships:

$$\beta_{00} = I + \frac{\omega^2 \rho_0 \alpha_{1,00}}{ik_z} \tag{6.53}$$

$$\beta_{mn} = \frac{\omega^2 \rho_0 \alpha_{1,mn}}{ik_{z,mn}}, \quad \text{at} \quad m \neq 0 \,\|\, n \neq 0 \tag{6.54}$$

$$\varepsilon_{mn} = \frac{\omega^2 \rho_{\text{cav}}\left[\alpha_{1,mn}e^{ik_{z,\text{cav},mn}(h_1+d)} - \alpha_{2,mn}e^{ik_{z,\text{cav},mn}h_1}\right]}{2k_{z,\text{cav},mn}\sin(k_{z,\text{cav},mn}d)} \tag{6.55}$$

$$\zeta_{mn} = \frac{\omega^2 \rho_{\text{cav}}\left[\alpha_{1,mn}e^{-ik_{z,\text{cav},mn}(h_1+d)} - \alpha_{2,mn}e^{-ik_{z,\text{cav},mn}h_1}\right]}{2k_{z,\text{cav},mn}\sin(k_{z,\text{cav},mn}d)} \tag{6.56}$$

$$\xi_{mn} = -\frac{\omega^2 \rho_0 \alpha_{2,mn}}{ik_{z,mn}}e^{ik_{z,mn}(h_1+h_2+d)} \tag{6.57}$$

6.2.4 Solution by Employing the Virtual Work Principle

Since the sandwich structure considered here is spatially periodic, the principle of virtual work [14–16] can be utilized to solve the theoretical formulations presented above and thence obtain the values of coefficients $\alpha_{1,mn}$ and $\alpha_{2,mn}$. As close relationships exist between the coefficients of panel displacements (i.e., $\alpha_{1,mn}$ and $\alpha_{2,mn}$) and those of sound pressure (i.e., β_{mn}, ε_{mn}, ζ_{mn}, and ξ_{mn}), the sound pressures can be straightforwardly obtained once the former is determined. To calculate the virtual work done by imposing the virtual displacements

$$\delta w_j^* = \delta \alpha_{j,kl} e^{-i(\alpha_k x+\beta_l y)} \quad (j=1,2) \tag{6.58}$$

on the sandwich, only one periodic element needs to be considered. The principle of virtual work states that the virtual work of the system stemming from the virtual displacements should be zero, from which the equilibrium equation of system can be established as detailed below.

6.2 Sound Transmission Through Absorptive Sandwich Structure

6.2.4.1 Virtual Work of Panel Elements

The virtual work contributed solely by one periodic element of each plate can be represented as

$$\delta\Pi_{p1} = \int_0^{l_x}\int_0^{l_y}\left[D_1\nabla^4 w_1 + m_1\frac{\partial^2 w_1}{\partial t^2} - p_i(x,y,0) + p_{cav}(x,y,h_1)\right]\cdot\delta w_1 dxdy$$

$$= \left\{\left[D_1(\alpha_k^2+\beta_l^2)^2 - m_1\omega^2\right]\alpha_{1,kl} - \frac{\omega^2\rho_0\alpha_{1,kl}}{ik_{z,kl}}\right.$$

$$\left.+\frac{\omega^2\rho_{cav}[\alpha_{1,kl}\cos(k_{z,cav,kl}d) - \alpha_{2,kl}]}{k_{z,cav,kl}\sin(k_{z,cav,kl}d)}\right\}\cdot l_x l_y \cdot \delta\alpha_{1,kl}$$

$$-\int_0^{l_x}\int_0^{l_y} 2Ie^{-i(k_x x+k_y y)}e^{i(\alpha_k x+\beta_l y)}dxdy \cdot \delta\alpha_{1,kl} \tag{6.59}$$

$$\delta\Pi_{p2} = \int_0^{l_x}\int_0^{l_y}\left[D_2\nabla^4 w_2 + m_2\frac{\partial^2 w_2}{\partial t^2} - p_{cav}(x,y,h_1+d)\right.$$

$$\left.+p_t(x,y,h_1+h_2+d)\right]\cdot\delta w_2 dxdy$$

$$= \left\{\left[D_2(\alpha_k^2+\beta_l^2)^2 - m_2\omega^2\right]\alpha_{2,kl} - \frac{\omega^2\rho_0\alpha_{2,kl}}{ik_{z,kl}}\right.$$

$$\left.-\frac{\omega^2\rho_{cav}[\alpha_{1,kl} - \alpha_{2,kl}\cos(k_{z,cav,kl}d)]}{k_{z,cav,kl}\sin(k_{z,cav,kl}d)}\right\}\cdot l_x l_y \cdot \delta\alpha_{2,kl} \tag{6.60}$$

6.2.4.2 Virtual Work of x-Wise Stiffeners

The virtual work contribution from the tensional force, bending moment, and torsional moment at the interface between the x-wise stiffeners (aligned with $y=0$) and upper or bottom panel is given by [18]

$$\delta\Pi_{x1} = -\int_0^{l_x}\left[Q_x^+(x,0) + \frac{\partial}{\partial x}M_x^+(x,0) + \frac{\partial}{\partial y}M_{Tx}^+(x,0)\right]\cdot\delta\alpha_{1,kl}e^{i\alpha_k x}dx$$

$$= \sum_{n=-\infty}^{+\infty}\left[\begin{array}{c}R_{Q1}\alpha_{1,kn} - R_{Q2}\alpha_{2,kn} + i\alpha_k^3(-R_{M1}\alpha_{1,kn}+R_{M2}\alpha_{2,kn}) \\ +i\alpha_k\beta_l\beta_n(-R_{T1}\alpha_{1,kn}+R_{T2}\alpha_{2,kn})\end{array}\right]\cdot l_x\delta\alpha_{1,kl}$$

$$\tag{6.61}$$

$$\delta\Pi_{x2} = \int_0^{l_x}\left[Q_x^-(x,0) + \frac{\partial}{\partial x}M_x^-(x,0) + \frac{\partial}{\partial y}M_{Tx}^-(x,0)\right]\cdot\delta\alpha_{2,kl}e^{i\alpha_k x}dx$$

$$= \sum_{n=-\infty}^{+\infty}\left[\begin{array}{c}-R_{Q2}\alpha_{1,kn}+R_{Q1}\alpha_{2,kn}-i\alpha_k^3(-R_{M2}\alpha_{1,kn}+R_{M1}\alpha_{2,kn}) \\ -i\alpha_k\beta_l\beta_n(-R_{T2}\alpha_{1,kn}+R_{T1}\alpha_{2,kn})\end{array}\right]\cdot l_x\delta\alpha_{2,kl}$$

$$\tag{6.62}$$

where $\dfrac{\partial M_{Tx}^{+}(x,0)}{\partial y} = \dfrac{\partial M_{Tx}^{+}(x,y)}{\partial y}\bigg|_{y=0}$ and $\dfrac{\partial M_{Tx}^{-}(x,0)}{\partial y} = \dfrac{\partial M_{Tx}^{-}(x,y)}{\partial y}\bigg|_{y=0}$.

6.2.4.3 Virtual Work of y-Wise Stiffeners

Likewise, the virtual work done by the tensional force, bending moment, and torsional moment at the interface between the y-wise stiffeners (aligned with $x = 0$) and upper or bottom panel is [18]

$$\delta\Pi_{y1} = -\int_0^{l_y} \left[Q_y^+(0,y) + \dfrac{\partial}{\partial y} M_y^+(0,y) + \dfrac{\partial}{\partial x} M_{Ty}^+(0,y) \right] \cdot \delta\alpha_{1,kl} e^{i\beta_l y} dy$$

$$= \sum_{m=-\infty}^{+\infty} \left[\begin{array}{l} R_{Q3}\alpha_{1,ml} - R_{Q4}\alpha_{2,ml} + i\beta_l^3 \left(-R_{M3}\alpha_{1,ml} + R_{M4}\alpha_{2,ml} \right) \\ +i\alpha_k\alpha_m\beta_l \left(-R_{T3}\alpha_{1,ml} + R_{T4}\alpha_{2,ml} \right) \end{array} \right] \cdot l_y \delta\alpha_{1,kl}$$

(6.63)

$$\delta\Pi_{y2} = \int_0^{l_y} \left[Q_y^-(0,y) + \dfrac{\partial}{\partial y} M_y^-(0,y) + \dfrac{\partial}{\partial x} M_{Ty}^-(0,y) \right] \cdot \delta\alpha_{2,kl} e^{i\beta_l y} dy$$

$$= \sum_{m=-\infty}^{+\infty} \left[\begin{array}{l} -R_{Q4}\alpha_{1,ml} + R_{Q3}\alpha_{2,ml} - i\beta_l^3 \left(-R_{M4}\alpha_{1,ml} + R_{M3}\alpha_{2,ml} \right) \\ -i\alpha_k\alpha_m\beta_l \left(-R_{T4}\alpha_{1,ml} + R_{T3}\alpha_{2,ml} \right) \end{array} \right] \cdot l_y \delta\alpha_{2,kl}$$

(6.64)

where $\dfrac{\partial M_{Ty}^{+}(0,y)}{\partial x} = \dfrac{\partial M_{Ty}^{+}(x,y)}{\partial x}\bigg|_{x=0}$ and $\dfrac{\partial M_{Ty}^{-}(0,y)}{\partial x} = \dfrac{\partial M_{Ty}^{-}(x,y)}{\partial x}\bigg|_{x=0}$.

6.2.4.4 Resultant Equations for Structure Motions

It follows from the virtual work principle that

$$\delta\Pi_{p1} + \delta\Pi_{x1} + \delta\Pi_{y1} = 0, \quad \delta\Pi_{p2} + \delta\Pi_{x2} + \delta\Pi_{y2} = 0 \quad (6.65)$$

Substituting Eqs. (6.59), (6.60), (6.61), (6.62), (6.63), and (6.64) into Eq. (6.65) and noticing that the virtual displacement is arbitrary, one obtains

$$\left\{ \left[D_1(\alpha_k^2 + \beta_l^2)^2 - m_1\omega^2 \right]\alpha_{1,kl} - \dfrac{\omega^2 \rho_0 \alpha_{1,kl}}{ik_{z,kl}} + \dfrac{\omega^2 \rho_{cav}\left[\alpha_{1,kl}\cos(k_{z,cav,kl}d) - \alpha_{2,kl}\right]}{k_{z,cav,kl}\sin(k_{z,cav,kl}d)} \right\} \cdot l_x l_y$$

$$+ \sum_{n=-\infty}^{+\infty} \left[R_{Q1}\alpha_{1,kn} - R_{Q2}\alpha_{2,kn} + i\alpha_k^3 \left(R_{M2}\alpha_{2,kn} - R_{M1}\alpha_{1,kn} \right) \right.$$

$$\left. + i\alpha_k\beta_n\beta_l \left(R_{T2}\alpha_{2,kn} - R_{T1}\alpha_{1,kn} \right) \right] \cdot l_x$$

6.2 Sound Transmission Through Absorptive Sandwich Structure

$$+ \sum_{m=-\infty}^{+\infty} \left[R_{Q3}\alpha_{1,ml} - R_{Q4}\alpha_{2,ml} + i\beta_l^3 \left(R_{M4}\alpha_{2,ml} - R_{M3}\alpha_{1,ml} \right) \right.$$

$$\left. + i\alpha_m\alpha_k\beta_l \left(R_{T4}\alpha_{2,ml} - R_{T3}\alpha_{1,ml} \right) \right] \cdot l_y$$

$$= \begin{cases} 2Il_x l_y & \text{when } k = 0 \text{ \& } l = 0 \\ 0 & \text{when } k \neq 0 \,||\, l \neq 0 \end{cases} \tag{6.66}$$

$$\left\{ \left[D_2\left(\alpha_k^2 + \beta_l^2\right)^2 - m_2\omega^2 \right]\alpha_{2,kl} - \frac{\omega^2 \rho_0 \alpha_{2,kl}}{ik_{z,kl}} - \frac{\omega^2 \rho_{\text{cav}}\left[\alpha_{1,kl} - \alpha_{2,kl}\cos\left(k_{z,\text{cav},kl}d\right)\right]}{k_{z,\text{cav},kl}\sin\left(k_{z,\text{cav},kl}d\right)} \right\} \cdot l_x l_y$$

$$+ \sum_{n=-\infty}^{+\infty} \left[-R_{Q2}\alpha_{1,kn} + R_{Q1}\alpha_{2,kn} - i\alpha_k^3 \left(R_{M1}\alpha_{2,kn} - R_{M2}\alpha_{1,kn} \right) \right.$$

$$\left. - i\alpha_k\beta_n\beta_l \left(R_{T1}\alpha_{2,kn} - R_{T2}\alpha_{1,kn} \right) \right] \cdot l_x$$

$$+ \sum_{m=-\infty}^{+\infty} \left[-R_{Q4}\alpha_{1,ml} + R_{Q3}\alpha_{2,ml} - i\beta_l^3 \left(R_{M3}\alpha_{2,ml} - R_{M4}\alpha_{1,ml} \right) \right.$$

$$\left. - i\alpha_m\alpha_k\beta_l \left(R_{T3}\alpha_{2,ml} - R_{T4}\alpha_{1,ml} \right) \right] \cdot l_y = 0 \tag{6.67}$$

It should be mentioned that the consideration of virtual work in any other periodic element of the sandwich structure would have yielded an identical set of equations.

In order to separate the variables $\alpha_{1,kl}$ and $\alpha_{2,kl}$, Eqs. (6.66) and (6.67) are rewritten as

$$\left[D_1\left(\alpha_k^2 + \beta_l^2\right)^2 - m_1\omega^2 - \frac{\omega^2 \rho_0}{ik_{z,kl}} + \frac{\omega^2 \rho_{\text{cav}}\cos\left(k_{z,\text{cav},kl}d\right)}{k_{z,\text{cav},kl}\sin\left(k_{z,\text{cav},kl}d\right)} \right] \cdot l_x l_y \alpha_{1,kl}$$

$$- \frac{\omega^2 \rho_{\text{cav}}}{k_{z,\text{cav},kl}\sin\left(k_{z,\text{cav},kl}d\right)} \cdot l_x l_y \alpha_{2,kl}$$

$$+ \sum_{n=-\infty}^{+\infty} \left[R_{Q1} - i\alpha_k^3 R_{M1} - i\beta_l \alpha_k \beta_n R_{T1} \right] \cdot l_x \alpha_{1,kn}$$

$$+ \sum_{n=-\infty}^{+\infty} \left[-R_{Q2} + i\alpha_k^3 R_{M2} + i\beta_l \alpha_k \beta_n R_{T2} \right] \cdot l_x \alpha_{2,kn}$$

$$+ \sum_{m=-\infty}^{+\infty} \left[R_{Q3} - i\beta_l^3 R_{M3} - i\alpha_k \alpha_m \beta_l R_{T3} \right] \cdot l_y \alpha_{1,ml}$$

$$+ \sum_{m=-\infty}^{+\infty} \left[-R_{Q4} + i\beta_l^3 R_{M4} + i\alpha_k \alpha_m \beta_l R_{T4} \right] \cdot l_y \alpha_{2,ml}$$

$$= \begin{cases} 2Il_x l_y & \text{when } k = 0 \text{ \& } l = 0 \\ 0 & \text{when } k \neq 0 \,||\, l \neq 0 \end{cases} \tag{6.68}$$

$$\left[D_2(\alpha_k^2 + \beta_l^2)^2 - m_2\omega^2 - \frac{\omega^2 \rho_0}{ik_{z,kl}} + \frac{\omega^2 \rho_{cav}\cos(k_{z,cav,kl}d)}{k_{z,cav,kl}\sin(k_{z,cav,kl}d)}\right] \cdot l_x l_y \alpha_{2,kl}$$

$$- \frac{\omega^2 \rho_{cav}}{k_{z,cav,kl}\sin(k_{z,cav,kl}d)} \cdot l_x l_y \alpha_{1,kl}$$

$$+ \sum_{n=-\infty}^{+\infty} \left[-R_{Q2} + i\alpha_k^3 R_{M2} + i\beta_l \alpha_k \beta_n R_{T2}\right] \cdot l_x \alpha_{1,kn}$$

$$+ \sum_{n=-\infty}^{+\infty} \left[R_{Q1} - i\alpha_k^3 R_{M1} - i\beta_l \alpha_k \beta_n R_{T1}\right] \cdot l_x \alpha_{2,kn}$$

$$+ \sum_{m=-\infty}^{+\infty} \left[-R_{Q4} + i\beta_l^3 R_{M4} + i\alpha_k \alpha_m \beta_l R_{T4}\right] \cdot l_y \alpha_{1,ml}$$

$$+ \sum_{m=-\infty}^{+\infty} \left[R_{Q3} - i\beta_l^3 R_{M3} - i\alpha_k \alpha_m \beta_l R_{T3}\right] \cdot l_y \alpha_{2,ml} = 0 \quad (6.69)$$

The infinite set of coupled algebraic simultaneous equation system of Eq. (6.68) and (6.69) can be simplified as a finite set of equations by applying a truncated series of the assumed modes, insofar as the solution converges. In the present study, the sum indices (m, n) are restricted to have finite values, i.e., $m = -\hat{k}$ to $\hat{k}$ and $n = -\hat{l}$ to $\hat{l}$. Upon necessary algebraic manipulations (Appendix C), the resultant equation system that contains a finite number [i.e., $2KL$, where $K = 2\hat{k} + 1$, $L = 2\hat{l} + 1$] of unknowns can be expressed in matrix notation as

$$\begin{bmatrix} T_{11,kl} & T_{12,kl} \\ T_{21,kl} & T_{22,kl} \end{bmatrix}_{2KL \times 2KL} \begin{Bmatrix} \alpha_{1,kl} \\ \alpha_{2,kl} \end{Bmatrix}_{2KL \times 1} = \begin{Bmatrix} F_{kl} \\ 0 \end{Bmatrix}_{2KL \times 1} \quad (6.70)$$

Solving Eq. (6.70), one can obtain the vibration displacements of the two faceplates, with which the acoustic pressures in different fields are readily determined. As an assessment of sound energy penetrating through the structure, the transmission coefficient is defined here as the ratio of the transmitted sound power to the incident sound power [16–18]:

$$\tau(\varphi, \theta) = \frac{\sum_{m=-\infty}^{+\infty} \sum_{n=-\infty}^{+\infty} |\xi_{mn}|^2 \text{Re}(k_{z,mn})}{|I|^2 k_z} \quad (6.71)$$

which is a function of sound incident angles φ and θ. The diffuse sound transmission coefficient is taken in an averaged form over all possible incident angles [1] as

$$\tau_{\text{diff}} = \frac{\int_0^{\pi/4} \int_0^{\varphi_{\text{lim}}} \tau(\varphi, \theta) \sin\varphi \cos\varphi d\varphi d\theta}{\int_0^{\pi/4} \int_0^{\varphi_{\text{lim}}} \sin\varphi \cos\varphi d\varphi d\theta} \quad (6.72)$$

6.2 Sound Transmission Through Absorptive Sandwich Structure

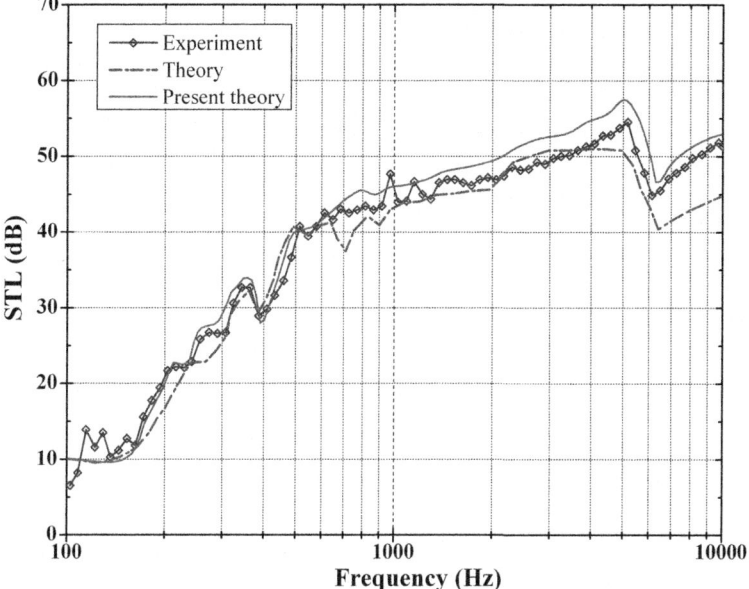

Fig. 6.11 Diffuse sound transmission loss (*STL*) plotted as a function of incident frequency: comparison between present model predictions with experimental measurements and theoretical results of Legault and Atalla [17] (With permission from Acoustical Society of America)

Finally, sound transmission loss (STL) is customarily defined in decibel scale [8, 9] as

$$\text{STL} = 10 \, \log_{10}\left(\frac{1}{\tau(\varphi,\theta)}\right) \tag{6.73}$$

which, intuitively, may be taken as a measure of the effectiveness of the sandwich structure in insulating the convective sound energy penetration.

6.2.5 Model Validation

For validation, the present model predictions are compared with the theoretical results and experimental measurements of Legault and Atalla [17] for 1D periodically rib-stiffened sandwich structures, as shown in Fig. 6.11. Since our model is developed for *orthogonally* rib-stiffened sandwich structures, it can be favorably degraded to the 1D case.

As can be seen from Fig. 6.11, overall agreement is achieved. Especially, the experimentally observed two significant resonance dips have been well captured by the present model, i.e., the first dip at approximately 400 Hz arising from the passband characteristics of periodic structures and the other due to coincidence resonance [17] at the critical frequency of approximately 6,200 Hz.

Since the diffuse sound transmission characteristics of periodically rib-stiffened sandwich structure have been well studied by Legault and Atalla [17], the focus of the present study thus turns to the normal sound incident case so as to explore more physical details. Moreover, in view of the fact that the sound transmission behavior of the sandwich structure at high frequency is complex with dense peaks and dips, the frequency range of 10–2,000 Hz is considered in subsequent analysis. As a result, the coincidence resonance dip is far beyond the considered frequency range which, even if calculated, would merge with the dense dips at high frequencies and impossible to distinguish.

6.2.6 Effects of Fluid-Structure Coupling on Sound Transmission

Since the absorption of sound by fibrous materials (i.e., fiberglass) is characterized using the frequency-dependent dynamic density and bulk modulus, the fluid-structure coupling effects are exploited below. For comparison, the air cavity case and the vacuum case are also considered to highlight the influence of the cavity-filling fiberglass.

Figures 6.12, 6.13, and 6.14 plot the STL as a function of frequency for the three different cases, with the periodic spacings selected as $l_x = l_y = 0.3$ m, 0.4 m, and 0.5 m, respectively. Previous theoretical studies often ignore the effects of fluid-structure coupling, assuming that sound transmission is dominated by the structure-borne path and that via the fluid-borne path is negligible. However, the results of Figs. 6.12, 6.13, and 6.14 demonstrate that fluid-structure coupling can alter the STL peaks and dips in the low-frequency range and thus affect sound transmission, especially when the rib-stiffeners are sparsely distributed (e.g., $l_x = l_y = 0.5$ m). In such cases, the response of the face panels is significantly affected by the fluid media confined in the cavities, since most of the panel surfaces are in contact with the fluid and the rib-stiffeners may only exert a local effect near the conjunctions. In other words, when the rib-stiffener separation is sufficiently large, fluid-structure coupling plays a role comparable to that of the rib-stiffeners. Under such conditions, the fluid-structure coupling effect can no longer be assumed negligible particularly in the stiffness-controlled low-frequency range.

As can be observed from the results of Figs. 6.12, 6.13, and 6.14, the three cases considered differ mainly in the low-frequency range where the fluid media confined in the partitioned cavities act on the face panels through fluid-structure coupling effect and work like pumping. That is, similar to elastic springs, the fluid media have equivalent stiffness, which affects significantly sound transmission in the stiffness-controlled low-frequency range while has almost no influence in the mass-controlled high-frequency range. As the stiffener separation is increased, the

6.2 Sound Transmission Through Absorptive Sandwich Structure

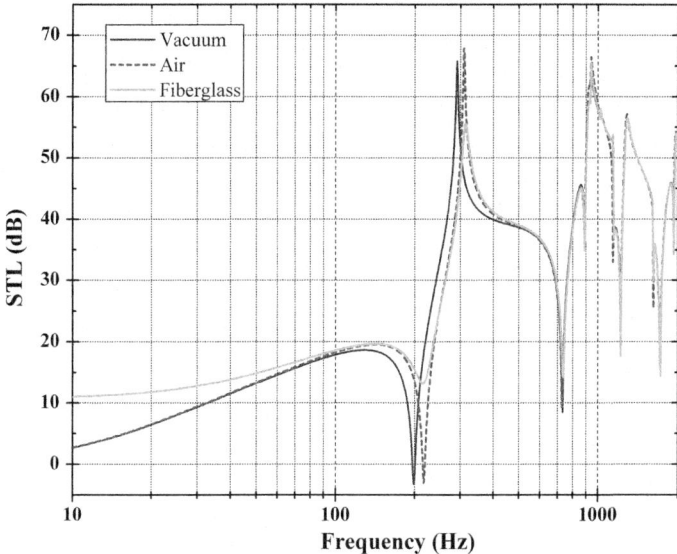

Fig. 6.12 Sound transmission loss (*STL*) plotted as a function of incident frequency for stiffener separations $l_x = l_y = 0.3$ m: comparison among three different kinds of orthogonally rib-stiffened sandwiches with cavities filled separately with vacuum, air, and fiberglass (With permission from Acoustical Society of America)

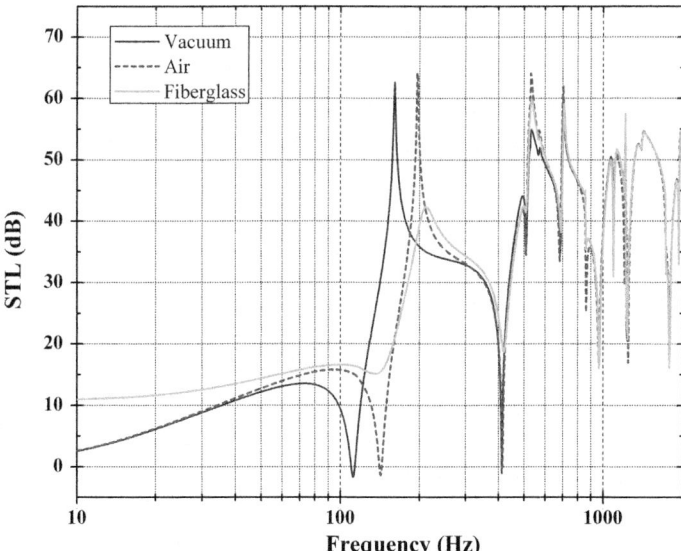

Fig. 6.13 Sound transmission loss (*STL*) plotted as a function of incident frequency for stiffener separations $l_x = l_y = 0.4$ m: comparison among three different kinds of orthogonally rib-stiffened sandwiches with cavities filled separately with vacuum, air, and fiberglass (With permission from Acoustical Society of America)

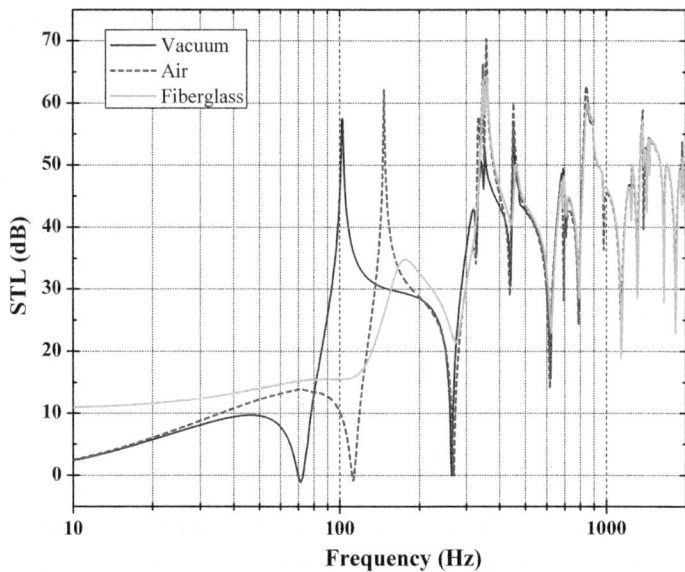

Fig. 6.14 Sound transmission loss (*STL*) plotted as a function of incident frequency for stiffener separations $l_x = l_y = 0.5$ m: comparison among three different kinds of orthogonally rib-stiffened sandwiches with cavities filled separately with vacuum, air, and fiberglass (With permission from Acoustical Society of America)

equivalent stiffness of the fluid media plays an increasingly important role in the transmission of sound, as the surface area of the face panels dominated by fluid media increases, while that controlled by the rib-stiffeners decreases. With this duly considered, it is then understandable that the divergences only exist in the low-frequency range, enlarging with increasing stiffener separation, as demonstrated in Figs. 6.12, 6.13, and 6.14.

6.2.7 Sound Transmission Loss Combined with Bending Stiffness and Structure Mass: Optimal Design of Sandwich

Due to high stiffness-to-weight ratio, sandwich structures have been widely applied in aeronautics and aerospace engineering, often providing acceptable sound insulation capability. To draw general guidelines for the practical engineering design of these weight-sensitive structures, an optimal design scheme for multifunctional sandwiches is presented and implemented below, combining low structure mass with high-stiffness and superior sound insulation requirements.

6.2 Sound Transmission Through Absorptive Sandwich Structure

Since both the face panels and rib-stiffeners considered in the present study have thin thickness compared with other geometrical dimensions, such as core depth d and periodic spacing l_x (or l_y), the most important structural geometry ratio only leaves the nondimensional variable l_x/d (or l_y/d). For simplicity, assuming that $l_x = l_y$, one then only needs to consider one variable (l/d) to seek for the optimal design of the sandwich for combined high STL, large bending stiffness, and low structure mass. Although diffuse sound transmission loss may be of more interest for practical engineering, the normal sound incident case is considered here to save computational efforts. To this end, several dimensionless parameters should be defined.

The first dimensionless parameter introduced is the normalized mass of the sandwich (i.e., ratio of the mass for one unit cell to that of the panel material filling the whole volume of the unit cell):

$$\overline{M} = \frac{\rho\left[(h_1 + h_2)\, l_x l_y + \left(l_x t_x + l_y t_y\right) d\right] + \rho_{\text{cav}} l_x l_y d}{\rho l_x l_y\, (d + h_1 + h_2)} \tag{6.74}$$

The above expression has accounted for the cavity-filling fiberglass. For the air cavity case, one only needs to eliminate the fiberglass term.

For external load bearing, the bending stiffness of the orthogonally rib-stiffened sandwich structure is important, given by [42, 43]

$$D_x = \frac{2Eh^3}{12} + \frac{Eh(d+h)^2}{2} + \frac{Ed^3}{12}\frac{t_x}{l_y}, \quad D_y = \frac{2Eh^3}{12} + \frac{Eh(d+h)^2}{2} + \frac{Ed^3}{12}\frac{t_y}{l_x} \tag{6.75}$$

which can be normalized as

$$\overline{D}_x = \frac{D_x}{Ed^3}, \quad \overline{D}_y = \frac{D_y}{Ed^3} \tag{6.76}$$

For $t_x = t_y$ and $l_x = l_y$ as in the present study, $\overline{D}_x = \overline{D}_y$, and hence, one can use only one symbol $\overline{D}$ to represent both $\overline{D}_x$ and $\overline{D}_y$. It should be pointed out that the fiberglass is loosely filled into the partitioned cavity and not bonded to the panels/rib-stiffeners and hence has no contribution to the structural rigidity $\overline{D}$.

Incorporating the above-defined dimensionless parameters $\overline{M}$, $\overline{D}$, and the sound insulation index STL, one may define an integrated index for optimal design toward high stiffness-to-mass ratio and superior sound isolation capability as [44]

$$\gamma_{\text{SDM}} = \frac{\text{STL} \times \overline{D}}{\overline{M}} \tag{6.77}$$

The larger the integrated index γ_{SDM} is, the more superior the combined acoustic and structural performance of the sandwich will be.

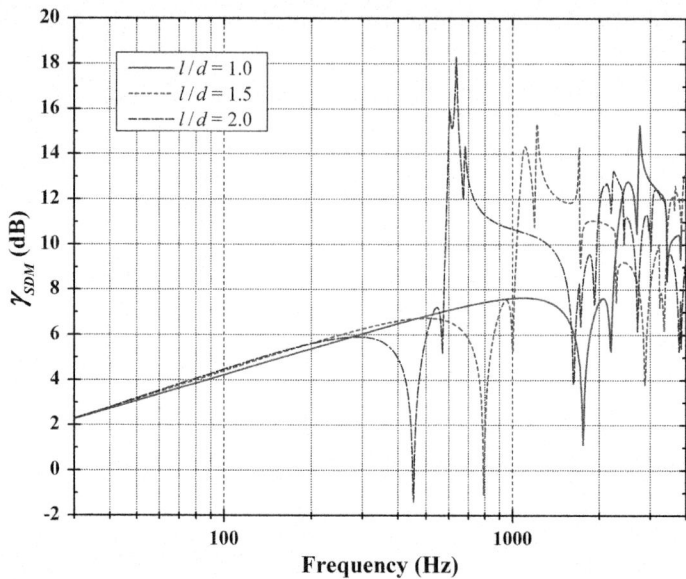

Fig. 6.15 Tendency plot of γ_{SDM} versus frequency for orthogonally rib-stiffened sandwich structures filled by air with selected structural geometry ratios: normal sound incident case (With permission from Acoustical Society of America)

Figures 6.15 and 6.16 show the tendency plots of γ_{SDM} versus frequency for orthogonally rib-stiffened sandwich structures having cavities filled separately with air and fiberglass. The influence of the key geometry ratio l/d on the integrated index γ_{SDM} is explored by comparing three typical cases, i.e., $l/d = 1.0$, 1.5, and 2.0. It is observed from Figs. 6.15 and 6.16 that while l/d has negligible influence on γ_{SDM} at low frequencies (<300 Hz), it causes significant changes of γ_{SDM} at relatively high frequencies. This implies that the integrated performance of the sandwich including mass, stiffness, and STL can be designed and optimized by varying the key structural geometry ratio l/d. Generally speaking, a larger l/d will help the structure to achieve a higher integrated index γ_{SDM}.

Relative to Fig. 6.15, the corresponding curves in Fig. 6.16 have slight alterations, resulting from the inclusion of fiberglass that induces changes in the parameters $\overline{M}$ and STL. In terms of the present optimal algorithm of Eq. (6.77), the inclusion of fiberglass does not appear to present additional benefits for the integrated performance. However, in accordance with different engineering requirements, the weight of the three parameters STL, $\overline{D}$, and $\overline{M}$ can be alternatively selected, and thus, different optimal designs may be achieved.

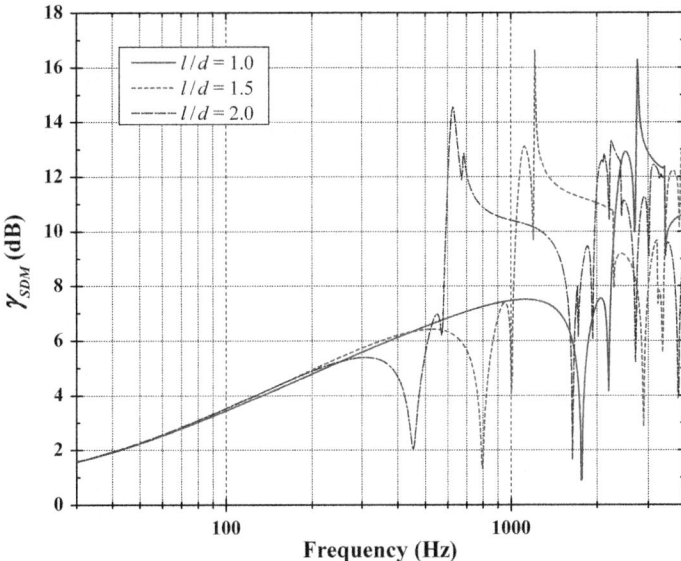

Fig. 6.16 Tendency plot of γ_{SDM} versus frequency for orthogonally rib-stiffened sandwich structures filled with fiberglass for selected structural geometry ratios: normal sound incident case (With permission from Acoustical Society of America)

6.2.8 Conclusions

Focusing on lightweight composite sandwich structures commonly used as aircraft fuselages, we propose a theoretical model to tackle with the sound transmission problem of infinite orthogonally rib-stiffened sandwich structures with fiberglass filled within the partitioned cavities. The process of sound penetration across the fiberglass is characterized by adopting the equivalent fluid model. The effects of fluid-structure coupling are also fully included by enforcing velocity continuity conditions at fluid-panel interfaces. The space-harmonic approach and the virtual work principle are applied to solve the resultant governing equations of the whole system. For validity check, the model predictions are compared with existing theoretical and experimental results for a simplified version of the sandwich structure, with good agreements achieved.

The model is subsequently applied to examine the influence of the cavity-filling fiberglass on sound transmission across the whole structure. It is demonstrated that the fluid-structure coupling effects should be taken into account in any theoretical attempt, especially when they play a role comparable with that of the rib-stiffeners when the rib-stiffener separations are sufficiently large. The inclusion of the fiberglass leads to remarkable changes of the STL versus frequency curves at low frequencies.

As a highlight of this research, an integrated optimal algorithm toward lightweight, high-stiffness, and superior sound insulation capability is proposed. With one key structural geometry ratio selected as the variable, a preliminary optimal design of the sandwich structure is carried out. It is found that the structural geometry ratio plays a significant role in the integrated mechanical and acoustic performance of the structure, providing therefore fundamental insight into the multifunctional design of the structure.

Appendices

Appendix A

Taking into account the inertial effects (due to stiffener mass) and applying both Hooke's law and Newton's second law, one can express the tensional forces arising from the rib-stiffeners as [18]

$$Q_x^+ = -R_{Q1}w_1 + R_{Q2}w_2, \quad Q_x^- = -R_{Q2}w_1 + R_{Q1}w_2 \quad (6.A.1)$$

$$Q_y^+ = -R_{Q3}w_1 + R_{Q4}w_2, \quad Q_y^- = -R_{Q4}w_1 + R_{Q3}w_2 \quad (6.A.2)$$

$$R_{Q1} = \frac{K_x(K_x - m_x\omega^2)}{(2K_x - m_x\omega^2)}, \quad R_{Q2} = \frac{K_x^2}{(2K_x - m_x\omega^2)} \quad (6.A.3)$$

$$R_{Q3} = \frac{K_y(K_y - m_y\omega^2)}{(2K_y - m_y\omega^2)}, \quad R_{Q4} = \frac{K_y^2}{(2K_y - m_y\omega^2)} \quad (6.A.4)$$

where ω is the circle frequency and (K_x, K_y) are the tensional stiffness of half the rib-stiffeners per unit length.

Likewise, the bending moments of the rib-stiffeners can be expressed as [18]

$$M_x^+ = R_{M1}\frac{\partial^2 w_1}{\partial x^2} - R_{M2}\frac{\partial^2 w_2}{\partial x^2}, \quad M_x^- = R_{M2}\frac{\partial^2 w_1}{\partial x^2} - R_{M1}\frac{\partial^2 w_2}{\partial x^2} \quad (6.A.5)$$

$$M_y^+ = R_{M3}\frac{\partial^2 w_1}{\partial y^2} - R_{M4}\frac{\partial^2 w_2}{\partial y^2}, \quad M_y^- = R_{M4}\frac{\partial^2 w_1}{\partial y^2} - R_{M3}\frac{\partial^2 w_2}{\partial y^2} \quad (6.A.6)$$

$$R_{M1} = \frac{E_x I_x^*(E_x I_x^* - \rho_x I_x \omega^2)}{(2E_x I_x^* - \rho_x I_x \omega^2)}, \quad R_{M2} = \frac{E_x^2 I_x^{*2}}{(2E_x I_x^* - \rho_x I_x \omega^2)} \quad (6.A.7)$$

$$R_{M3} = \frac{E_y I_y^* \left(E_y I_y^* - \rho_y I_y \omega^2\right)}{\left(2E_y I_y^* - \rho_y I_y \omega^2\right)}, \quad R_{M4} = \frac{E_y^2 I_y^{*2}}{\left(2E_y I_y^* - \rho_y I_y \omega^2\right)} \qquad (6.A.8)$$

where $(E_x I_x^*, E_y I_y^*)$ are the bending stiffness of half the rib-stiffeners and (ρ_x, ρ_y) and (I_x, I_y) are mass density and polar moment of inertia for the rib-stiffeners, with subscripts x and y indicating the direction of the stiffeners.

In a similar manner, the torsional moments of the rib-stiffeners are obtained as [18]

$$M_{Tx}^+ = R_{T1}\frac{\partial^2 w_1}{\partial x \partial y} - R_{T2}\frac{\partial^2 w_2}{\partial x \partial y}, \quad M_{Tx}^- = R_{T2}\frac{\partial^2 w_1}{\partial x \partial y} - R_{T1}\frac{\partial^2 w_2}{\partial x \partial y} \qquad (6.A.9)$$

$$M_{Ty}^+ = R_{T3}\frac{\partial^2 w_1}{\partial y \partial x} - R_{T4}\frac{\partial^2 w_2}{\partial y \partial x}, \quad M_{Ty}^- = R_{T4}\frac{\partial^2 w_1}{\partial y \partial x} - R_{T3}\frac{\partial^2 w_2}{\partial y \partial x} \qquad (6.A.10)$$

$$R_{T1} = \frac{G_x J_x^* \left(G_x J_x^* - \rho_x J_x \omega^2\right)}{\left(2G_x J_x^* - \rho_x J_x \omega^2\right)}, \quad R_{T2} = \frac{G_x^2 J_x^{*2}}{\left(2G_x J_x^* - \rho_x J_x \omega^2\right)} \qquad (6.A.11)$$

$$R_{T3} = \frac{G_y J_y^* \left(G_y J_y^* - \rho_y J_y \omega^2\right)}{\left(2G_y J_y^* - \rho_y J_y \omega^2\right)}, \quad R_{T4} = \frac{G_y^2 J_y^{*2}}{\left(2G_y J_y^* - \rho_y J_y \omega^2\right)} \qquad (6.A.12)$$

where $(G_x J_x^*, G_y J_y^*)$ are the torsional stiffness of half the rib-stiffeners and (J_x, J_y) are the torsional moment of inertia for the rib-stiffeners.

The Fourier transforms of the tensional forces, bending moments, and torsional moments are listed below:

1. Fourier transforms of tensional forces

$$\tilde{Q}_x^+ = R_{Q2}\tilde{w}_2(a, \beta_n) - R_{Q1}\tilde{w}_1(a, \beta_n), \quad \tilde{Q}_x^- = R_{Q1}\tilde{w}_2(a, \beta_n) - R_{Q2}\tilde{w}_1(a, \beta_n) \qquad (6.A.13)$$

$$\tilde{Q}_y^+ = R_{Q4}\tilde{w}_2(a_m, \beta) - R_{Q3}\tilde{w}_1(a_m, \beta), \quad \tilde{Q}_y^- = R_{Q3}\tilde{w}_2(a_m, \beta) - R_{Q4}\tilde{w}_1(a_m, \beta) \qquad (6.A.14)$$

2. Fourier transforms of bending moments

$$\tilde{M}_x^+ = \alpha^2 \left[R_{M2}\tilde{w}_2(a, \beta_n) - R_{M1}\tilde{w}_1(a, \beta_n)\right],$$
$$\tilde{M}_x^- = \alpha^2 \left[R_{M1}\tilde{w}_2(a, \beta_n) - R_{M2}\tilde{w}_1(a, \beta_n)\right] \qquad (6.A.15)$$

$$\tilde{M}_y^+ = \beta^2 \left[R_{M4}\tilde{w}_2(a_m, \beta) - R_{M3}\tilde{w}_1(a_m, \beta)\right],$$
$$\tilde{M}_y^- = \beta^2 \left[R_{M3}\tilde{w}_2(a_m, \beta) - R_{M4}\tilde{w}_1(a_m, \beta)\right] \qquad (6.A.16)$$

3. Fourier transforms of torsional moments

$$\tilde{M}_{Tx}^+ = \alpha\beta_n \left[R_{T2}\tilde{w}_2(a,\beta_n) - R_{T1}\tilde{w}_1(a,\beta_n)\right],$$
$$\tilde{M}_{Tx}^- = \alpha\beta_n \left[R_{T1}\tilde{w}_2(a,\beta_n) - R_{T2}\tilde{w}_1(a,\beta_n)\right] \quad (6.A.17)$$

$$\tilde{M}_{Ty}^+ = \alpha_m\beta \left[R_{T4}\tilde{w}_2(a_m,\beta) - R_{T3}\tilde{w}_1(a_m,\beta)\right],$$
$$\tilde{M}_{Ty}^- = \alpha_m\beta \left[R_{T3}\tilde{w}_2(a_m,\beta) - R_{T4}\tilde{w}_1(a_m,\beta)\right] \quad (6.A.18)$$

Appendix B

Adopting the similar procedure of Takahashi's beam model [11] for rib-stiffeners, taking the inertial effects of the rib-stiffeners into consideration, and applying Hooke's law and the Newton's second law, one can obtain the tensional forces of the rib-stiffeners as [18]

$$Q_x^+ = -\frac{K_x\left(K_x - m_x\omega^2\right)}{2K_x - m_x\omega^2}w_1 + \frac{K_x^2}{2K_x - m_x\omega^2}w_2 \quad (6.B.1)$$

$$Q_x^- = -\frac{K_x^2}{2K_x - m_x\omega^2}w_1 + \frac{K_x\left(K_x - m_x\omega^2\right)}{2K_x - m_x\omega^2}w_2 \quad (6.B.2)$$

$$Q_y^+ = -\frac{K_y\left(K_y - m_y\omega^2\right)}{2K_y - m_y\omega^2}w_1 + \frac{K_y^2}{2K_y - m_y\omega^2}w_2 \quad (6.B.3)$$

$$Q_y^- = -\frac{K_y^2}{2K_y - m_y\omega^2}w_1 + \frac{K_y\left(K_y - m_y\omega^2\right)}{2K_y - m_y\omega^2}w_2 \quad (6.B.4)$$

where ω is the circular frequency, (K_x, K_y) are the tensional stiffness of half rib-stiffeners per unit length, and (m_x, m_y) are the line mass density of the x- and y-wise stiffeners, respectively.

Likewise, the bending moments of the rib-stiffeners can be expressed as [18]

$$M_x^+ = \frac{E_xI_x^*\left(E_xI_x^* - \rho_xI_x\omega^2\right)}{2E_xI_x^* - \rho_xI_x\omega^2}\frac{\partial^2 w_1}{\partial x^2} - \frac{E_x^2I_x^{*2}}{2E_xI_x^* - \rho_xI_x\omega^2}\frac{\partial^2 w_2}{\partial x^2} \quad (6.B.5)$$

$$M_x^- = \frac{E_x^2I_x^{*2}}{2E_xI_x^* - \rho_xI_x\omega^2}\frac{\partial^2 w_1}{\partial x^2} - \frac{E_xI_x^*\left(E_xI_x^* - \rho_xI_x\omega^2\right)}{2E_xI_x^* - \rho_xI_x\omega^2}\frac{\partial^2 w_2}{\partial x^2} \quad (6.B.6)$$

$$M_y^+ = \frac{E_yI_y^*\left(E_yI_y^* - \rho_yI_y\omega^2\right)}{2E_yI_y^* - \rho_yI_y\omega^2}\frac{\partial^2 w_1}{\partial y^2} - \frac{E_y^2I_y^{*2}}{2E_yI_y^* - \rho_yI_y\omega^2}\frac{\partial^2 w_2}{\partial y^2} \quad (6.B.7)$$

Appendices

$$M_y^- = \frac{E_y^2 I_y^{*2}}{2E_y I_y^* - \rho_y I_y \omega^2} \frac{\partial^2 w_1}{\partial y^2} - \frac{E_y I_y^* \left(E_y I_y^* - \rho_y I_y \omega^2\right)}{2E_y I_y^* - \rho_y I_y \omega^2} \frac{\partial^2 w_2}{\partial y^2} \quad (6.B.8)$$

where $(E_x I_x^*, E_y I_y^*)$ are the bending stiffness of half rib-stiffeners per unit length and (ρ_x, ρ_y), (I_x, I_y) are the mass density and polar moment of inertia of the stiffeners, respectively, with subscripts x and y indicating the corresponding orientations of the stiffeners.

In a similar scheme, the torsional moments of the rib-stiffeners are given by [18]

$$M_{Tx}^+ = \frac{G_x J_x^* \left(G_x J_x^* - \rho_x J_x \omega^2\right)}{2G_x J_x^* - \rho_x J_x \omega^2} \frac{\partial^2 w_1}{\partial x \partial y} - \frac{G_x^2 J_x^{*2}}{2G_x J_x^* - \rho_x J_x \omega^2} \frac{\partial^2 w_2}{\partial x \partial y} \quad (6.B.9)$$

$$M_{Tx}^- = \frac{G_x^2 J_x^{*2}}{2G_x J_x^* - \rho_x J_x \omega^2} \frac{\partial^2 w_1}{\partial x \partial y} - \frac{G_x J_x^* \left(G_x J_x^* - \rho_x J_x \omega^2\right)}{2G_x J_x^* - \rho_x J_x \omega^2} \frac{\partial^2 w_2}{\partial x \partial y} \quad (6.B.10)$$

$$M_{Ty}^+ = \frac{G_y J_y^* \left(G_y J_y^* - \rho_y J_y \omega^2\right)}{2G_y J_y^* - \rho_y J_y \omega^2} \frac{\partial^2 w_1}{\partial y \partial x} - \frac{G_y^2 J_y^{*2}}{2G_y J_y^* - \rho_y J_y \omega^2} \frac{\partial^2 w_2}{\partial y \partial x} \quad (6.B.11)$$

$$M_{Ty}^- = \frac{G_y^2 J_y^{*2}}{2G_y J_y^* - \rho_y J_y \omega^2} \frac{\partial^2 w_1}{\partial y \partial x} - \frac{G_y J_y^* \left(G_y J_y^* - \rho_y J_y \omega^2\right)}{2G_y J_y^* - \rho_y J_y \omega^2} \frac{\partial^2 w_2}{\partial y \partial x} \quad (6.B.12)$$

where $(G_x J_x^*, G_y J_y^*)$ are the torsional stiffness of half rib-stiffeners per unit length and (J_x, J_y) are the torsional moments of inertia of the stiffeners.

To simplify Eqs. (6.B.1), (6.B.2), (6.B.3), (6.B.4), (6.B.5), (6.B.6), (6.B.7), (6.B.8), (6.B.9), (6.B.10), (6.B.11), and (6.B.12), the following sets of specified characteristics are utilized to replace the coefficients of the general displacements:

1. Replacement of tensional force coefficients

$$R_{Q1} = \frac{K_x \left(K_x - m_x \omega^2\right)}{2K_x - m_x \omega^2}, \quad R_{Q2} = \frac{K_x^2}{2K_x - m_x \omega^2} \quad (6.B.13)$$

$$R_{Q3} = \frac{K_y \left(K_y - m_y \omega^2\right)}{2K_y - m_y \omega^2}, \quad R_{Q4} = \frac{K_y^2}{2K_y - m_y \omega^2} \quad (6.B.14)$$

2. Replacement of bending moment coefficients

$$R_{M1} = \frac{E_x I_x^* \left(E_x I_x^* - \rho_x I_x \omega^2\right)}{2E_x I_x^* - \rho_x I_x \omega^2}, \quad R_{M2} = \frac{E_x^2 I_x^{*2}}{2E_x I_x^* - \rho_x I_x \omega^2} \quad (6.B.15)$$

$$R_{M3} = \frac{E_y I_y^* \left(E_y I_y^* - \rho_y I_y \omega^2\right)}{2E_y I_y^* - \rho_y I_y \omega^2}, \quad R_{M4} = \frac{E_y^2 I_y^{*2}}{2E_y I_y^* - \rho_y I_y \omega^2} \quad (6.B.16)$$

3. Replacement of torsional moment coefficients

$$R_{T1} = \frac{G_x J_x^* \left(G_x J_x^* - \rho_x J_x \omega^2\right)}{2 G_x J_x^* - \rho_x J_x \omega^2}, \quad R_{T2} = \frac{G_x^2 J_x^{*2}}{2 G_x J_x^* - \rho_x J_x \omega^2} \quad (6.B.17)$$

$$R_{T3} = \frac{G_y J_y^* \left(G_y J_y^* - \rho_y J_y \omega^2\right)}{2 G_y J_y^* - \rho_y J_y \omega^2}, \quad R_{T4} = \frac{G_y^2 J_y^{*2}}{2 G_y J_y^* - \rho_y J_y \omega^2} \quad (6.B.18)$$

In terms of space-harmonic series, the expressions of the tensional forces, bending moments, and torsional moments can be simplified as follows:

1. Tensional forces

$$Q_x^+ = \sum_{m=-\infty}^{+\infty} \sum_{n=-\infty}^{+\infty} \left(-R_{Q1}\alpha_{1,mn} + R_{Q2}\alpha_{2,mn}\right) e^{-i(\alpha_m x + \beta_n y)} \quad (6.B.19)$$

$$Q_x^- = \sum_{m=-\infty}^{+\infty} \sum_{n=-\infty}^{+\infty} \left(-R_{Q2}\alpha_{1,mn} + R_{Q1}\alpha_{2,mn}\right) e^{-i(\alpha_m x + \beta_n y)} \quad (6.B.20)$$

$$Q_y^+ = \sum_{m=-\infty}^{+\infty} \sum_{n=-\infty}^{+\infty} \left(-R_{Q3}\alpha_{1,mn} + R_{Q4}\alpha_{2,mn}\right) e^{-i(\alpha_m x + \beta_n y)} \quad (6.B.21)$$

$$Q_y^- = \sum_{m=-\infty}^{+\infty} \sum_{n=-\infty}^{+\infty} \left(-R_{Q4}\alpha_{1,mn} + R_{Q3}\alpha_{2,mn}\right) e^{-i(\alpha_m x + \beta_n y)} \quad (6.B.22)$$

2. Bending moments

$$M_x^+ = \sum_{m=-\infty}^{+\infty} \sum_{n=-\infty}^{+\infty} \left(-R_{M1}\alpha_{1,mn} + R_{M2}\alpha_{2,mn}\right) \alpha_m^2 e^{-i(\alpha_m x + \beta_n y)} \quad (6.B.23)$$

$$M_x^- = \sum_{m=-\infty}^{+\infty} \sum_{n=-\infty}^{+\infty} \left(-R_{M2}\alpha_{1,mn} + R_{M1}\alpha_{2,mn}\right) \alpha_m^2 e^{-i(\alpha_m x + \beta_n y)} \quad (6.B.24)$$

$$M_y^+ = \sum_{m=-\infty}^{+\infty} \sum_{n=-\infty}^{+\infty} \left(-R_{M3}\alpha_{1,mn} + R_{M4}\alpha_{2,mn}\right) \beta_n^2 e^{-i(\alpha_m x + \beta_n y)} \quad (6.B.25)$$

$$M_y^- = \sum_{m=-\infty}^{+\infty} \sum_{n=-\infty}^{+\infty} \left(-R_{M4}\alpha_{1,mn} + R_{M3}\alpha_{2,mn}\right) \beta_n^2 e^{-i(\alpha_m x + \beta_n y)} \quad (6.B.26)$$

3. Torsional moments

$$M_{Tx}^+ = \sum_{m=-\infty}^{+\infty} \sum_{n=-\infty}^{+\infty} (-R_{T1}\alpha_{1,mn} + R_{T2}\alpha_{2,mn}) \alpha_m \beta_n e^{-i(\alpha_m x + \beta_n y)} \quad (6.B.27)$$

$$M_{Tx}^- = \sum_{m=-\infty}^{+\infty} \sum_{n=-\infty}^{+\infty} (-R_{T2}\alpha_{1,mn} + R_{T1}\alpha_{2,mn}) \alpha_m \beta_n e^{-i(\alpha_m x + \beta_n y)} \quad (6.B.28)$$

$$M_{Ty}^+ = \sum_{m=-\infty}^{+\infty} \sum_{n=-\infty}^{+\infty} (-R_{T3}\alpha_{1,mn} + R_{T4}\alpha_{2,mn}) \alpha_m \beta_n e^{-i(\alpha_m x + \beta_n y)} \quad (6.B.29)$$

$$M_{Ty}^- = \sum_{m=-\infty}^{+\infty} \sum_{n=-\infty}^{+\infty} (-R_{T4}\alpha_{1,mn} + R_{T3}\alpha_{2,mn}) \alpha_m \beta_n e^{-i(\alpha_m x + \beta_n y)} \quad (6.B.30)$$

Appendix C

The deflection coefficients of the two face panels are

$$\{\alpha_{1,kl}\} = [\alpha_{1,11}\ \alpha_{1,21}\ \cdots\ \alpha_{1,K1}\ \alpha_{1,12}\ \alpha_{1,22}\ \cdots\ \alpha_{1,K2}\ \cdots\ \alpha_{1,KL}]^T_{KL \times 1} \quad (6.C.1)$$

$$\{\alpha_{2,kl}\} = [\alpha_{2,11}\ \alpha_{2,21}\ \cdots\ \alpha_{2,K1}\ \alpha_{2,12}\ \alpha_{2,22}\ \cdots\ \alpha_{2,K2}\ \cdots\ \alpha_{2,KL}]^T_{KL \times 1} \quad (6.C.2)$$

The right-hand side of Eq. (6.70) represents the generalized force, that is,

$$\{F_{kl}\} = [F_{11}\ F_{21}\ \cdots\ F_{K1}\ F_{12}\ F_{22}\ \cdots\ F_{K2}\ \cdots\ F_{KL}]^T_{KL \times 1} \quad (6.C.3)$$

where

$$F_{kl} = \begin{cases} 2Il_xl_y & \text{at } k = \frac{K+1}{2}\ \&\ l = \frac{L+1}{2} \\ 0 & \text{at } k \neq \frac{K+1}{2}\ \|\ l \neq \frac{L+1}{2} \end{cases} \quad (6.C.4)$$

$$\lambda_{kl}^{11,1} = \left[D_1(\alpha_k^2 + \beta_l^2)^2 - m_1\omega^2 - \frac{\omega^2 \rho_0}{ik_{z,kl}} + \frac{\omega^2 \rho_{cav} \cos(k_{z,cav,kl}d)}{k_{z,cav,kl} \sin(k_{z,cav,kl}d)}\right] \cdot l_x l_y \quad (6.C.5)$$

$$T_{11,1} = \text{diag}[\lambda_{11}^{11,1}\ \lambda_{21}^{11,1}\ \cdots\ \lambda_{K1}^{11,1}\ \lambda_{12}^{11,1}\ \lambda_{22}^{11,1}\ \cdots\ \lambda_{K2}^{11,1}\ \cdots\ \lambda_{KL}^{11,1}]_{KL \times KL} \quad (6.C.6)$$

$$\lambda_{KL}^{11,2} = l_x \times \text{diag}[R_{Q1} - i\alpha_1^3 R_{M1}\ R_{Q1} - i\alpha_2^3 R_{M1}\ \cdots\ R_{Q1} - i\alpha_K^3 R_{M1}]_{K \times L} \quad (6.C.7)$$

$$T_{11,2} = \begin{bmatrix} \lambda_{KL}^{11,2} & \lambda_{KL}^{11,2} & \cdots & \lambda_{KL}^{11,2} \\ \lambda_{KL}^{11,2} & \lambda_{KL}^{11,2} & \cdots & \lambda_{KL}^{11,2} \\ \vdots & \vdots & \ddots & \vdots \\ \lambda_{KL}^{11,2} & \lambda_{KL}^{11,2} & \cdots & \lambda_{KL}^{11,2} \end{bmatrix}_{KL \times KL} \tag{6.C.8}$$

$$\lambda_{Kl,n}^{11,3} = l_x \times \text{diag}\begin{bmatrix} -i\alpha_1 \beta_l \beta_n R_{T1} & -i\alpha_2 \beta_l \beta_n R_{T1} & \cdots & -i\alpha_K \beta_l \beta_n R_{T1} \end{bmatrix}_{K \times L} \tag{6.C.9}$$

$$T_{11,3} = \begin{bmatrix} \lambda_{K1,1}^{11,3} & \lambda_{K1,2}^{11,3} & \cdots & \lambda_{K1,L}^{11,3} \\ \lambda_{K2,1}^{11,3} & \lambda_{K2,2}^{11,3} & \cdots & \lambda_{K2,L}^{11,3} \\ \vdots & \vdots & \ddots & \vdots \\ \lambda_{KL,1}^{11,3} & \lambda_{KL,2}^{11,3} & \cdots & \lambda_{KL,L}^{11,3} \end{bmatrix}_{KL \times KL} \tag{6.C.10}$$

$$\lambda_{Kl}^{11,4} = l_y \times \begin{bmatrix} R_{Q3} - i\beta_l^3 R_{M3} & R_{Q3} - i\beta_l^3 R_{M3} & \cdots & R_{Q3} - i\beta_l^3 R_{M3} \\ R_{Q3} - i\beta_l^3 R_{M3} & R_{Q3} - i\beta_l^3 R_{M3} & \cdots & R_{Q3} - i\beta_l^3 R_{M3} \\ \vdots & \vdots & \ddots & \vdots \\ R_{Q3} - i\beta_l^3 R_{M3} & R_{Q3} - i\beta_l^3 R_{M3} & \cdots & R_{Q3} - i\beta_l^3 R_{M3} \end{bmatrix}_{K \times L} \tag{6.C.11}$$

$$T_{11,4} = \text{diag}\begin{bmatrix} \lambda_{K1}^{11,4} & \lambda_{K2}^{11,4} & \cdots & \lambda_{KL}^{11,4} \end{bmatrix}_{KL \times KL} \tag{6.C.12}$$

$$\lambda_{Kl}^{11,5} = l_y \times \begin{bmatrix} -i\alpha_1 \beta_l \alpha_1 R_{T3} & -i\alpha_1 \beta_l \alpha_2 R_{T3} & \cdots & -i\alpha_1 \beta_l \alpha_K R_{T3} \\ -i\alpha_2 \beta_l \alpha_1 R_{T3} & -i\alpha_2 \beta_l \alpha_2 R_{T3} & \cdots & -i\alpha_2 \beta_l \alpha_K R_{T3} \\ \vdots & \vdots & \ddots & \vdots \\ -i\alpha_K \beta_l \alpha_1 R_{T3} & -i\alpha_K \beta_l \alpha_2 R_{T3} & \cdots & -i\alpha_K \beta_l \alpha_K R_{T3} \end{bmatrix}_{K \times L} \tag{6.C.13}$$

$$T_{11,5} = \text{diag}\begin{bmatrix} \lambda_{K1}^{11,5} & \lambda_{K2}^{11,5} & \cdots & \lambda_{KL}^{11,5} \end{bmatrix}_{KL \times KL} \tag{6.C.14}$$

$$\lambda_{kl}^{12,1} = -\frac{\omega^2 \rho_{cav}}{k_{z,cav,kl} \sin(k_{z,cav,kl} d)} \cdot l_x l_y \tag{6.C.15}$$

$$T_{12,1} = \text{diag}\begin{bmatrix} \lambda_{11}^{12,1} & \lambda_{21}^{12,1} & \cdots & \lambda_{K1}^{12,1} & \lambda_{12}^{12,1} & \lambda_{22}^{12,1} & \cdots & \lambda_{K2}^{12,1} & \cdots & \lambda_{KL}^{12,1} \end{bmatrix}_{KL \times KL} \tag{6.C.16}$$

$$\lambda_{KL}^{12,2} = l_x \times \text{diag}\begin{bmatrix} -R_{Q2} + i\alpha_1^3 R_{M2} & -R_{Q2} + i\alpha_2^3 R_{M2} & \cdots & -R_{Q2} + i\alpha_K^3 R_{M2} \end{bmatrix}_{K \times L} \tag{6.C.17}$$

$$T_{12,2} = \begin{bmatrix} \lambda_{KL}^{12,2} & \lambda_{KL}^{12,2} & \cdots & \lambda_{KL}^{12,2} \\ \lambda_{KL}^{12,2} & \lambda_{KL}^{12,2} & \cdots & \lambda_{KL}^{12,2} \\ \vdots & \vdots & \ddots & \vdots \\ \lambda_{KL}^{12,2} & \lambda_{KL}^{12,2} & \cdots & \lambda_{KL}^{12,2} \end{bmatrix}_{KL \times KL} \tag{6.C.18}$$

$$\lambda_{Kl,n}^{12,3} = l_x \times \text{diag}\begin{bmatrix} i\alpha_1 \beta_l \beta_n R_{T2} & i\alpha_2 \beta_l \beta_n R_{T2} & \cdots & i\alpha_K \beta_l \beta_n R_{T2} \end{bmatrix}_{K \times L} \tag{6.C.19}$$

$$T_{12,3} = \begin{bmatrix} \lambda_{K1,1}^{12,3} & \lambda_{K1,2}^{12,3} & \cdots & \lambda_{K1,L}^{12,3} \\ \lambda_{K2,1}^{12,3} & \lambda_{K2,2}^{12,3} & \cdots & \lambda_{K2,L}^{12,3} \\ \vdots & \vdots & \ddots & \vdots \\ \lambda_{KL,1}^{12,3} & \lambda_{KL,2}^{12,3} & \cdots & \lambda_{KL,L}^{12,3} \end{bmatrix}_{KL \times KL} \tag{6.C.20}$$

$$\lambda_{Kl}^{12,4} = l_y \times \begin{bmatrix} -R_{Q4} + i\beta_l^3 R_{M4} & -R_{Q4} + i\beta_l^3 R_{M4} & \cdots & -R_{Q4} + i\beta_l^3 R_{M4} \\ -R_{Q4} + i\beta_l^3 R_{M4} & -R_{Q4} + i\beta_l^3 R_{M4} & \cdots & -R_{Q4} + i\beta_l^3 R_{M4} \\ \vdots & \vdots & \ddots & \vdots \\ -R_{Q4} + i\beta_l^3 R_{M4} & -R_{Q4} + i\beta_l^3 R_{M4} & \cdots & -R_{Q4} + i\beta_l^3 R_{M4} \end{bmatrix}_{K \times L} \tag{6.C.21}$$

$$T_{12,4} = \text{diag}\begin{bmatrix} \lambda_{K1}^{12,4} & \lambda_{K2}^{12,4} & \cdots & \lambda_{KL}^{12,4} \end{bmatrix}_{KL \times KL} \tag{6.C.22}$$

$$\lambda_{Kl}^{12,5} = l_y \times \begin{bmatrix} i\alpha_1 \beta_l \alpha_1 R_{T4} & i\alpha_1 \beta_l \alpha_2 R_{T4} & \cdots & i\alpha_1 \beta_l \alpha_K R_{T4} \\ i\alpha_2 \beta_l \alpha_1 R_{T4} & i\alpha_2 \beta_l \alpha_2 R_{T4} & \cdots & i\alpha_2 \beta_l \alpha_K R_{T4} \\ \vdots & \vdots & \ddots & \vdots \\ i\alpha_K \beta_l \alpha_1 R_{T4} & i\alpha_K \beta_l \alpha_2 R_{T4} & \cdots & i\alpha_K \beta_l \alpha_K R_{T4} \end{bmatrix}_{K \times L} \tag{6.C.23}$$

$$T_{12,5} = \text{diag}\begin{bmatrix} \lambda_{K1}^{12,5} & \lambda_{K2}^{12,5} & \cdots & \lambda_{KL}^{12,5} \end{bmatrix}_{KL \times KL} \tag{6.C.24}$$

$$\lambda_{kl}^{21,1} = -\frac{\omega^2 \rho_{\text{cav}}}{k_{z,\text{cav},kl} \sin(k_{z,\text{cav},kl} d)} \cdot l_x l_y \tag{6.C.25}$$

$$T_{21,1} = \text{diag}\begin{bmatrix} \lambda_{11}^{21,1} & \lambda_{21}^{21,1} & \cdots & \lambda_{K1}^{21,1} & \lambda_{12}^{21,1} & \lambda_{22}^{21,1} & \cdots & \lambda_{K2}^{21,1} & \cdots & \lambda_{KL}^{21,1} \end{bmatrix}_{KL \times KL} \tag{6.C.26}$$

$$\lambda_{KL}^{21,2} = l_x \times \text{diag}\begin{bmatrix} -R_{Q2} + i\alpha_1^3 R_{M2} & -R_{Q2} + i\alpha_2^3 R_{M2} & \cdots & -R_{Q2} + i\alpha_K^3 R_{M2} \end{bmatrix}_{K \times L} \tag{6.C.27}$$

$$T_{21,2} = \begin{bmatrix} \lambda_{KL}^{21,2} & \lambda_{KL}^{21,2} & \cdots & \lambda_{KL}^{21,2} \\ \lambda_{KL}^{21,2} & \lambda_{KL}^{21,2} & \cdots & \lambda_{KL}^{21,2} \\ \vdots & \vdots & \ddots & \vdots \\ \lambda_{KL}^{21,2} & \lambda_{KL}^{21,2} & \cdots & \lambda_{KL}^{21,2} \end{bmatrix}_{KL \times KL} \tag{6.C.28}$$

$$\lambda_{Kl,n}^{21,3} = l_x \times \text{diag}\begin{bmatrix} i\alpha_1\beta_l\beta_n R_{T2} & i\alpha_2\beta_l\beta_n R_{T2} & \cdots & i\alpha_K\beta_l\beta_n R_{T2} \end{bmatrix}_{K \times L} \tag{6.C.29}$$

$$T_{21,3} = \begin{bmatrix} \lambda_{K1,1}^{21,3} & \lambda_{K1,2}^{21,3} & \cdots & \lambda_{K1,L}^{21,3} \\ \lambda_{K2,1}^{21,3} & \lambda_{K2,2}^{21,3} & \cdots & \lambda_{K2,L}^{21,3} \\ \vdots & \vdots & \ddots & \vdots \\ \lambda_{KL,1}^{21,3} & \lambda_{KL,2}^{21,3} & \cdots & \lambda_{KL,L}^{21,3} \end{bmatrix}_{KL \times KL} \tag{6.C.30}$$

$$\lambda_{Kl}^{21,4} = l_y \times \begin{bmatrix} -R_{Q4} + i\beta_l^3 R_{M4} & -R_{Q4} + i\beta_l^3 R_{M4} & \cdots & -R_{Q4} + i\beta_l^3 R_{M4} \\ -R_{Q4} + i\beta_l^3 R_{M4} & -R_{Q4} + i\beta_l^3 R_{M4} & \cdots & -R_{Q4} + i\beta_l^3 R_{M4} \\ \vdots & \vdots & \ddots & \vdots \\ -R_{Q4} + i\beta_l^3 R_{M4} & -R_{Q4} + i\beta_l^3 R_{M4} & \cdots & -R_{Q4} + i\beta_l^3 R_{M4} \end{bmatrix}_{K \times L}$$
$$\tag{6.C.31}$$

$$T_{21,4} = \text{diag}\begin{bmatrix} \lambda_{K1}^{21,4} & \lambda_{K2}^{21,4} & \cdots & \lambda_{KL}^{21,4} \end{bmatrix}_{KL \times KL} \tag{6.C.32}$$

$$\lambda_{Kl}^{21,5} = l_y \times \begin{bmatrix} i\alpha_1\beta_l\alpha_1 R_{T4} & i\alpha_1\beta_l\alpha_2 R_{T4} & \cdots & i\alpha_1\beta_l\alpha_K R_{T4} \\ i\alpha_2\beta_l\alpha_1 R_{T4} & i\alpha_2\beta_l\alpha_2 R_{T4} & \cdots & i\alpha_2\beta_l\alpha_K R_{T4} \\ \vdots & \vdots & \ddots & \vdots \\ i\alpha_K\beta_l\alpha_1 R_{T4} & i\alpha_K\beta_l\alpha_2 R_{T4} & \cdots & i\alpha_K\beta_l\alpha_K R_{T4} \end{bmatrix}_{K \times L}$$
$$\tag{6.C.33}$$

$$T_{21,5} = \text{diag}\begin{bmatrix} \lambda_{K1}^{21,5} & \lambda_{K2}^{21,5} & \cdots & \lambda_{KL}^{21,5} \end{bmatrix}_{KL \times KL} \tag{6.C.34}$$

$$\lambda_{kl}^{22,1} = \left[D_2(\alpha_k^2 + \beta_l^2)^2 - m_2\omega^2 - \frac{\omega^2 \rho_0}{ik_{z,kl}} + \frac{\omega^2 \rho_{\text{cav}} \cos(k_{z,\text{cav},kl} d)}{k_{z,\text{cav},kl} \sin(k_{z,\text{cav},kl} d)} \right] \cdot l_x l_y \tag{6.C.35}$$

$$T_{22,1} = \text{diag}\begin{bmatrix} \lambda_{11}^{22,1} & \lambda_{21}^{22,1} & \cdots & \lambda_{K1}^{22,1} & \lambda_{12}^{22,1} & \lambda_{22}^{22,1} & \cdots & \lambda_{K2}^{22,1} & \cdots & \lambda_{KL}^{22,1} \end{bmatrix}_{KL \times KL} \tag{6.C.36}$$

$$\lambda_{KL}^{22,2} = l_x \times \text{diag}\begin{bmatrix} R_{Q1} - i\alpha_1^3 R_{M1} & R_{Q1} - i\alpha_2^3 R_{M1} & \cdots & R_{Q1} - i\alpha_K^3 R_{M1} \end{bmatrix}_{K \times L} \tag{6.C.37}$$

$$T_{22,2} = \begin{bmatrix} \lambda_{KL}^{22,2} & \lambda_{KL}^{22,2} & \cdots & \lambda_{KL}^{22,2} \\ \lambda_{KL}^{22,2} & \lambda_{KL}^{22,2} & \cdots & \lambda_{KL}^{22,2} \\ \vdots & \vdots & \ddots & \vdots \\ \lambda_{KL}^{22,2} & \lambda_{KL}^{22,2} & \cdots & \lambda_{KL}^{22,2} \end{bmatrix}_{KL \times KL} \quad (6.C.38)$$

$$\lambda_{Kl,n}^{22,3} = l_x \times \text{diag}\begin{bmatrix} -i\alpha_1 \beta_l \beta_n R_{T1} & -i\alpha_2 \beta_l \beta_n R_{T1} & \cdots & -i\alpha_K \beta_l \beta_n R_{T1} \end{bmatrix}_{K \times L} \quad (6.C.39)$$

$$T_{22,3} = \begin{bmatrix} \lambda_{K1,1}^{22,3} & \lambda_{K1,2}^{22,3} & \cdots & \lambda_{K1,L}^{22,3} \\ \lambda_{K2,1}^{22,3} & \lambda_{K2,2}^{22,3} & \cdots & \lambda_{K2,L}^{22,3} \\ \vdots & \vdots & \ddots & \vdots \\ \lambda_{KL,1}^{22,3} & \lambda_{KL,2}^{22,3} & \cdots & \lambda_{KL,L}^{22,3} \end{bmatrix}_{KL \times KL} \quad (6.C.40)$$

$$\lambda_{Kl}^{22,4} = l_y \times \begin{bmatrix} R_{Q3} - i\beta_l^3 R_{M3} & R_{Q3} - i\beta_l^3 R_{M3} & \cdots & R_{Q3} - i\beta_l^3 R_{M3} \\ R_{Q3} - i\beta_l^3 R_{M3} & R_{Q3} - i\beta_l^3 R_{M3} & \cdots & R_{Q3} - i\beta_l^3 R_{M3} \\ \vdots & \vdots & \ddots & \vdots \\ R_{Q3} - i\beta_l^3 R_{M3} & R_{Q3} - i\beta_l^3 R_{M3} & \cdots & R_{Q3} - i\beta_l^3 R_{M3} \end{bmatrix}_{K \times L} \quad (6.C.41)$$

$$T_{22,4} = \text{diag}\begin{bmatrix} \lambda_{K1}^{22,4} & \lambda_{K2}^{22,4} & \cdots & \lambda_{KL}^{22,4} \end{bmatrix}_{KL \times KL} \quad (6.C.42)$$

$$\lambda_{Kl}^{22,5} = l_y \times \begin{bmatrix} -i\alpha_1 \beta_l \alpha_1 R_{T3} & -i\alpha_1 \beta_l \alpha_2 R_{T3} & \cdots & -i\alpha_1 \beta_l \alpha_K R_{T3} \\ -i\alpha_2 \beta_l \alpha_1 R_{T3} & -i\alpha_2 \beta_l \alpha_2 R_{T3} & \cdots & -i\alpha_2 \beta_l \alpha_K R_{T3} \\ \vdots & \vdots & \ddots & \vdots \\ -i\alpha_K \beta_l \alpha_1 R_{T3} & -i\alpha_K \beta_l \alpha_2 R_{T3} & \cdots & -i\alpha_K \beta_l \alpha_K R_{T3} \end{bmatrix}_{K \times L} \quad (6.C.43)$$

$$T_{22,5} = \text{diag}\begin{bmatrix} \lambda_{K1}^{22,5} & \lambda_{K2}^{22,5} & \cdots & \lambda_{KL}^{22,5} \end{bmatrix}_{KL \times KL} \quad (6.C.44)$$

Using the definition of the sub-matrices presented above, one obtains

$$T_{11} = T_{11,1} + T_{11,2} + T_{11,3} + T_{11,4} + T_{11,5}, \quad T_{22} = T_{22,1} + T_{22,2} + T_{22,3} + T_{22,4} + T_{22,5} \quad (6.C.45)$$

$$T_{12} = T_{12,1} + T_{12,2} + T_{12,3} + T_{12,4} + T_{12,5}, \quad T_{21} = T_{21,1} + T_{21,2} + T_{21,3} + T_{21,4} + T_{21,5} \quad (6.C.46)$$

References

1. Xin FX, Lu TJ, Chen CQ (2010) Sound transmission through simply supported finite double-panel partitions with enclosed air cavity. J Vib Acoust 132(1):011008:011001–011011
2. Mace BR (1980) Sound radiation from a plate reinforced by two sets of parallel stiffeners. J Sound Vib 71(3):435–441
3. Mace BR (1981) Sound radiation from fluid loaded orthogonally stiffened plates. J Sound Vib 79(3):439–452
4. Yin XW, Gu XJ, Cui HF et al (2007) Acoustic radiation from a laminated composite plate reinforced by doubly periodic parallel stiffeners. J Sound Vib 306(3–5):877–889
5. Xin FX, Lu TJ, Chen CQ (2009) Dynamic response and acoustic radiation of double-leaf metallic panel partition under sound excitation. Comput Mater Sci 46(3):728–732
6. Maury C, Gardonio P, Elliott SJ (2001) Active control of the flow-induced noise transmitted through a panel. AIAA J 39(10):1860–1867
7. Xin FX, Lu TJ, Chen CQ (2009) External mean flow influence on noise transmission through double-leaf aeroelastic plates. AIAA J 47(8):1939–1951
8. Xin FX, Lu TJ (2009) Analytical and experimental investigation on transmission loss of clamped double panels: implication of boundary effects. J Acoust Soc Am 125(3): 1506–1517
9. Xin FX, Lu TJ, Chen CQ (2008) Vibroacoustic behavior of clamp mounted double-panel partition with enclosure air cavity. J Acoust Soc Am 124(6):3604–3612
10. Lin G-F, Garrelick JM (1977) Sound transmission through periodically framed parallel plates. J Acoust Soc Am 61(4):1014–1018
11. Takahashi D (1983) Sound radiation from periodically connected double-plate structures. J Sound Vib 90(4):541–557
12. Trochidis A, Kalaroutis A (1986) Sound transmission through double partitions with cavity absorption. J Sound Vib 107(2):321–327
13. Alba J, Ramis J, Sanchez-Morcillo VJ (2004) Improvement of the prediction of transmission loss of double partitions with cavity absorption by minimization techniques. J Sound Vib 273(4–5):793–804
14. Mead DJ, Pujara KK (1971) Space-harmonic analysis of periodically supported beams: response to convected random loading. J Sound Vib 14(4):525–532
15. Lee JH, Kim J (2002) Analysis of sound transmission through periodically stiffened panels by space-harmonic expansion method. J Sound Vib 251(2):349–366
16. Wang J, Lu TJ, Woodhouse J et al (2005) Sound transmission through lightweight double-leaf partitions: theoretical modelling. J Sound Vib 286(4–5):817–847
17. Legault J, Atalla N (2009) Numerical and experimental investigation of the effect of structural links on the sound transmission of a lightweight double panel structure. J Sound Vib 324(3–5):712–732
18. Xin FX, Lu TJ (2010) Analytical modeling of fluid loaded orthogonally rib-stiffened sandwich structures: sound transmission. J Mech Phys Solids 58(9):1374–1396
19. Panneton R, Atalla N (1996) Numerical prediction of sound transmission through finite multilayer systems with poroelastic materials. J Acoust Soc Am 100(1):346–354
20. Sgard FC, Atalla N, Nicolas J (2000) A numerical model for the low frequency diffuse field sound transmission loss of double-wall sound barriers with elastic porous linings. J Acoust Soc Am 108(6):2865–2872
21. Brown SM, Niedzielski J, Spalding GR (1978) Effect of sound-absorptive facings on partition airborne-sound transmission loss. J Acoust Soc Am 63(6):1851–1856
22. Allard J-F, Champoux Y (1992) New empirical equations for sound propagation in rigid frame fibrous materials. J Acoust Soc Am 91(6):3346–3353
23. Rumerman ML (1975) Vibration and wave propagation in ribbed plates. J Acoust Soc Am 57(2):370–373
24. Lu TJ, Hess A, Ashby MF (1999) Sound absorption in metallic foams. J Appl Phys 85(11):7528–7539

25. Wang XL, Lu TJ (1999) Optimized acoustic properties of cellular solids. J Acoust Soc Am 106(2):756–765
26. Lu TJ, Chen F, He D (2000) Sound absorption of cellular metals with semiopen cells. J Acoust Soc Am 108(4):1697–1709
27. Lu TJ, Kepets M, Dowling A (2008) Acoustic properties of sintered FeCrAlY foams with open cells (I): static flow resistance. Sci China Series E Technol Sci 51(11):1803–1811
28. Lu TJ, Kepets M, Dowling A (2008) Acoustic properties of sintered FeCrAlY foams with open cells (II): sound attenuation. Sci China Series E Technol Sci 51(11):1812–1837
29. Zwikker C, Kosten CW (1949) Sound absorbing materials. Elsevier, New York
30. Morse PM, Ingard KU (1968) Theoretical acoustics. McGraw-Hill, New York
31. Biot MA (1956) Theory of propagation of elastic waves in a fluid-saturated porous solid. I. Low-frequency range. J Acoust Soc Am 28(2):168–178
32. Biot MA (1956) Theory of propagation of elastic waves in a fluid-saturated porous solid. II. Higher frequency range. J Acoust Soc Am 28(2):179–191
33. Johnson DL, Koplik J, Dashen R (1987) Theory of dynamic permeability and tortuosity in fluid-saturated porous-media. J Fluid Mech 176:379–402
34. Xin FX, Lu TJ (2010) Analytical modeling of sound transmission across finite aeroelastic panels in convected fluids. J Acoust Soc Am 128(3):1097–1107
35. Mace BR (1980) Periodically stiffened fluid-loaded plates, I: Response to convected harmonic pressure and free wave propagation. J Sound Vib 73(4):473–486
36. Legault J, Atalla N (2010) Sound transmission through a double panel structure periodically coupled with vibration insulators. J Sound Vib 329(15):3082–3100
37. Mead DJ (1970) Free wave propagation in periodically supported, infinite beams. J Sound Vib 11(2):181–197
38. Xin FX, Lu TJ (2010) Sound radiation of orthogonally rib-stiffened sandwich structures with cavity absorption. Compos Sci Technol 70(15):2198–2206
39. Delany ME, Bazley EN (1970) Acoustical properties of fibrous absorbent materials. Appl Acoust 3(2):105–116
40. Brillouin L (1953) Wave propagation in periodic structures. Dover, New York
41. Graham WR (1995) High-frequency vibration and acoustic radiation of fluid-loaded plates. Philos Trans R Soc Phys Eng Sci (1990–1995) 352(1698):1–43
42. Kolsters H, Wennhage P (2009) Optimisation of laser-welded sandwich panels with multiple design constraints. Mar Struct 22(2):154–171
43. Lok TS, Cheng QH (1999) Elastic deflection of thin-walled sandwich panel. J Sandw Struct Mater 1(4):279–298
44. Xin FX, Lu TJ (2010) Effects of core topology on sound insulation performance of lightweight all-metallic sandwich panels. Mat Manuf Process 26(9):1213–1221